Herman Franz Mark

Otto Vogl
Chairman of Symposium

Front row from left to right:
Karl Schlögl, Takeo Saegusa, Linus Pauling, Otto Vogl, Herman Mark, Hans Mark, Ephraim Katchalski-Katzir, Paul Flory, John McTague.

Second row from left to right:
Charles Overberger, Günther Wilke, Piero Pino, Eli Pearce, Ellis Fields, George Smets, C. Bamford, James O'Donnell, Edward Teller.

Third row from left to right:
Bengt Rånby, Guy Donaruma, Helmut Ringsdorf, Herbert Morawetz, Naoya Ogata, Norbert Bikales, Kurt Komarek, Bruce Merrifield, Murray Goodman.

Polymer Science in the Next Decade

Polymer Science in the Next Decade
Trends, Opportunities, Promises

An International Symposium Honoring Herman F. Mark on His 90th Birthday

Edited by

Otto Vogl

Polytechnic University
Brooklyn, New York

Edmund H. Immergut

Consulting Editor
Brooklyn, New York

Published by John Wiley & Sons, Inc.
New York · Chichester · Brisbane · Toronto · Singapore

Library of Congress Cataloging-in-Publication Data

Polymer science in the next decade, trends, opportunities, promises.

1. Polymers and polymerization—Congresses.
2. Mark, H. F. (Herman Franz), 1895–
I. Vogl, Otto, 1927– . II. Immergut, Ed.
III. Mark, H. F. (Herman Franz), 1895–
QD380.P644 1987 547.7 87-14280
ISBN 0-471-63239-2

Printed in the United States of America

ORGANIZING COMMITTEE

Otto Vogl, Chairman

Sheldon Atlas	Eli Pearce
L. Guy Donaruma	George J. Schmitt
Murray Goodman	Jane C. Vogl

SPONSORING ORGANIZATIONS:

Polytechnic University
John Wiley & Sons, Inc.
New York Academy of Science
American Chemical Society
Division of Polymer Chemistry Inc., ACS
New York Section, ACS

CONTRIBUTING ORGANIZATIONS

Allied Corporation Foundation
The Dow Chemical Company Foundation
E. I. Du Pont De Nemours and Company
General Electric Company
National Patent Development Corporation
PPG Industries Incorporated
National Science Foundation

SPONSORS

Air Products and Chemicals, Incorporated
Allied Chemical Company
American Cyanamid Company
American Hoechst, Incorporated
ARCO Chemical Company
Beatrice Chemicals, Incorporated
Carl Hanser Verlag
Celanese Research Company
CIBA-GEIGY Corporation
Dow Chemical Company
Dow Corning Corporation
E. I. Du Pont De Nemours and Company
Exxon Research and Engineering Company
Fuji Film Limited
General Electric Company
General Motors Research Laboratories
GTE Laboratories, Incorporated
Hercules, Incorporated
Himont, Incorporated
IBM Corporation
International Paper Company
Lockheed Electronics, Incorporated
Milton Harris
Marcel Dekker, Incorporated
Milliken Research Corporation
Mobay Chemical Corporation
Monsanto Polymer Products Company
3M Corporation
Pfizer, Incorporated
Philip Morris Incorporated
Phillips Petroleum Company
Polaroid Foundation, Incorporated
Research Corporation
Rohm and Haas Company
Toray Industries, Incorporated
Union Carbide Corporation
Uniroyal, Incorporated
Xerox Corporation

ADVISORY COUNCIL

PREFACE

An International Symposium entitled "Polymer Science in the Next Decade" was held at the Plaza Hotel in New York City, May 6 and 7, 1985. This Symposium emphasized trends in the development of polymer science in the future.

With this event Professor Herman F. Mark, who is often called the "Father of Polymer Science in the United States," was honored on his 90th birthday. The subject of the Symposium emphasized the future of polymer science, an appropriate theme to reflect the visionary thinking of Professor Mark. Nearly 500 people attended the Symposium and 550 the banquet, which was the highlight of the event.

A few years ago, under the auspices of the National Academy of Sciences and the National Research Council a report was prepared entitled "Polymer Science and Engineering: Needs, Opportunities and Challenges." This document was meant to provide guidance for the development of polymer science in the next decade and beyond. Polymer Science and Engineering developed in periods; major developments in polymer science occurred in organic chemistry of polymers in the 1950s, in the understanding of the solid state properties of polymers in the 1960s, in polymer processing, the optimization of polymer properties and the development of polymer composites in the 1970s. The present decade is dominated by the development of functional and specialty polymers and the tailor making of specific new compositions. As a consequence, overlapping areas are developing between polymer science and the life sciences, medicine and biology; polymers are now also used in agriculture, the development and preservation of energy, in information storage and retrieval, and in health care and nutrition.

It is, consequently, timely that we look one decade ahead toward the development of ideas and directions as we approach the turn of the 21st Century.

Professor Mark, who has done so much in the field of polymer science, began his career in Europe; his early work on structure/property relationship in Polymer Science distinguished him as an eminent scientist and scholar. When he came to Brooklyn, he set up the Polymer Research Institute at the Polytechnic Institute of Brooklyn (now New York), the first of its kind in the world. This initiative and his publication activities became the spawning ground of much of the polymer industry in the United States. Over the years, many notable academicians came to the Polytechnic Institute of Brooklyn to do research or to teach. The Institute provided the training for the professors who then set up their own polymer institutions and/or the legions of scientists who became leaders in research and management in industry and government.

Otto Vogl
Edmund H. Immergut

HERMAN F. MARK

Herman F. Mark was born in Vienna, Austria on May 3, 1895. Over the past 70 years, he has contributed greatly to the scientific communities: he played a leading role in the field of Polymer Science and is often referred to as the "Father of Polymer Science in the United States." He continues his active schedule as a world traveler, is a frequent lecturer at symposia, seminars and in informal discussions.

Dr. Mark has published over 500 papers and review articles and more than 20 books on various topics of polymer science. He was instrumental in initiating publications on polymer science at Interscience Publishing, now John Wiley and Sons, Inc. He founded the Journal of Polymer Science, currently published as four separate periodicals; he also organized the Encyclopedia of Polymer Science and Technology and a book series on High Polymers.

Not only as a scientist, editor, and publisher is Herman Mark well known, but also for his exceptional personality; he is respected equally as a teacher, lecturer, and disseminator of polymer science. He is highly regarded as a compassionate human being, known for his humor, sensitivity, and wit.

At the Polytechnic Institute of Brooklyn, (now known as the Polytechnic Institute of New York), he established the unique and internationally famous Polymer Research Institute; he also started the first program leading to a degree in polymer science. He was also instrumental in the founding of the Weizman Institute of Science, initially organized at the Polytechnic Institute of Brooklyn.

Herman F. Mark received his Ph.D. in 1921 in Organic Chemistry at the University of Vienna. He then went to the Institute of Fiber Chemistry of the Kaiser Wilhelm Institute in Berlin-Dahlem. In this period, he accomplished some of his most important work on the Structure of Cellulose by X-ray diffraction studies.

In 1927, he joined the BASF in Ludwigshafen as a research chemist and rose to the position of Assistant Research Director. In 1932 he was appointed Professor and Director of the Physical-Chemical Institute of the University of Vienna; he joined the International Paper Company in 1938 and the Polytechnic Institute of Brooklyn in 1940. He served at *Brooklyn Poly* on the faculty until 1965, ultimately as Dean of Science.

Herman F. Mark's accomplishments are too numerous to mention in detail. They have been recognized in many ways: 20 honorary doctoral degrees, 45 membership and honorary memberships of academies and professional societies, 40 medals and awards, including the National Medal of Science and, most recently, the Grand Decoration of Honor in Gold with Star for Services to the Republic of Austria.

To many of us, Professor Herman F. Mark is affectionally known as the "Geheimrat," a title bestowed on exceptional persons that have been leaders of the community and society, and that have achieved special status among their friends.

We honor our "Geheimrat" on the occasion of his 90th birthday by bringing together an exceptional scientific community of his friends, colleagues, and admirers with an exceptional theme — with futuristic vision — "Polymer Science in the Next Decade."

We expect that the scientists and scholars involved in polymer science present at this symposium will again be inspired by our nonagenarian, the visionary Herman Mark. The speakers of the Symposium are expected to point the direction of polymer science and technology as it moves into the next decade, the next century, and the next millennium.

Otto Vogl
Jane C. Vogl

Contents

OPENING REMARKS

OTTO VOGL

Ladies and gentlemen, welcome to the International Symposium, "Polymer Science in the Next Decade." This Symposium has brought together scientists from all over the world to recognize the importance of polymer science, and also to recognize the importance of trying to predict the future development of polymer science. At the same time the scientific community has come together to honor one of the giants of polymer science, Professor Herman F. Mark, on his 90th birthday.

Professor Mark has been active in chemistry, physical chemistry, and polymer science for over sixty-five years, starting from his early days at the University of Vienna where in 1921 he received his Ph.D. in Organic Chemistry with Professor Schlenck as his advisor. He then went to the Kaiser Wilhelm Institute in Berlin and BASF in Ludwigshafen, ultimately as the Assistant Director of Research. From 1932 to 1938 he was Professor at the University of Vienna and the Director of the Institute of Physical Chemistry. In 1938 he emigrated to the New World and went initially to the International Paper Company in Canada. In 1940, he came to the United States and to the Polytechnic Institute of New York, which was then the Polytechnic Institute of Brooklyn.

Throughout the two days of this Symposium, we will hear prospects, promises, and trends in polymer science. In some cases, especially in the historic part of the discussions and deliberations, mention will be made of Professor Mark's contribution as a scientist, teacher, editor, but most importantly as *the* communicator of polymer science. Herman Mark was always involved in all aspects of polymer science and technology; he was and is the disseminator of knowledge, the initiator of new developments, and the nucleating agent for new ideas. It is the foresight of Herman Mark and his scientific conviction, his personality, and his capability to see into the future that has inspired us to name this symposium "Polymer Science in the Next Decade," or if we wish to call it even more specifically "Polymer Science as it Approaches the 21st Century." It is the objective of this meeting, to develop ideas for the future of polymer science and provide guidance to achieve those ideas.

Polymer science in the last 40 years can be divided roughly into a number of epochs of developments starting with physical chemistry. Organic chemistry was the primary driving force in the 1950s, polymer physics in the 1960s, polymer engineering in the 1970s, and now the most important thrusts in the 1980s are functional and specialty polymers, tailor-made for specific uses. In this present period of forging ahead, the individual disciplines of polymer science are strongly interacting with the life sciences: as indicated by the use of polymers in biological and medical applications, in agriculture and nutrition, but also the use of polymers in energy development and conservation, in communication and information storage and retrieval, and for the general benefit and well being of mankind.

I have asked the speakers of this Symposium to stress the future developments of their disciplines in polymer science, to try to foresee what might be the status of their part of polymer science in the next decade, to try to forecast new developments, new thrusts, and new directions, for courageous researchers, inventors, and innovators.

As you look at the program, you will see that after the introduction and the talks on science policy and the professional society, we will have three very eminent scientists who will discuss various aspects of polymer science. The first two lectures in the afternoon are still involved with the interface of polymer science with medicine and biology. The first speaker will be Professor Katzir, who, together with Mark, was responsible for the creation of the Weizmann Institute, which was founded in Brooklyn and existed there for several months.

The last two lectures on Tuesday afternoon follow Professor Mark's early scientific development: his Ph.D. at the University of Vienna; the talk given by the Director of the Organic Institute of the University of Vienna who is also the President of the Austrian Chemical Society. This talk will be followed by a lecture by the Vice President of the Max-Planck Society, which is the successor of the Kaiser Wilhelm Society, which had the responsibility for administrating the Kaiser Wilhelm Institutes between the two world wars; it was there that Herman Mark had his first significant scientific successes.

The European and world communities in polymer science will pay tribute to the future of polymer science and Herman Mark on Tuesday morning, with talks on photochemistry, stereoselective polymerization, and new composite materials. Finally, on Tuesday afternoon, we will have talks by selected students and colleagues of Herman Mark of the early period of the Polymer Research Institute of the Polytechnic Institute of Brooklyn, an Institute founded by Herman Mark in 1942.

In the early and middle 1940s Herman Mark became very much involved with the publication of journals and books on Polymer Science with Interscience Publishers, now John Wiley and Sons, Incorporated. The *Journal of Polymer Science* was founded, which now has four parts. Herman Mark edited books on high polymers and brought out the *Encyclopedia of Polymer Science and Technology* which is now in its second edition.

It was the Polytechnic Institute of Brooklyn and the Polymer Research Institute which served as the operating basis of Professor Mark, from where he made his enormous contribution to polymer science in general and to the scientific and intellectual community of the world.

Herman F. Mark's accomplishments are too numerous to mention in detail. They include the publication of about 600 papers and articles, and about 20 books. These accomplishments were recognized by: 20 honorary doctoral degrees, 45 memberships and honorary memberships of academies and professional societies, and 40 medals and awards, including the National Medal of Science.

THOMAS KLESTIL

Ambassador of Austria in Washington, DC

Ladies and gentlemen, we are assembled here tonight to honor one of the greatest scientific pioneers of our time. This morning I had already the great privilege to convey to Professor Herman Mark on behalf of the Austrian Institute of Science and Research, and on my own behalf, the best wishes of the Austrian government on his 90th birthday. Professor Herman Mark's extraordinary and most distinguished career has earned him a unique place in the scientific community as the "founding father" of polymer science. His international reputation as an innovator and teacher is unparalleled and his academic achievements are recognized not only in this country, but practically all over the world. His outstanding work is the result of a rare combination of genius, knowledge, diligence, and dedication.

It is with particular pride and at the same time with special gratitude that we claim Professor Mark as one of the great sons of Austria. Among the many Austrians and former Austrians who had emigrated from their native country and settled in the United States Professor Mark is not only one of the most famous but also one whose human and intellectual qualities became most apparent in the face of adversity. Beyond all the vicissitudes of time and the versatility of man that shaped his life Professor Mark's genius has triumphed. He has changed our perception of the physical world and increased our mastery of it. The impact of his work has demonstrated the power of scientific innovation on the life of mankind.

The more we recognize the interaction between science, technology, and society the more we become aware of how imperative it is to reconcile scientific and technological progress with human aspirations and social development.

Modern science is changing man's view of himself. We no longer see ourselves as merely observers, but the movers and shapers of this modern world, as free and responsible agents of change. Science and technology have become a distinct human experience, an adventure and a challenge to create a better world. In this sense, scientific innovation and technological progress have to become an enactment of human rationality in history, a portrayal of vision of our future life, and a choice of means for moving toward it. The human imperative, man's responsibility toward history, thus becomes the commanding criterion with which we may rein the rapid, exponential advances of science and technology—but it is also this human imperative that makes these advances possible.

Professor Mark's life and scientific achievements bear witness to that dual impact of great innovators on mankind. They also testify to his human greatness and compassion which found such overwhelming recognition by the many honors he received during his long and so productive life. The presence of so many of his friends, colleagues, and admirers here tonight show the extraordinary degree to which Professor Mark has succeeded to combine his pursuit of scientific excellence with the pursuit of human goals and values.

It is for this reason that the President of Austria has decided to bestow upon Professor Mark one of the highest Austrian decorations, the "Knight Commander's Great Golden Cross First Class." It is a great honor and privilege for me to present to you, Professor Mark, this decoration. Please accept it as a special recognition of the admiration your native country and its people have for you and your tremendous scientific and human achievements.

FEDERAL SCIENCE AND TECHNOLOGY POLICY

JOHN P. McTAGUE
Office of Science and Technology, Washington, D.C. 20506

SYNOPSIS

During the first four years of the Reagan Administration the groundwork was laid for a U.S. science policy which will endure for many years. The basic principles of this policy are that basic research is vitally important, technical talent must be stimulated and nurtured, excellence in research and teaching must be encouraged, partnerships among industry, universities and government must be stimulated, the development of "leap frog" technologies should be pursued, and science policies and funding must be made consistent.

In the first four years of the Administration basic research moved from the smallest to the largest component in nondefense R&D expenditures. Government funded basic research in universities grew 26% in real terms during that period. New programs have been started which emphasize engineering education and research, and improved interactions between universities, Federal laboratories, and industry.

It is a pleasure for me to be here today as the White House spokesman at this International Symposium on Polymer Science in the Next Decade, honoring Professor Mark. It is special to me because, as a chemistry graduate student doing research in polymers, I kept seeing the influence of his work just about everywhere. To quantify that observation, over the weekend I picked up my dog-eared copy of Professor Flory's 640-page *Principles of Polymer Chemistry*. The first reference to Herman Mark is on page 22, and the last is on page 615.

But those references are to the past. I have two tasks today, the first of which is to look to the future. What *is* the nation's science policy for the coming decade?

Before we try to chart a course for the future, it is only prudent to examine our past and present activities. Are they coherent? Do they follow a plan? In other words—do we have a science policy?

Sherlock Holmes once remarked that "It is a capital mistake to theorize before one has data." I hadn't realized how well he understood the Washington scene.

Taking his warning to heart, just about this time last year we took a broad look at the data to see if our deeds matched our words. Perhaps the most concise statement of the federal government's current science and technology policy is contained in Figure 1.

This focuses on the drastic changes in the allocation of nondefense research and development over the past four years. I want to emphasize that these

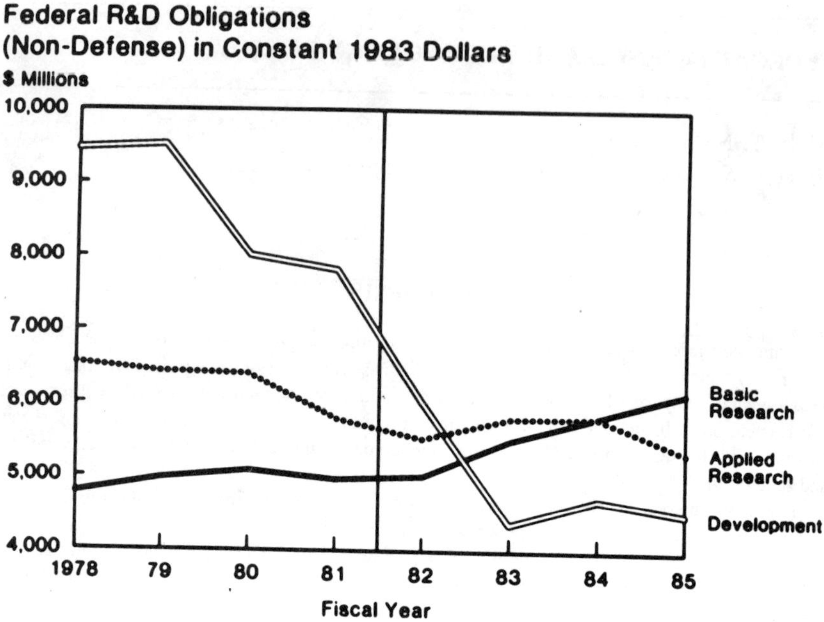

FIG. 1. Federal R&D obligations (nondefense) in constant 1983 dollars.

graphs show constant 1983 dollars, which means that the curves have been corrected for inflation and represent actual purchasing power.

Although it is not shown on this particular graph, the curve for the overall expenditures for civilian R&D during the recent period of constrained federal budgets is relatively flat. But essentially level funding actually masks two dramatic, countervailing trends—a steady *drop* in development and a steady *rise* in basic research. The Reagan administration has made particularly heavy reductions in those well-intentioned but incredibly expensive and misguided energy-related demonstration projects left over from the 1970s. For the most part, such investments are far more appropriately left to the private sector.

So when we look here at the three categories of funding in nondefense areas—basic research, applied research, and development—we see a startling shift in relative priorities. That is clearly *policy*. You can see how basic research, *for the first time ever,* has gone from the smallest fraction of that nondefense R&D to the *largest*. At the same time, development funding has dropped by nearly 40% (Fig. 2).

Finally, we see the growth in *basic research* funds that go *specifically* to universities and colleges. Here, clear policy is even more pronounced. Although it is not shown on this graph, there had been a consistent decline in basic research funding for universities all the way back to 1968. However, from 1981 to the level for 1985, this support for universities grows by 30%—again, in real terms.

To reiterate, over the past four years the administration has *de-emphasized* federal involvement in development projects, the phase of R&D closest to commercial production. At the same time it has substantially *increased* support of

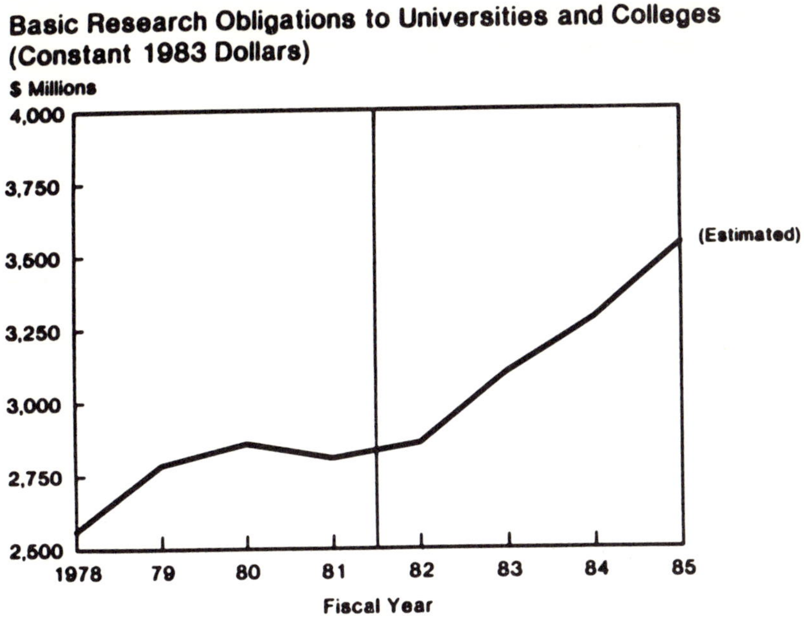

FIG. 2. Basic research obligations to universities and colleges (constant 1983 dollars).

basic research in just about all disciplines of physical and life science, as well as engineering.

In thinking about science policy, we have to keep firmly in mind that we're serving the *national* good, not the good of science, as we allocate taxpayer funds for R&D. Because one of the critical issues before government today is how well we're prepared—and preparing—for a world environment that looks considerably different in 1985 than it did only a decade ago.

The driving force behind so much re-evaluation and change in our approach to R&D is the sheer pervasiveness of science and technology in the world today. And one inescapable manifestation of that change is the degree to which industry rises or falls on its technological advantages—and the rise of fierce foreign competition for sales in the world market and especially in our own very large domestic market. Not very long ago U.S. industry did not have to worry about that competition. We dominated the introduction of new industrial technologies, which had always been the core of our commercial success, and we had always been able to develop and introduce them at pretty much our own pace. But anyone who drives a car, uses a copying machine, or owns a television set knows we don't have that luxury of self-pacing any more. Now it's a challenge to stay at the head of the pack, and that means much more aggressive use of our technological resources. The reason is plain enough: Technology and talent are virtually our only clear competitive advantages in a world where the dollar may be permanently overvalued, where foreign governments are subsidizing capital costs, and where foreign labor may hold an order-of-magnitude advantage over us.

In the decades after World War II, technology spinoffs from federal R&D,

particularly defense R&D, helped lay the foundation for some of today's most successful industries—such as computers, semiconductors and integrated circuits, and commercial jetliners. So in those days industry was strongly stimulated by and benefited from the federal role in R&D. But once again, competition has reared its inevitable head, and the tables have turned.

In the couple of past decades the commercial technology market has expanded tremendously. Today it is *industry, not government,* that's pushing hardest at the frontiers of technology. In fact, so convincingly have the tables turned that the federal government now relies heavily on *industry* to provide *it* with the technology it needs.

But at the *same* time there's been an equally striking *increase* in industry's reliance on two things it *can't* produce for itself and that *only* government can produce. *One* is the fundamental knowledge that ultimately drives the process of industrial innovation—tomorrow's advance in polymer science, for example. The other thing industry relies on government for is the new *technical talent* who can use that knowledge to respond to technological and industrial opportunities. So the more that industry has become the prime developer of new technology, the more it has become dependent on universities. Ironically, in the 1980s, the ivory towers are emerging as the bulwarks of industrial competitiveness.

As I said earlier, the Reagan administration recognized that changed relationship and took decisive steps over the past four years to adjust the federal role to this new reality. The administration, with strong bipartisan congressional cooperation, beefed up the support for the science and engineering on which industrial technology draws—basic research. At the same time, government has continued to get *out* of the development of technology that industry is far better qualified and motivated to do. The most striking example of how government can flounder—at great taxpayer expense—was the well-motivated but naive rush of the federal government into the development of new energy technologies in the late 1970s. For the billions and billions of dollars we spent during that time, our biggest return was a lesson in "Economics 1A"—the law of supply and demand as it operates in the marketplace.

I believe the record of the past four years speaks for itself, but in spite of the changes we've already brought about, there's much more to do. So let me outline the major concerns that have to characterize federal R&D programs in coming years.

First, government has a responsibility to help our colleges and universities create an environment for attracting and retaining faculty of the highest quality. In simple terms, that means improving the *climate for research* on campus so that ambitious scientists and engineers won't feel compelled to take jobs in industry if they want to be competitive in research. In the long run, an academic brain drain is devastating for universities and industry alike, because universities simply won't have enough faculty to teach students, particularly in the fastest moving areas of science and engineering. And without newly trained talent, industrial technology will surely dry up.

Of course, the best help of all for that academic climate has been the increased support for university research projects. But over the past few years we've taken important *additional* steps to improve that campus environment. These include the rejuvenated funding for research instrumentation, these include programs to improve access of university researchers to supercomputers, and these include the innovative new Presidential Young Investigator Program to attract and retain new faculty in critical disciplines—just to name a few. I should also mention that the White House Science Council is currently nearing the end of a year-long review of the health of American universities. Under the chairmanship of David Packard and Allan Bromley, this panel will be making a number of recommendations for further actions to strengthen universities for their long-term job of generating knowledge and talent to meet national needs and opportunities.

The *second* area of emphasis for government is support for the creation of knowledge. Moreover, we have to pay special attention to those areas of research that are of particular importance to industrial competitiveness. For example, today the newly emerging biotechnology industries exist because of decades of federally sponsored basic research in molecular biology. We should be making sure now that we're investing in comparable kinds of frontier research to fuel tomorrow's new technologies—whatever they may be.

Emphasis number *three* is some changes in government's traditional way of supporting university research almost exclusively on a narrow, project-by-project basis. That method, while very successful for progress within disciplinary boundaries, is severely limited in its usefulness to broad areas of industrially important topics. Working in partnership with both academia and industry, government should help establish means for university faculty and students to work on forefront industrial problems—which these days happen to be some of the most challenging scientific and technical problems as well. We have to find mechanisms for students to learn how to solve problems that have multidisciplinary dimensions. That is, real-world problems.

I'm particularly excited about the new program in the National Science Foundation to establish cross-disciplinary engineering research centers. In actuality, I think these centers will be the forerunners of university-based institutes covering a broad spectrum of engineering and science research. They'll operate in parallel with the strong disciplinary programs on campus, but they will be places where students can get the kind of experience in problem solving that just doesn't exist at universities today, and where industry can bring its perspectives to defining the kinds of theoretical developments that are likely to affect how science and technology can be used.

The interest of the universities in participating in this kind of program is incredibly high. In the first round of proposals for these engineering centers, NSF was inundated by the response—141 proposals from 106 different universities, totalling some $2 billion in requests.

The *fourth* major responsibility of government is to find better ways to stimulate the flow of ideas, expertise, and people between the federal laboratories, universities, and industry. And here I have to point out that one-sixth of the nation's scientists and engineers are employed in federal laboratories, which themselves have a combined annual budget of some 18 billion dollars. They are often superb and, frankly, underutilized resources. We darn well better be finding ways to get substantial industrial benefit out of a federal investment of that magnitude.

As I said, the administration has addressed all of those challenges over the past four years, and we've had many reassuring successes. The President's science and technology policy clearly identifies these objectives, and that gives us a means to target them as we work with the individual agencies in developing programs. But my continuing concern is that we should be doing more, that our piecemeal approach fails to truly capture the potential that a 55-billion-dollar federal R&D program holds out. We want to see a better balance in federal R&D spending between essential mission requirements and the strong science and technology needed to support American technological leadership.

This is a challenge to all of us, and it's one that I believe the science community, the universities, the administration, and the Congress can make great progress on in coming years. Perhaps more than at any time in the recent past, we have a firm sense of the challenges that face us, the priorities we must set in these times of budgetary restraint, and the actions we have to take to build a base for long-term growth and prosperity.

Dr. John P. McTague was appointed Deputy Director, Office of Science and Technology Policy, Executive Office of the President, on November 8, 1983. Following nomination by President Reagan he was confirmed unanimously by the U.S. Senate.

Dr. McTague was born in Jersey City, New Jersey, on November 28, 1938. Upon completion of high school in the New York area he entered Georgetown University and received his B.S. degree in chemistry with honors in 1960, and his Ph.D. in physical chemistry from Brown University in 1965.

From 1964 to 1970, Dr. McTague was a member of the technical staff at the North American Rockwell Science Center. He then became professor of chemistry and member of the Institute of Geophysics and Planetary Physics at the University of California at Los Angeles until 1982 at which time he was appointed Chairman of the National Synchrotron Light Source Department at Brookhaven National Laboratory. Dr. McTague was also adjunct professor of chemistry at Columbia University.

Dr. McTague is the author of over 80 experimental and theoretical papers in condensed matter physics and chemistry. He is a member of the American Chemical Society and a fellow of the American Physical Society, and has served as Associate Editor of the Journal of Chemical Physics. Dr. McTague has received senior fellowships from A. P. Sloan, John Simon Guggenheim, and NATO. In 1975 he was honored with the California Section Award of the American Chemical Society.

Dr. McTague is married, has four children and resides in Potomac, Maryland.

HERMAN MARK: HIS INFLUENCE IN THE AMERICAN CHEMICAL SOCIETY AND ON AMOCO'S POLYMER RESEARCH

ELLIS K. FIELDS
American Chemical Society, and Amoco Chemicals Corporation, Exploratory Research, Naperville, Illinois 60566

SYNOPSIS

Herman Mark and I first crossed paths at the American Chemical Society meeting in Baltimore in the fall of 1939. Both of us were looking for jobs. Mark already had a job as Research Director of Industrial Cellulose Research, Ltd., a subsidiary of the Canadian International Paper Company. I was looking for my first permanent job while serving as an Eli Lilly Postdoctorate Fellow at the University of Chicago with Kharasch. Mark was sought after eagerly by several companies and landed an adjunct professorship at Brooklyn Polytechnic. I went back to postdoc for two more years. During the war I worked on nerve gases, drugs to combat malaria and schistosomiasis, and many other defense-oriented projects. Mark developed an unsinkable, impregnable aircraft carrier of wood pulp and ice, named Habakkuk I. I joined Standard Oil of Indiana in 1950 and again crossed paths with Herman Mark, who was a consultant at Standard. It was in the 1950s that Mark became active in the American Chemical Society of which he had been a member since 1938. In addition to other posts he was elected Chairman of the Polymer Division (1954) and Chairman of the Cellulose Division (1958). Mark was awarded the American Chemical Society Awards in Polymer Chemistry (1965) and in Applied Polymer Science (1976). He has been on the Advisory Board of the ACS journal *Macromolecules* since 1980. In one of his periodic updates on "What's New in Polymer Science" Mark lectured to us at Standard Oil on the outstanding properties of polyimides synthesized at duPont from pyromellitic dianhydride and diamines. This inspired me to prepare new, better dianhydrides and polyimides, although Mark cautioned it would not be easy. Our first attempt, pyrolysis of phthalic anhydride, led us to a new synthesis of benzynes in the gas phase and a fruitful collaboration with Seymour Meyerson, a superb mass spectrometrist. We began the renaissance of pyrolysis for synthesis and mechanisms, still in vogue. Our second try led to formation of free radicals from relatively low-temperature decomposition of silver arylcarboxylates, an inexpensive, stable source of free radicals. We did make biphenyl tetracarboxylic anhydride by pyrolysis of silver trimellitate, but the yield was terrible. Herman Mark was right! It was not easy to beat pyromellitic dianhydride. Our third attempt led to the preparation of new dianhydrides that are stable to high temperatures and to hydrolysis by acid or base. These involve dimers of disubstituted maleic anhydrides that are formed photochemically or by modified Perkin condensations. Several examples are discussed. In all this research, extending over many decades, Herman Mark has been a source of inspiration and a shining light to me and to countless chemists all over the world.

THE AMERICAN CHEMICAL SOCIETY TOWARD THE TWENTY-FIRST CENTURY

I would like to tell you briefly where the American Chemical Society is today and where we hope to be in the future.

The year 1876 saw the founding of the American Chemical Society with 133 charter members. Over the years, the Society has flourished; today it has 135,000 members and assets of $150 million. In 1938, the U.S. Congress granted the Society a national charter, signed by Franklin Delano Roosevelt. It calls on the Society to encourage in the broadest and most liberal manner, the advancement of chemistry in all its branches. . . thereby fostering public welfare and education, aiding the development of our country's industries, and adding to the material prosperity and happiness of our people.

Each year, 3,000 members, both elected and appointed, devote their energies to the ACS through service as officers in the 180 local sections and 30 divisions, and as members of the Board of Directors and the national council. Other members serve on committees and on editorial boards for ACS publications. It is within these groups that policy, goals, and objectives are formulated. That so many people give so much of their time to Society affairs clearly indicates the value they place on what they help accomplish—not for themselves or their profession alone, but for everyone.

Day-to-day operation of the Society is entrusted to a staff of approximately 1,600 located principally in Washington, D.C. and Columbus, Ohio. News bureaus for the Society's weekly news magazine are maintained in five other cities in the United States and Europe. The Society is the key provider of chemical information to industry, government, and academe.

Chemistry is essential to all physical and life sciences. It is dynamic and ever-changing. Reflecting this are its many subdisciplines arising in recent times—nuclear chemistry, geochemistry, cosmochemistry, and biotechnology, to name a few.

The American Chemical Society is made up mainly of member chemists and chemical engineers, and it is to their professional enhancement that much of the Society's work is dedicated. The Society is the world's largest organization devoted to a single scientific discipline; its primary product is scientific information, and its principal activities are collecting and disseminating that information.

Chemical Abstracts Service (CAS), Columbus, Ohio, publishes *Chemical Abstracts* (CA), the world's foremost compendium of chemical knowledge. Since its founding in 1907, CA has published more than 10 million abstracts, and the vast expansion of chemical knowledge is shown in their compilation. It took 32 years to amass the first million, 33 years to reach five million in 1971, and only 13 years to arrive at 10 million in March 1984. Virtually all of the world's chemical knowledge is contained within the pages of CA and the accompanying computerized registry of chemical substances. Now most of this immense store of information can be searched and displayed through on-line computer terminals.

Improved international on-line access to CAS information, and to similar European data, resulted from a 1983 agreement between ACS and a West German scientific organization to link computers to CAS and those at Karlsruhe, using the same search software. This network is being expanded to other nations, notably Japan, eventually giving users around the world access to international scientific data banks via their telephone systems.

The ACS is part of the team chosen to develop and install a multimillion dollar computer system for the United States Patent and Trademark Office (USPTO). The system will automate patent examining and processing. By 1990, USPTO will be the most automated patent office in the world.

Papers published in the Society's prestigious journals form an indispensable foundation for future advances in chemistry and other fields. The worldwide pre-eminence of ACS primary journals is indicated by *Science Citation Index,* which annually ranks science journals by the number of references made to papers appearing in the journals during the year. The *Journal of the American Chemical Society* recently ranked first among chemistry research journals and second among all journals listed in the *Index*.

The Society initiates new journals to fill demonstrated needs. *Langmuir,* with its first edition in January 1985, is the newest of its 19 journals devoted to surface and colloid chemistry.

The increasing computerization at CAS also is evident in ACS journals. "ACS Journals Online" a primary journal data base service, has been offered since June 1983. It provides ready access to the approximately 9,000 research papers published annually by ACS.

The Society's official news-magazine, *Chemical & Engineering News* (*C&EN*), has the largest circulation of any chemistry publication. Each week C&EN reports on national and international events in industry, government, and academe bearing on the chemical enterprise. The Society's primary link with its membership, C&EN, also reports ACS news of importance to members.

In addition to its periodicals, ACS publishes a broad range of books, monographs, symposium series, reprint books, and other specialized volumes, with some 500 titles in print. Out-of-print books and all ACS journals and magazines are available on microfilm and microfiche.

Another large body of information on current research is presented annually at our two national meetings and eight regional meetings. Attendance at a national meeting frequently exceeds 10,000, with about 3,000 technical papers being presented. Approximately 3,000 more technical papers are presented annually at the regional meetings. Thus, this impressive total, added to that of the published papers, means that an average of 18,000 research reports is made public each year through ACS auspices.

Information designed to increase public understanding and appreciation of chemistry and chemists is brought to the attention of news media—and the public—in large part through the ACS News Service. Currently, News Service efforts result in some 6,500 articles in publications having a combined circulation exceeding half a billion. The News Service also arranges television and radio appearances by ACS officers and members. During a lecture tour in Louisiana and Texas I gave press and television interviews. In Miami and Los Angeles I participated on radio shows with telephone call-ins from listeners over several hours. If this exposure continues, I may abandon mass spectrometry and polyimide syntheses for a livelier career. Established in 1919, the News Service was the first effort by any scientific organization to help news reporters interpret technical stories for the public.

The Society's public information program now reaches millions of television viewers through a series of chemistry-related news features and public service announcements produced in ACS headquarters' facilities.

Each week 800 U.S. radio stations broadcast ACS programs, including "Dimensions in Science," the longest-running science program in radio history. ACS programs are heard weekly by mainly prime-time audiences of more than five million. They also are aired worldwide by the Armed Forces Network, Radio Free Europe, and the Voice of America.

ACS is the foremost supplier of educational programs to the international planetarium community. Its first show, "The Loneliness Factor," premiered in 1977, has attracted more viewers than any other program in the history of the planetarium medium. The current program, the fourth for the ACS, explores the history and chemistry of Halley's Comet and is in demand everywhere.

Among other ACS materials promoting understanding of science are films, filmstrips, and numerous specialized publications.

We have close relationships with Congress and key science-oriented federal agencies. The Society, through its Department of Government Relations and Science Policy, monitors federal action on such issues as science and technology, education, occupational safety and health, and environmental improvement.

The Society develops policy statements, presents testimony to Congress, publishes information pamphlets for nonspecialists, and maintains a registry of experts who can provide technical advice to the government. The goal of these activities, in keeping with the Society's congressionally mandated charter, is to serve the public interest.

We are involved intimately with the House Committee Task Force on National Science Policy. On March 26, 1985 the ACS presented a briefing to representatives and their staff on the Bhopal disaster with a panel of four experts from government, industry, and consultants.

The Society also seeks to improve the chemical community's understanding of new federal policies. *The Washington Alert,* an ACS newsletter for members, reports on congressional and regulatory actions, and forums at ACS meetings feature experts who present up-to-date information.

ACS conducts a Congressional Fellows Program designed to acquaint scientists firsthand with operations in the federal government's legislative branch, provide the government with increased scientific and technical expertise, and enhance awareness in the scientific and governmental communities of the value of such interactions.

In all of this work, the Society frequently interacts with other science and engineering societies on questions of science and technology policy.

We are deeply involved in the educational process. Programs begin at precollege levels, go on through college, and extend to lifelong continuing education for the practicing chemist or chemical engineer.

High school and college students can benefit from ACS career publications, including information on financial aid and the employment outlook. For those students who do not plan careers in science, ACS is developing "ChemCom," a modular curriculum presenting chemistry in a societal context. "Project SEED" provides summer jobs in research laboratories for disadvantaged high school students. Other activities for high schools include *Chem Matters,* a popular magazine for students, and teacher workshops on demonstrations and laboratory safety. Through its local chapters, the Society sponsors a variety of other high school-oriented programs. A new program helps students enter the annual International Chemistry Olympiad. About 100 ACS local sections were involved in examinations to select 25 students for participation in intensive courses at the Air Force Academy. Of these, the top four students will represent the United States in the Olympiads at Bratislava, Czechoslovakia this year.

At the college level, student involvement with ACS is open to undergraduates in chemistry and related sciences through affiliate chapters at many schools. With more than a third of all undergraduate chemistry students on their rolls, two-year colleges are offered ACS assistance in meeting the special needs of their students. ACS also helps both schools and employers develop cooperative education programs in chemistry and refers chemistry departments wishing to upgrade their curricula to highly qualified consultants.

The Society develops guidelines for evaluating undergraduate programs in chemistry, examines these programs at institutions seeking ACS approval, and publishes a list of schools meeting its standards. The biennial *ACS Directory of Graduate Research* gives potential graduate students detailed information on chemistry-related programs in the United States and Canada.

To meet the lifelong continuing education needs of its members, the Society offers courses on a wide variety of subjects. Included are in-person short courses, multisite teleconference courses, and audio, video, and computer courses. Grants help college teachers attend the short courses.

A major ACS-sponsored national study of U.S. chemistry education, completed in 1984, calls for sweeping improvements in the effectiveness of chemical education and also for increases in related financial and human resources. We have expanded our video conferencing courses, and will be conducting more teleconferencing of selected symposia from national meetings.

The Society compiles standard examinations in chemistry for use at both high school and college levels and organizes numerous educational conferences, symposia, and workshops.

The ACS conducts manpower and salary surveys to help chemists, chemical engineers, and others keep pace with these aspects of their profession. The survey results can be used by employers and educational institutions in manpower and curriculum planning.

The Society is responsive to its unemployed members and to members in positions for which they are overqualified. Assistance programs include employment clearinghouses at national meetings and a year-round clearinghouse at its headquarters. "ACS Professional Employment Guidelines" serve as recommendations for enhancing relations between employers and employees.

Recognition of chemists and chemical engineers is provided by the Society's national awards program and awards of ACS local sections and divisions. The national awards, renowned throughout the scientific world, include the Priestley Medal, highest award in U.S. chemistry. At the Miami ACS meeting I had the pleasure of presenting the Priestley Medal to Henry Taube. In addition to some 40 national ACS awards to individuals, more than 150 others are presented each year by local sections and divisions. These awards provide prizes ranging up to $15,000. Most are sponsored by industrial organizations.

Recognition of a different sort is provided by grants from the ACS-Petroleum Research Fund (PRF) in support of advanced scientific education and fundamental research. Commitments for ACS-PRF grants are now at the $9 million level annually, based on income from the PRF Trust. The fund was established in 1944 by seven oil companies as a charitable, scientific, and educational trust. More than 300 institutions now have active PRF grants.

Although primarily an American institution, the Society is becoming more international in scope every year. In recognition of this, a section was added to its constitution in 1979. "To foster the objectives specified in this Article, the Society shall cooperate with scientists internationally and shall be concerned with the worldwide application of chemistry to the needs of humanity." About 40% of the Society's general revenues now come from foreign sources—mainly through sales of *Chemical Abstracts* service products and ACS journals. Nearly 9% of ACS members are citizens of other nations.

ACS interaction with foreign scientific organizations is increasing. ACS initiated the biennial meetings of chemical society presidents, now attended regularly by more than half of the world's 65 leading societies. An increasing number of major ACS meetings are held jointly with foreign chemical societies. In addition, ACS officials meet with their counterparts in other countries and with foreign government officials involved in scientific matters.

Where is the American Chemical Society headed? In information, we will become more automated; journals and *Chemical Abstracts* clear back to first issues will be on-line and easily accessible. National meetings will continue as they have in the past, two per year. We are slotted in various cities through 1995 for national meetings. Teleconferencing of special symposia will increase, with possible international involvement. A new journal on energy and fuels is being researched thoroughly.

The 3rd North American Congress is definite for Toronto, June 1988, with Canada and Mexico. This will be a regular ACS national meeting, to be held somewhat later than the usual spring meeting.

Increasing the public's perception of chemists and chemistry is a goal to which we all are devoted. I'd like to mention just a few facets of this area. *ChemCom* tries to reach the community at large, nonscientific and mostly nontechnical. It explains what chemistry is and how it is integral with our entire life style. We have received sizeable sums from federal agencies in this work. Another priority is to establish a program of chemistry and science education from kindergarten through college. Let's make children aware that chemistry is fun and that it touches every aspect of their lives. A new office of pre-high school chemistry and science education has been established to coordinate efforts with other societies such as the AAAS.

Besides the Bhopal briefing, we have held several other briefings for congressional science and technology committees and staff. More are in the planning, to keep our legislators aware of the great resources of the ACS.

On the international front, we have signed protocols with the Soviet Union on joint research and exchange of chemists in the area of catalysis and the environment. Similar agreements are being drawn with the Chemical Society of China. The Pacific Basin Chemical Congress in Honolulu 1984 was so successful, I presently am proposing to Professors Tsuruta and Lemieux, presidents of the chemical societies of Japan and Canada, that we hold another such congress in 1989 or 1990, then again in 1994 or 1995. Both have responded favorably.

In conclusion, I believe these items suffice to show that the ACS is dynamic, constructive, progressive, and restless, ever seeking to fulfill and enlarge the missions of its Congressional charter. I thank Professor Vogl and the rest of the organizing committee for inviting me to this symposium and I thank Herman Mark for being such a shining example and inspiration to me and countless chemists in the ACS over many decades.

MY INDEBTEDNESS TO HERMAN MARK

LINUS PAULING
Linus Pauling Institute of Science and Medicine, 440 Page Mill Road, Palo Alto, California 94306

SYNOPSIS

Herman Mark was led to make his important contributions to the understanding of the nature of colloids by his work, mainly in the early 1920s, on the determination of the structure of crystals by the x-ray diffraction method. I discussed his crystal-structure investigation in my paper "Herman F. Mark and the Structure of Crystals," published in *Polymer Science Overview* in the ACS Symposium Series of the American Chemical Society, in 1981. He was one of the most active workers in the field between 1923 and 1928. He tied for third place, with fifty references quoted in the first volume of the Strukturbericht, which covered the period 1913 to 1928. In his x-ray work Mark showed a great range of interest. He and his collaborators studied elements, minerals, inorganic compounds, simple organic compounds, condensed gases, and macromolecular substances, and in addition they studied the physics of x-rays and of the diffraction phenomenon.

For part of this early period he was working in the Kaiser Wilhelm Institute of the Chemistry of Fibers, in Berlin-Dahlem. He had many collaborators who became well known: R. Brill, O. Hassel, J. R. Katz, K. H. Meyer, M. Polanyi, Leo Szilard, K. Weissenberg, and E. Wigner. His work on fibrous macromolecular substances began in 1925, with his publication, together with Katz, of a paper on cellulose. He continued to work on cellulose and also on rubber. He had developed excellent ideas about the nature of rubber and the explanation of its extensibility and elasticity. He also, with his collaborators, made good x-ray fiber diagrams of silk fibroin in 1920. He and K. H. Meyer suggested that silk fibroin contained polypeptide chains in the extended configuration. He was the principal supporter of the idea that colloidal substances contain macromolecules, and he strove vigorously to uphold this idea, which, of course, turned out to be correct.

Herman Mark is one of my best friends. It is now nearly sixty years since I met him and benefited by his advice. He is six years older, which means that when I was 25 years old he was 31, and he seemed to me almost to belong to an older generation. At any rate, he had had much more experience as a scientist, and I was happy to learn about his work and his ideas. I think that it is likely that his interest in cellulose and fibrous proteins, such as silk, stimulated me to begin to think about the ways in which polypeptide chains are folded.

My wife and I were in Germany for more than a year in 1926 and 1927. I had received a Guggenheim fellowship and was attending lectures by Professor Arnold Sommerfeld and attempting to apply quantum mechanics to chemical problems in the Institute for Theoretical Physics of Munich University. It was during this period that I met Herman Mark. Probably our discussions were mainly about the structure of crystals, a field in which I had been working and he was active.

My main indebtedness to him came as a result of my visit to I. G. Farbenindustrie in Ludwigshafen in the summer of 1930. My wife and I had returned to Europe that spring, in order that I might spend some time in Manchester, where W. Lawrence Bragg was investigating the structure of silicate minerals, and in Munich, to continue my study of quantum mechanics. My students and I had been using the x-ray diffraction method to determine the structure of minerals, including some of the silicate minerals, and I had formulated a set of principles determining the structure of minerals and other inorganic crystals in which the bonds have a considerable amount of ionic character. I had also begun to think about the possibility of introducing refinements into the chemical structure theory of organic compounds. One of these refinements was the concept of resonance of a molecule between two or more valence-bond structures. This concept had been formulated independently by John C. Slater, at the Massachusetts Institute of Technology, and myself. I recognized that it accounted in a rather simple way for many of the properties of organic compounds, such as benzene. Aromatic compounds do not show the striking properties of unsaturation that are shown by aliphatic compounds containing double bonds. The theory of resonance immediately suggested that the resonance between the two Kekulé structures of benzene would stabilize the molecule to such an extent as to make it more resistant to hydrogenation than an unsaturated molecule such as cyclohexene.

The theory also permitted predictions to be made about the structure of the benzene molecule, which had not yet been determined experimentally. The theory led to the prediction that the benzene ring is planar, with the carbon–carbon bonds equivalent; that is, it is a planar symmetrical hexagonal ring molecule. The theory also permitted the prediction to be made that the carbon–carbon bond lengths would be intermediate between those for a single bond, as in the diamond crystal or in ethane and other saturated hydrocarbon molecules, and a double bond, as in ethylene.

At that time the power of the x-ray diffraction method of determining the structure of crystals was limited. Roscoe G. Dickinson in the California Institute of Technology had in 1922, with his student Albert Raymond, determined the crystal structure of hexamethylene tetramine and had evaluated the bond lengths and bond angles, but very little successful work on determining the crystal structure of organic compounds had been reported by 1930, because the number of parameters determining this structure was usually too great to permit their evaluation. I thought that much progress could be made in organic structural chemistry if a way could be found to determine the interatomic distances and bond angles in organic molecules.

When I visited Professor Mark in his laboratory in Ludwigshafen in 1930, he showed me the apparatus that he and his associate R. Wierl had built and which was being used for determining the structure of gas molecules by the diffraction of a beam of electrons, accelerated by a potential difference of about 50,000 volts. The task of determining the structure of an organic molecule from the electron diffraction pattern of the gas was far simpler than that of determining the structure of the crystal by x-ray diffraction, because the structure of the crystal is complicated by the way in which the molecules are related to one another, whereas the diffraction pattern of the gas molecule depends only on the interatomic distances in the molecule itself. Mark and Wierl had studied carbon tetrachloride, benzene, and a number of other substances with this technique, which they themselves had invented. I was overwhelmed by the recognition of the possibilities of applying this technique in the determination of the structure not only of organic molecules but also of inorganic ones.

I immediately asked Dr. Mark if he would object to my building an apparatus of this sort in Pasadena and applying it in the study of the molecular structure of organic and inorganic substances. He said that he had no objection; in fact, he urged me to move along this line, saying that he was not intending to apply the technique in any extensive way, because his job was to work on somewhat different problems. Moreover, he gave me the drawings for the apparatus, in order to facilitate our work.

When I returned to Pasadena in the fall of 1930 I found a young graduate student, just arrived at the California Institute of Technology to work for his Ph.D. degree, and interested in molecular structure. I asked him, Lawrence Brockway, if he would be interested in building this apparatus and determining the structures of some molecules by the electron diffraction method. With the help and advice of Richard Badger, a member of the staff of the Division of Chemistry and Chemical Engineering, he and the instrument makers in the Gates and Crellin Laboratories constructed the apparatus, and soon the workers in our laboratory began to determine the structures of many gas molecules by this technique. The information about interatomic distances and bond angles gathered in this way has been of great value in connection with the development of the modern theory of the nature of the chemical bond.

Herman Mark has made many contributions to science. I feel that his invention of the electron diffraction method of determining the structure of gas molecules is one of the most important of his contributions. It is, at any rate, the one that has had the greatest influence on my own career, and I continue to be grateful to him.

Linus Pauling was born in Portland, Oregon, on 28 February 1901, and was educated in Oregon (B.S. in Chemical Engineering, Oregon State College, 1922) and California (Ph.D., California Institute of Technology, 1925). He was a member of the teaching staff of the California Institute of Technology from 1922 to November 1963, Research Professor of the Physical and Biological Sciences in the Center for the Study of Democratic Institutions, Santa Barbara, California, 1963 to 1967, Professor of Chemistry in the University of California, San Diego, 1967 to 1969, Professor of Chemistry in Stanford University, 1969 to 1973 (now Professor Emeritus), and Research Professor, Linus Pauling Institute of Science and Medicine, 1973 to present. He was George Eastman Professor at Oxford University in 1948 and has been a visiting professor in the University of California, Cornell University, University of Illinois, Massachusetts Institute of Technology, Harvard, Princeton, Madras, and several other universities and colleges.

Much of his scientific work has dealt with the nature of the chemical bond. It has included experimental studies on the structure of crystals by x-ray diffraction and the structure of gas molecules by electron diffraction, the study of the magnetic properties of substances, the investigation of the nature of serological systems and the structure of antibodies, the structure of proteins, the molecular basis of general anesthesia, and the role of abnormal molecules in causing disease, especially abnormal hemoglobins in relation to sickle-cell anemia and other hereditary hemolytic anemias, and abnormal enzymes in relation to mental disease. In addition, he has carried on theoretical studies, especially the application of quantum mechanics to the structure of molecules and the nature of the chemical bond, the extension of the theory of valence to include metals and intermetallic compounds, and the development of a theory of the structure of atomic nuclei and the nature of the process of nuclear fission. During recent years much of his work has been on the application of chemistry to biological and medical problems.

Professor Pauling was awarded the Nobel Prize in Chemistry for 1954 for his research on the nature of the chemical bond and its application to the elucidation of the structure of complex substances. His contributions to chemistry have been recognized also by several other awards, including the American Chemical Society Award in Pure Chemistry, the Nichols Medal, the Gibbs Medal, the Richards Medal, the Gilbert Newton Lewis medal, the Avogadro Medal, the Pasteur Medal, the Pierre Fermat Medal, the Sabatier Medal, the Davy Medal of the Royal Society, The Linus Pauling Medal of the Puget Sound and Oregon Sections of the American Chemical Society, and the Priestley Medal. In 1967 he received the Roebling Medal of the Mineralogical Society of America. On September 18, 1975 he was awarded the National Medal of Science for 1974 by President Ford. In February 1978 the Presidium of the Academy of the U.S.S.R. awarded him the 1977 Lomonosov Gold Medal for his work in chemistry and biochemistry, and in April 1979 he received the Chemical Sciences Award of the National Academy of Sciences, U.S.A.

His discoveries in the field of medicine led to the award to him of the Thomas Addis Medal of the National Nephrosis Foundation, the Phillips Medal for Contributions to Internal Medicine by the American College of Physicians, the Gold Medal of the Rudolph Virchow Medical Society of New York, the Gold Medal of the French Academy of Medicine, the Vermeil Medal of the City of Paris, the Modern Medicine Award for Distinguished Achievement, the Eliasberg and Goedel Medallions in Anesthesiology, and the Rachel Carson Memorial Award. He was the first recipient of the Dr. Martin Luther King, Jr. Medical Achievement Award, given for pioneering work in determining the cause of sickle-cell anemia.

He has been given honorary doctorates by thirty-six universities, including Chicago, Princeton, Yale, Cambridge, Oxford, London, Paris, Toulouse, Montpellier, Liege, Melbourne, Cracow, and Berlin. He was President of the American Chemical Society for 1949 and Vice-President of the American Philosophical Society from 1951 to 1954. He is a foreign member of the Royal Society of London, Associé etranger of the French Academy of Sciences, and an honorary member of the academies of science of Norway, U.S.S.R., India, Italy, Belgium, Portugal, Poland, Austria, Yugoslavia, Romania, and several other countries.

In 1948 he was given the Presidential Medal for Merit by President Truman "for exceptionally meritorious conduct in the performance of outstanding services to the United States from October, 1940 to June, 1946." He is Grand Officer of the Order of Merit of the Italian Republic and recipient of the Medal of the Senate of the Republic of Chile.

On October 10, 1963 he was awarded the Nobel Peace Prize for 1962. He has also received The International Lenin Peace Prize, the Ghandi Peace Prize, the Grotius Medal for Contributions to International Law, the Janice Holland Peace Award (jointly with Ava Helen Pauling), and the Women Strike for Peace Annual Award (May, 1982). In 1961 he was chosen Humanist of the Year by the American Humanist Association. He received the Gold Medal of the National Institute of Social Sciences in 1979, the Vollum Award, and the 1978 Award of Merit of the Decalogue Society of Lawyers.

He has published over 600 scientific papers, about 200 articles on social and political questions, especially about peace, and several books, including *The Structure of Line Spectra* (with S. Goudsmit); *Introduction to Quantum Mechanics* (with E. B. Wilson, Jr.); *The Nature of the Chemical Bond; General Chemistry*; *College Chemistry*; *No More War!*; *The Architecture of Molecules* (with Roger Hayward); *Science and World Peace*; *Vitamin C and the Common Cold* (which received the Phi Beta Kappa prize for best scientific book of the year); *Orthomolecular Psychiatry: Treatment of Schizophrenia*, co-edited with Dr. David Hawkins; *Chemistry* (with Peter Pauling); *Vitamin C, the Common Cold, and the Flu*; and *Cancer and Vitamin C* (with Ewan Cameron).

Professor Pauling in 1923 married Ava Helen Miller, also a native of Oregon. He and his wife have four children, fifteen grandchildren, and five great-grandchildren. Their home is Deer Flat Ranch, Salmon Creek, Big Sur, California. Mrs. Pauling died on 7 December 1981.

CHAIN MOLECULES AT INTERFACES

PAUL J. FLORY
Department of Chemistry, Stanford University, Stanford, California 94305

The presence of an interface imposes severe constraints on the configurations of chain molecules. They are required to be highly ordered in the surface layer, the nature of the order being dependent upon the type of interface. In the interphase region, this order is progressively dissipated with distance from the interface; the isotropy of the fluid in the interior of the phase is approached more or less rapidly depending on the flexibility of the chains and the operative constraints for the given type of interface. These constraints are twofold. First, the space must be discretely filled and, second, the continuity of the molecular chains must be preserved. The analytical study of interfaces became feasible with the ability to describe the configurations of chain molecules in the bulk.

Various types of interphases populated by chain molecules may be treated according to statistical mechanical methods based on appropriate lattice models. In monolayers and bilayers consisting of amphiphilic molecules [1, 3, 4] the initial flux of chains normal to the surface is large owing to the high density of polar head groups at the interface. This flux is dissipated by chain ends, the presence of which diminishes the flux into subsequent layers. Reduction of the flux allows greater randomness and, hence, promotes further incidence of chain ends which, in turn, allows an additional reduction in the flux in the next layer, etc. The concentration of chain ends increases rapidly with depth. Consequently, it is greatest in the midplane of a bilayer. The associated increase in configurational freedom with depth is the source of the well known disorder gradient in monolayers and bilayers.

In the case of cylindrical or spherical micelles and vesicles [2] [5] the lattice model must be adapted to the curvilinear geometry of the micelle whereby the number of lattice cells decreases in accordance with the decrease in area of successive layers. The chain ends are deduced to be nonuniformly distributed, with maximal incidence approximately midway between the center of the micelle and the outer surface. The order parameter changes progressively with distance from the center in a manner dependent on the flexibility of the chain and on the density in the outermost layer. Theoretical predictions concerning the structure of micelles are well confirmed by experiments and by molecular dynamic simulations.

The interphase in lamellar semicrystalline polymers [6, 7] is a case of special interest. Here the flux of chains protruding from the crystalline phase is very large. For long molecular chains, it can be dissipated only by reversals in the directions of the chains that enable them to return to the crystal interface and to be "absorbed" therein. Attainment of isotropy requires dissipation of approximately 70% of the chains through re-entry into the same lamella from which they emerged. Some adjacent re-entry of chains is required in the initial layer. However, the preponderence of the re-entry occurs through longer trajectories which return at a point in the interface appreciably removed from the point of exit. In the case of polyethylene, the thickness of the interphase is predicted to be 10–12Å, in agreement with experiments. The observed inclination of chain axes from the normal to the interface markedly reduces the incidence of regular folding required in the initial layer.

The theory is applicable also to polymer melts bounded by a hard wall [7]. Chain segments in the initial layer must be aligned parallel to the surface. According to theory this initial orientation is rapidly dissipated with distance from the interface. The concepts that have been developed here have been applied to disorder in *n*-alkane crystals and are currently being applied to mixtures of incompatible polymer blends.

REFERENCES

[1] K. A. Dill and P. J. Flory, *Proc. Natl. Acad. Sci. USA, 77,* 3115 (1980).
[2] K. A. Dill and P. J. Flory, *Proc. Natl. Acad. Sci. USA, 78,* 676 (1981).
[3] K. A. Dill and R. S. Cantor, *Macromolecules, 17,* 380 (1984).
[4] R. S. Cantor and K. A. Dill, *Macromolecules, 17,* 384 (1984).
[5] K. A. Dill, D. E. Koppel, R. S. Cantor, J. D. Dill, D. Bendedouch, and S. H. Chen, *Nature, 309,* 42 (1984).
[6] P. J. Flory, D. Y. Yoon, and K. A. Dill, *Macromolecules, 17,* 862 (1984).
[7] D. Y. Yoon and P. J. Flory, *Macromolecules, 17,* 868 (1984).

Paul J. Flory was born on 19 June 1910 in Sterling, Illinois, the son of Ezra and Martha (Brumbaugh) Flory. He attended Manchester College in Indiana where he received the B.Sc. degree in 1931. He was awarded the Ph.D. degree in physical chemistry from The Ohio State University in 1934. Paul Flory and Emily Catherine Tabor were married in 1936. They have three children, Susan Springer, Melinda Groom and Paul John, Jr., and five grandchildren.

Flory was engaged in research in industry before, during, and for several years following World War II. From 1934 to 1938 he was employed by the Du Pont Company in Wilmington, Delaware, where he was privileged to work in the group headed by Wallace H. Carothers whose investigations led to the first synthetic fiber, nylon. From 1938–1940 he was a Research Associate in the Basic Science Laboratory of the University of Cincinnati. He was employed subsequently at the ESSO (now Exxon) research laboratory (1940–1943) and at the Goodyear research laboratory in Akron (1943–1948). In 1948 he accepted a professorship at Cornell University, where he remained until 1957. From 1957 to 1961 Flory was Executive Director of Research at the Mellon Institute in Pittsburgh. In 1961 he joined the faculty of the Department of Chemistry at Stanford University. From 1965 to 1975 he held the J. G. Jackson-C. J. Wood professorship at Stanford. He is now Professor Emeritus.

Professor Flory's early investigations in the field of polymers, commencing in 1935, were concerned with the kinetics of polycondensation and of the chain reactions whereby vinyl monomers are polymerized. Based on the principle of equal reactivity of functional groups established by these investigations, he derived the statistical distributions of molecular species in various types of polymers including non-linear polymers. His work on systems in the latter category accounts for gelation due to the formation of networks. At about the same time (1942), he and, independently, the late Maurice L. Huggins founded the well-known theory of solutions of macromolecules. Flory extended these investigations to the treatment of crystallization of polymers and of rubber elasticity, including the swelling of networks. In 1949 he treated the effect of excluded volume in an isolated polymer chain, and showed it to be non-asymptotic with chain length. He also introduced the concept of the Theta point and demonstrated its universal significance. In the mid-1950s he applied lattice methods to rigid rods and to semi-flexible chains, with emphasis on spontaneous formation of ordered phases (e.g., liquid crystals). Commencing in 1964 he developed rigorous, realistic and versatile methods for analysis of the spatial configurations of polymer chains and their associated properties. This is an area of continuing interest to Professor Flory and his collaborators. Their current work includes also comprehensive theoretical treatment of rubber elasticity and, drawing on earlier work he published in the 1950 s, investigations on liquid crystallinity in polymers.

Flory is the author of two books, "Principles of Polymer Chemistry", (1953) and "Statistical Mechanics of Chain Molecules" (1969). He has published 350 scientific papers and has been granted some 20 patents. In 1974 he received the Nobel Prize for Chemistry and in 1975 was awarded the National Medal of Science.

ENGINEERING NEW PROTEINS

M. F. PERUTZ
Laboratory of Molecular Biology, Cambridge CB2 2QH, England, U.K.

SYNOPSIS

Recent advances in genetic manipulations have given us the tools for making new enzymes: the only thing lacking is the theoretical knowledge needed to design them. In the absence of such knowledge, we had better find out how nature evolves new enzymatic activities and try to follow her example. Bacteria learn to live on new carbon sources if put under selective pressure. This learning process consists of random mutation and natural selection. Do enzymes evolve new activities as a result of amino acid replacements, each generating a small change in chemical affinity or by a few amino acid substitutions in key positions producing large changes in chemical affinity?

Hemoglobins of different animal species respond to very different chemical stimuli. Comparison of amino acid sequences of related enzymes suggest that new enzymes evolve by duplication of the gene of an existing enzyme, followed by mutation of one of the twin genes.

The combination of enzyme biochemistry and protein crystallography has given us an insight into the probable catalytic mechanism of several enzymes. Recent advances in genetic manipulation have made it possible to measure the contributions of individual amino acid residues to the catalytic mechanism.

Chemists seem to have succeeded in modifying the catalytic activity of enzymes by genetic manipulation, but no one so far has engineered a new activity. However, the exploration of enzymatic mechanisms made possible by genetic manipulation is preparing the ground for that next big step. It will take imagination and a great deal of luck.

In the late 1960s, I set an examination paper in applied molecular biology for some future century. It ran like this:

1. Design an enzyme to catalyze the methylation of *n*-propanol at C_2.
2. Formulate the template required for the catalytic synthesis of an ordered copolymer from the monomers isoprene (1), vinylchloride (2), and ethylene (3), such that they are arranged in the sequence 1, 2, 3, 1, 2, 3, etc.

How near are we now to being able to answer such questions? Recent advances in genetic manipulation have provided the tools, but our theoretical knowledge still lags behind our technical capabilities.

All chemical reactions in living cells are catalyzed by enzymes. Ususally each simple reaction or each step in a complex one is catalyzed by an enzyme existing specifically for that one purpose, so that living cells contain thousands of different enzymes. All enzymes are proteins: large molecules made of amino acids strung together into long chains called polypeptides, which are coiled and folded into compact globular molecules. Each enzyme may contain from tens to thousands of amino acids and have a unique, specific, and largely rigid structure. That structure is determined by the sequence of the twenty different kinds of amino acid residues along its chain or chains; the amino acid sequence in turn is specified by genes. Many enzymes also require for their function cofactors such as metals, pigments, nucleotides, or carbohydrates.

The genetic message is written in a four-letter alphabet of nucleotides along a linear scroll of DNA. Each amino acid is coded for by a deoxynucleotide triplet. Of the 64 possible triplets 61 can code for amino acids and three code for the termination of the chain. Though genes determine enzymes, enzymes are not made by them directly. To make proteins the genetic message is first transcribed from one of the two strands of DNA onto a single strand of RNA. This messenger RNA then serves as a template for the synthesis of polypeptide chains in a complex organelle called a ribosome. When the correct sequence of amino acids has been assembled, the newly made chain folds up spontaneously to its specific three-dimensional structure. For example, to destroy bacterial membranes, nature has evolved a chain of 129 amino acids that fold up spontaneously to form the enzyme lysozyme, discovered by Alexander Fleming in his nasal mucus when suffering from a cold.

How does it work? Lysozyme was the first enzyme structure to be solved 20 years ago in a brilliant piece of x-ray analysis by D. C. (now Sir David) Phillips and his colleagues in Sir Lawrence Bragg's laboratory at the Royal Institution in London. When his pupils had solved a structure Bragg used to ask them: "What is it trying to tell you?" Stimulated by his question and an imminent Royal Society discussion, Phillips and Charles Vernon found the answer in one afternoon's intensive work [1, 2]. Lysozyme was known to split polysaccharide chains in bacterial membranes between the carbon atom of one sugar ring and the oxygen that links it to a carbon in the next ring. The structure showed the enzyme to be made of two domains separated by a cleft where it clamps the polysaccharide chain and brings the bond to be split between jaws of chemical pliers. The jaws are made up of an aspartic acid on one side and a glutamic acid on the other side of the polysaccharide chain (Fig. 1). The aspartate carries a negative charge that pushes an electron from the carbon of the first ring toward the bridging oxygen, turning the carbon into a positively charged carbenium ion, and thus making it susceptible to attack by a water molecule or hydroxyl ion. The glutamate donates a proton to the leaving oxygen, forming an OH group. In 1948 Linus Pauling said: "I think that enzymes are molecules that are complementary in structure to the activated complexes of the reactions that they catalyse, that is, to the molecular configuration that is intermediate between the reacting substances and the products of the reaction." The structure of lysozyme and of many other enzymes solved since have borne out his prediction.

Hydrolysis of polysaccharides also takes place in aqueous solution, but would normally require much stronger acids then aspartic and glutamic acid. Why then can they split the polysaccharide at a high rate when they are part of an enzyme? In aqueous solution, ions are solvated by shells of water molecules whose dipoles try to compensate the ions' charge, thus screening acids from the molecules they are trying to attack. To overcome this problem, chemists use organic solvents of low dielectric constant in which strong electrical interactions between the reactants can take place. Enzymes have a largely nonpolar interior which provides the living cell with the organic solvent used by the chemists. The molecule to be attacked, known as the substrate, is drawn into this interior where the polar groups of the enzyme can interact powerfully with it in the absence of screening water molecules. There is another important factor that favors reactions in enzymes over those taking place in solution: interaction between two molecules colliding in solution is weakened by their loss of rotational and translational entropy. On the other hand, once a molecule has combined with an enzyme it is rigidly fixed and no further loss of entropy slows its reactions. This entropic factor can speed the reaction many thousandfold over its rate in solution [3].

Until last year, theories of enzyme catalysis could be tested only by classical chemistry, but genetic engineering has opened the way to testing them by altering the structure of enzymes at will. If the amino acid sequence of a protein is known, its gene can be isolated, not only from the genomes of bacteria or yeast, but even from among the tens of thousands of human genes. Once isolated, the gene can be introduced into *E. Scherichia coli* or yeast or frog eggs and transcribed into messenger RNA which acts as a template for the synthesis of the enzyme. By suitable tailoring of the prefix that is attached to the gene and by a variety of other tricks, micro-organisms have been made to produce such enzymes in high yield. Once synthesis of the wild-type enzyme has been made to work, mutations can be introduced at will. For example, substitution of a single nucleotide base can change an aspartate to an asparagine which has a similar structure, and can still form hydrogen bonds, but lacks a negative charge.

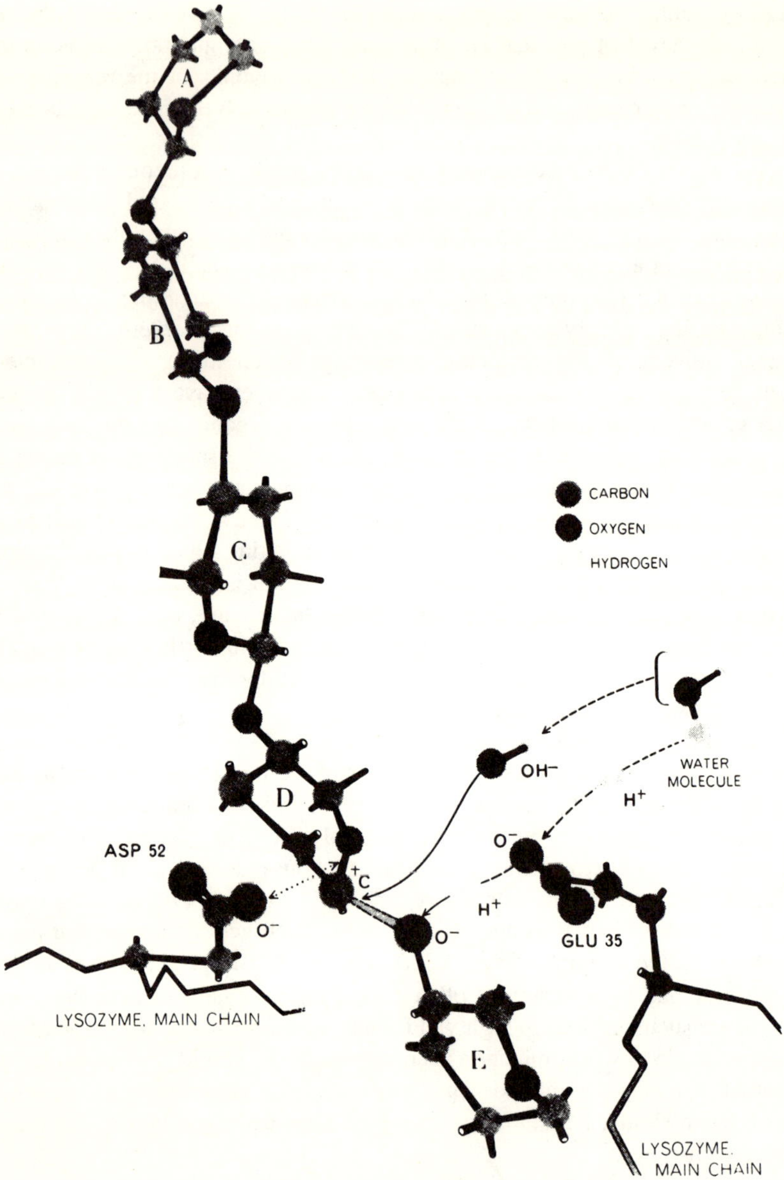

FIG. 1. Active site of lysozyme showing a polysaccharide chain between the "jaws" of aspartic acid 52 and glutamic acid 35, and with an OH^- attacking the carbenium ion (^+C), splitting its bond with O^-. The glutamic acid donates a proton to the leaving O^-. (Courtesy *Scientific American*).

My colleagues, Homme Hellinga and Phil Evans, have used this method to test the reaction mechanism of phosphofructokinase, an enzyme in the glycolytic pathway that transfers a phosphate from adenosine triphosphate (ATP) to fructose-6-phosphate, making fructose 1,6-phosphate [4]. Its active site is tailored so as to juxtapose the 1-OH of the fructose and the gamma phosphate of ATP (Fig. 2). To form what Pauling called the activated complex or what we would now call the transition state of the substrates, the 1-OH has to mount a nucleophilic attack on the phosphate, leading to the formation of a five-coordinated phosphorus atom with three oxygens in a plane, and one oxygen on either side forming the apex of a trigonal pyramid. Hellinga and Evans guessed that the neighboring aspartate might activate this reaction by its negative charge drawing the proton away from the 1-OH, thus making it more reactive. The aspartate could have a powerful electrostatic effect because it is shut off from water and only 3 Å away from the hydroxyl [5, 6]. By a single base substitution in the phosphofructokinase gene, Hellinga and Evans changed the aspartate to an asparagine, which is uncharged, with the result that the *E. coli* extract no longer exhibited phosphofructokinase activity. By similar methods, other biochemists have proved the vital parts played by buried aspartates in the reactions catalyzed by the enzymes dihydrofolate reductase [7] and chymotrypsin [8].

The most extensive exploration of the active site of an enzyme has been done by A. R. Fersht and D. M. Blow and their colleagues at Imperial College, London and G. Winter and his colleagues at the MRC Laboratory of Molecular Biology in Cambridge [9,10]. They have used tyrosyl-t-RNA-synthetase, an

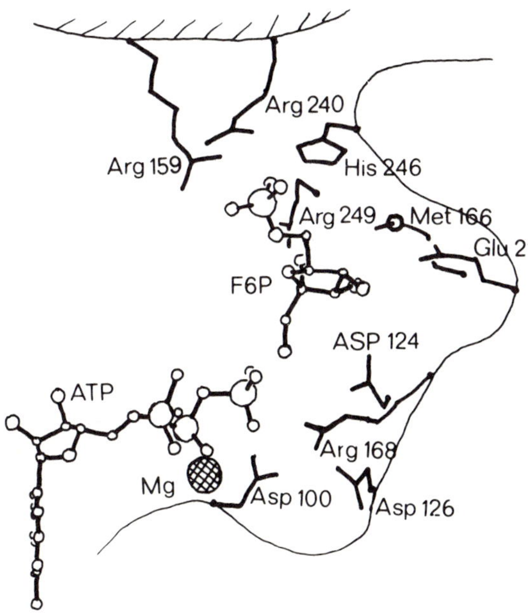

FIG. 2. Active site of fructophosphokinase, showing the substrates fructose-6-phosphate (F6P) and adenosine triphosphate (ATP) with the surrounding amino acids. Aspartic acid 124 polarizes the 1-OH of the fructose to facilitate its attack on the gamma phosphate of ATP. (Courtesy *Nature*.)

enzyme catlyzing two preliminary steps in the incorporation of tyrosine into growing polypeptide chains: the first step is the transfer of adenosine monophosphate from ATP to the carboxylate of tyrosine, forming the energy-rich tyrosine adenylate, and the second step is the transfer of the tryrosine to tyrosyl-t-RNA. X-ray analysis has given a detailed picture of the active site that catalyzes the first step, while the site for the second step is still obscure (Figs. 3 and 4). The structure of the enzyme with tyrosyl adenylate in the active site shows a multitude of contacts between enzyme and substrate, at least eight of them involving hydrogen bonds. How much does each of these bonds contribute to the stabilization of the transition state? The energy of a hydrogen bond between uncharged NH and CO groups is 21 kJ/mol, or about one tenth of the energy of a covalent carbon–carbon bond, but a hydrogen bond in an enzyme-substrate complex contributes less energy, because each of the groups that combine to form a hydrogen bond in the complex also forms hydrogen bonds with water when enzyme and substrate are free. Hence the free energy that stabilizes the transition state is contributed by the difference between the energy of the bond linking enzyme and substrate together and the energy of the bonds that linked water to each of them before they were joined. For example, the tyrosyl hydroxyl of the tyrosyladenylate forms hydrogen bonds with an aspartate (176) and a tyrosyl hydroxyl (34) of the enzyme. Fersht and his

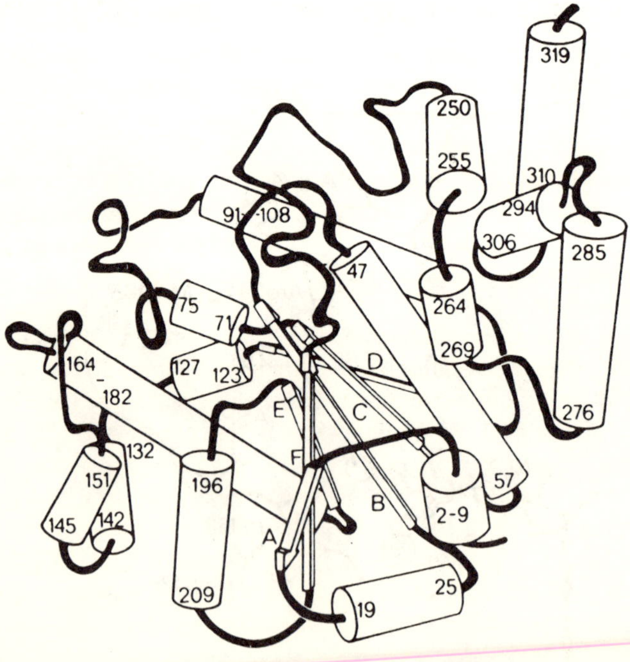

FIG. 3. Three-dimensional structure of tyrosyl-t-RNA-synthetase, showing the fold of its polypeptide chain of 319 amino acid residues. The cylinders represent α-helices and the straight bars β-pleated sheets. The active site, where tyrosyl adenylate is bound, lies between helices 47–57, 164–182 and the pleated sheet. (Courtesy *Angewandte Chemie*.)

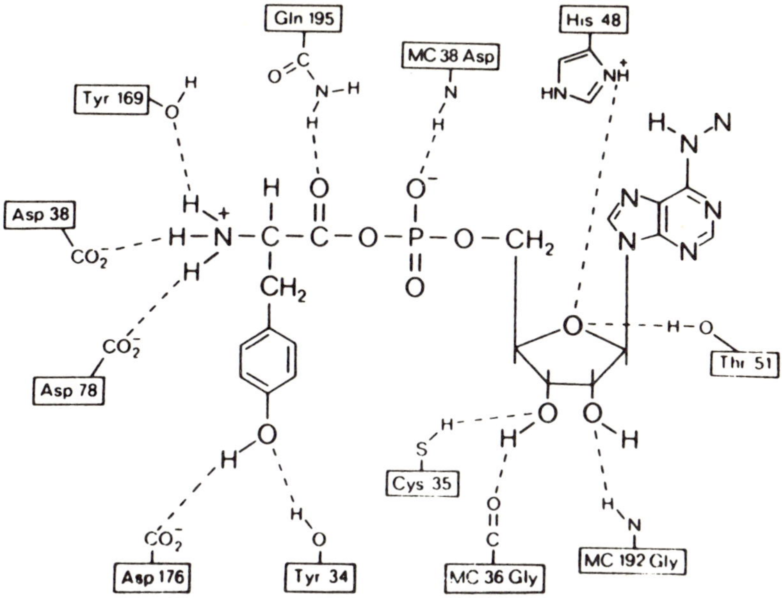

FIG. 4. Active site binding tyrosyl adenylate by means of hydrogen bonds between polar amino acid side chains of the enzyme and polar groups of the substrate. (Courtesy *Angwandte Chemie*.)

colleagues [9,10] measured the energy contributed by those bonds by replacing each of the tyrosines in turn, and both of them together, by phenylalanine. This amino acid differs from tyrosine only by the absence of the hydrogen bond-forming OH. The results showed that the bond from the substrate's OH to the negatively charged aspartate is worth at least 15–20 kJ/mol, an energy corresponding to a thousandfold acceleration in the reaction rate. On the other hand, the energy contributed by hydrogen bonds between uncharged donors and acceptors, such as that between the tyrosine of the substrate and tyrosine 34 of the enzyme, generally amounts to only 2–6 kJ/mol and provides no more than a 10-fold acceleration.

One of the amino acid substitutions made by the team actually increased the efficiency of the enzyme. Histidine 48 and threonine 51 form part of an α-helix facing the active site. Prolines are misfits in α-helices and cause them to bend. Replacement of threonine 51 by proline was therefore expected to alter the contact of histidine 48 with the oxygen of the ribose. When this mutant enzyme was made, it was found to have an improved affinity for ATP and a nearly three times higher free energy of interaction of histidine 48 with the transition state of the substrate. The construction of this mutant of tyrosyl-t-RNA synthetase that is a more powerful catalyst than the wide-type enzyme was a spectacular feat, and raises hopes that this might be the first step toward the construction of new enzymes.

However, the design of radically new enzymes is still utopian. Despite many years of research by able scientists, the rules that determine how a polypeptide chain of a given amino acid sequence folds up in space are still unknown, so

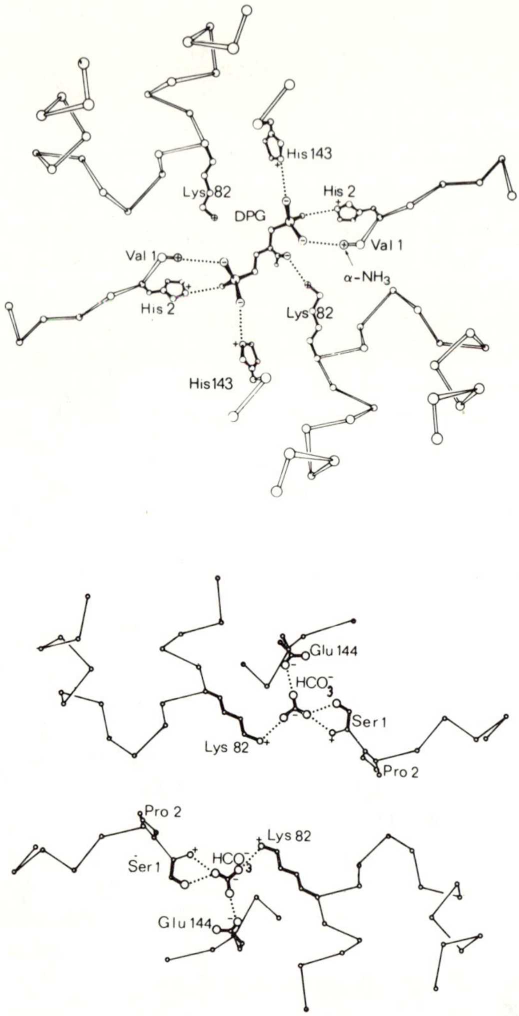

FIG. 5. Top: The allosteric regulator 2,3-diphosphoglycerate (DPG) which has the formula

$$\begin{array}{c} {}^{-}\mathrm{O} \quad \mathrm{O} \\ \diagdown \;\; \diagup \\ \mathrm{C} \\ | \\ {}^{=}\mathrm{HO_3P{-}O{-}CH_2{-}CH{-}O{-}PO_3H^{=}}, \end{array}$$

bound between positively charged amino acid residues of human hemoglobin. In crocodilians, valine 1 is replaced by serine, histidine 2 by proline, histidine 143 by alanine, and lysine 144 (side chain not shown) by glutamic acid. These substitutions suffice to convert the single diphosphoglycerate binding site to a pair of bicarbonate binding sites (bottom). The small circles connected by thin lines represent the α-carbon atoms and amides of the main polypeptide chain. The amino acid side chains binding diphosphoglycerate or bicarbonate are shown in bold lines, hydrogen bonds as dotted lines. In the bottom figure small circles represent carbons, medium circles nitrogen, large circles oxygen.

that we cannot specify a sequence that will make a polypeptide chain form a chosen catalytic site. Failing such knowledge, the best we can do is to find out how Nature generates new enzymatic activities and follow her example.

I have tried to discover Nature's ways by studying the adaptation of vertebrate hemoglobins to different environments [11] and the adaptation of an enzyme to high temperatures in thermophile bacteria [12]. All vertebrate hemoglobins carry oxygen from the lungs to the tissues, but their response to other chemical stimuli varies widely among different animal classes and phyla. For example, the oxygen affinity of mammalian hemoglobins is regulated by an organic phosphate which facilitates the release of oxygen in the tissues. In crocodilians, on the other hand, Christian Bauer at Regensburg discovered that regulation is taken over by bicarbonate ions, so that there is a direct reciprocating action between oxygen and CO_2, the end product of oxidative metabolism. When Bauer told me of his finding, I wondered if such a drastic change in chemical affinity had been brought about by major structural changes in the hemoglobin molecule, involving many amino acid substitutions, or by a few amino acid substitutions in key positions. To my surprise, I found that there was no need to invoke any major structural changes: no more than four amino acid substitutions were sufficient to convert the binding site for one organic phosphate molecule in human hemoglobin into two binding sites for bicarbonate ions in the hemoglobin of Caiman crocodilus. K. Nagai, C. Poyart, and I are now testing these ideas by directed mutagenesis of human hemoglobins made in *E. coli* bacteria [13,14].

M. Brunori and his colleagues in Rome discovered that trout have two different hemoglobins, one whose oxygen affinity is regulated by hydrogen ions and organic phosphates and another whose oxygen affinity is insensitive to these regulators. Comparison of their amino acid sequences showed that no more than five amino acid substitutions are needed to switch these regulatory functions on or off.

Most vertebrate proteins are denatured above 55°C, but proteins of thermophile bacteria that live in hot springs can stand much higher temperatures. H. Raidt and I discovered that the thermal stability of the enzyme ferredoxin in *Bacterium stearothermophilus* arises from only 10 amino acid substitutions which create six additional hydrogen bonds between oppositely charged side chains on the surface of the enzyme, while its internal structure remains unchanged [12].

Patricia Clarke at University College, London, attacked the problem in a different way [15]. She used that most adaptable of micro-organisms, *Pseudomonas aeroginosa,* and put it under selective pressure designed to make it grow on new carbon sources. Her wild-type strain grew well on acetamide, but not on longer chain amides. A single-point mutation that substituted a serine for a phenylalanine in one of the bacterium's enzymes produced a strain that grew on butyramide. A further single mutation produced a strain that grew on phenylacetamide but had lost the ability to grow on acetamide. When B. S. Hartley and others at Imperial College, London, grew *Klebsiella aerogenes* on the sugar xylitol instead of its "natural" sugar source ribitol, this led to a point mutation in the gene for the enzyme ribitol dehydrogenase that increased the enzyme's affinity for xylitol [16].

All my examples go to show that new ways of regulating the activity of proteins or new enzymatic activities can arise as a result of a small number of mutations, leading to a few amino acid substitutions in key positions, while the overall structure of the protein remains undisturbed. What Nature does by random mutations and natural selection, chemists ought to be able to accomplish by design. But rather than designing a new enzyme from scratch, success is likely to come from detailed analysis of the structure and mechanism of an existing enzyme such as Fersht and his colleagues [9,10] have carried out, followed by a few amino acid substitutions in key positions which change the enzyme's substrate specificity or, for example, make it heat resistant. Since it would be a relatively simple matter to introduce a few extra hydrogen bonds between side chains carrying opposite charges on the surface of an enzyme, heat resistance would probably be easiest to engineer. Joe Kraut, John Abelson, and their colleagues at the University of California in San Diego have in fact already tried to do this by an alternative method [17]. They cross-linked two segments of polypeptide chain in dihydrofolate reductase by an S-S bridge and obtained a thermostable, active enzyme. Subtilisin, an enzyme widely used in industry and as an addition to soap powders, is readily inactivated by oxidation. Wells and others found that this was due to auto-oxidation of a single methionine residue. When this was replaced by alanine, the enzyme still retained activity, but was no longer auto-oxidizable [18].

REFERENCES

[1] A discussion on the structure and function of lysozyme. *Proc. Roy. Soc. Lond. B, 167,* 349–449 (1967).

[2] D. C. Phillips, "Three-Dimensional Structure of an Enzyme Molecule," *Sci. Am.*, November, 78 (1966).

[3] A. Fersht, *Enzyme Structure and Mechanism,* 2nd ed., W. H. Freeman, New York, 1985.

[4] H. Hellinga and P. R. Evans, to be published.

[5] P. R. Evans and P. J. Hudson, "Structure and Control of Phosphofructokinase from *B. stearothermophilus," Nature (Lond.), 279,* 500–504 (1979).

[6] P. R. Evans, G. W. Farrants, and P. J. Hudson, "Structure and Mechanism of Phosphofructokinase from *B. stearothermophilus,*" *Phil. Trans. Roy. Soc. Lond. B, 293,* 53–62 (1981).

[7] J. E. Villafranca, E. E. Howell, and J. Kraut, "Directed Mutagenesis of Dihydrofolate Reductase," *Phil. Trans. Roy. Soc. Lond. B,* in press.

[8] E. T. Kaiser, "Better Enzymes by Design,"*Nature, Lond., 313,* 630–631 (1985).

[9] A. B. Fersht, J-P. Shi, J. Knill-Jones, D. M. Lowe, A. J. Wilkinson, D. M. Blow, P. Brick, P. Carter, M. M. Y. Waye, and G. Winter, "Hydrogen Bonding and Biological Specificity Analysed by Protein Engineering," *Nature (Lond), 314,* 235–238 (1985).

[10] A. R. Fersht, J-P. Shi, A. J. Wilkinson, D. M. Blow, P. Carter, M. M. Y. Waye, and G. P. Winter, "Analysis of Enzyme Structure and Activity by Protein Engineering," *Angewandte Chemie, 23,* 467–473 (1984).

[11] M. F. Perutz, "Species Adaptation in a Protein Molecule," *Adv. Protein Chem., 36,* 213–245 (1984).

[12] M. F. Perutz and H. Raidt, "Stereochemical Basis of Heat Stability in Bacterial Ferrodoxins and in Hemoglobin A2,"*Nature (Lond.), 255,* 256–259 (1975).

[13] K. Nagai and H. C. Thøgerson, "Generation of β-Globin by Sequence-Specific Proteolysis of a Hybrid Protein Produced in *Escherichia coli*," *Nature (Lond.)*, *309*, 810–812 (1984).
[14] K. Nagai, M. F. Perutz, and C. Poyart, "Oxygen Binding Properties of Human Hemoglobin Mutants Synthesized in *Escherichia coli*," *Proc. Nat. Acad. Sci. USA*, in press.
[15] P. H. Clarke, "Experiments in Microbial Evolution: New Enzymes, New Metabolic Activities," *Proc. Roy. Soc. Lond. B*, *207*, 385–404 (1980).
[16] B. S. Hartley, B. M. Altosaar, J. M. Dothie, and M. S. Neuberger, in *Structure-Function Relationship of Proteins*, R. Markham and R. W. Horne, Eds. North Holland, Amsterdam, 1976, p. 191.
[17] J. Kraut, private communication.
[18] J. A. Wells, D. B. Powers, T. P. Graycor, and D. A. Estell, "Subtilisin: Alteration of Substrate Specificity and Oxidative Stability by Site-Directed Mutagenesis," *Phil. Trans. Roy. Soc. Lond. B*, in press.

CHEMICALLY ACTIVE AND BIOLOGICALLY SPECIFIC POLYMERS

EPHRAIM KATCHALSKI-KATZIR,* MEIR WILCHEK, and AVRAHAM PATCHORNIK
The Weizmann Institute of Science, Rehovot, 76100 Israel

INTRODUCTION

Biopolymers, in general, and proteins, in particular, show high specificity when interacting with low or high molecular weight compounds. Enzymes display marked catalytic activity when brought into contact with their corresponding substrates. In contrast, classical synthetic polymers, such as polyethylene and polystyrene, lack the biological specificity and the chemical reactivity which characterize the native biopolymers.

An interest in synthetic polymers as well as in biopolymers prompted the authors and their collaborators to undertake the transformation of inert polymers, native or synthetic, into polymer derivatives displaying catalytic activity, biological specificity, or pronounced chemical activity. E. K. -K. concentrated on the preparation and characterization of polymer–enzyme conjugates, namely, immobilized enzymes; M. W. prepared and characterized polymer-ligand conjugates for affinity chromatography; and A. P. prepared chemically active polymers and tested their usefulness as polymer reagents in various chemical reactions.

The following contains a brief description of the preparation, properties, and uses of immobilized enzymes in the laboratory and in industry, the application of the affinity chromatography technique in the isolation and purification of proteins and peptides, and the chemical and physical characteristics which account for the usefulness of polymeric reagents in organic synthesis.

IMMOBILIZED ENZYMES

From the theoretical point of view, the interest in immobilized enzymes in general and enzyme-polymer conjugates in particular derives from the fact that many of the cellular enzymes are embedded within cell membranes or organelles. In this form they act as heterogeneous catalysts, whose activity and mode of action can be elucidated by studying the properties of enzymes artificially bound to well-characterized carriers or embedded in natural or synthetic membranes.

Biochemists are well acquainted with enzymes such as hydrolases, isomerases, transferases, lyases, ligases, and oxidoreductases which catalyze a great

*Also at Tel-Aviv University, Ramat Aviv, Israel.

variety of chemical reactions. The availability of different enzyme-polymer conjugates capable of acting as specific heterogeneous catalysts is thus likely to be of considerable interest to biochemists, organic chemists, and chemical engineers.

Enzymes can be immobilized by gel entrapment, by microencapsulation, by physical or ionic adsorption, by covalent binding to inorganic or organic carriers, or by whole cell immobilization. Of particular interest is the large number of chemical reactions developed for the covalent binding of enzymes via their nonessential functional groups to inorganic carriers such as glass, ceramics, and iron, to natural polymers such as cellulose and Sepharose, and to synthetic polymers such as nylon, polyacrylamide, and other vinyl polymers and copolymers possessing reactive chemical groups [1–5]. Glass and ceramics, for example, can be readily transformed into chemically reactive carriers by treatment with γ-aminopropyltriethoxy silane, followed by coupling with bifunctional reagents such as glutaraldehyde, phosgene, and *p*-nitrochlorobenzene, according to the scheme outlined below. The polyfunctional carriers thus obtained were found to react readily with different enzymes, yielding covalently bound insoluble catalysts which were highly reactive [5].

Polysaccharides such as cellulose, Sepharose, and agarose can be readily activated by cyanogen bromide according to the well-known procedure of Axén, Porath, and Ernback [6, 7]. A reactive intermediate is formed which readily couples with proteins via their amino groups. Relatively stable enzyme-polysaccharide conjugates can thus be obtained.

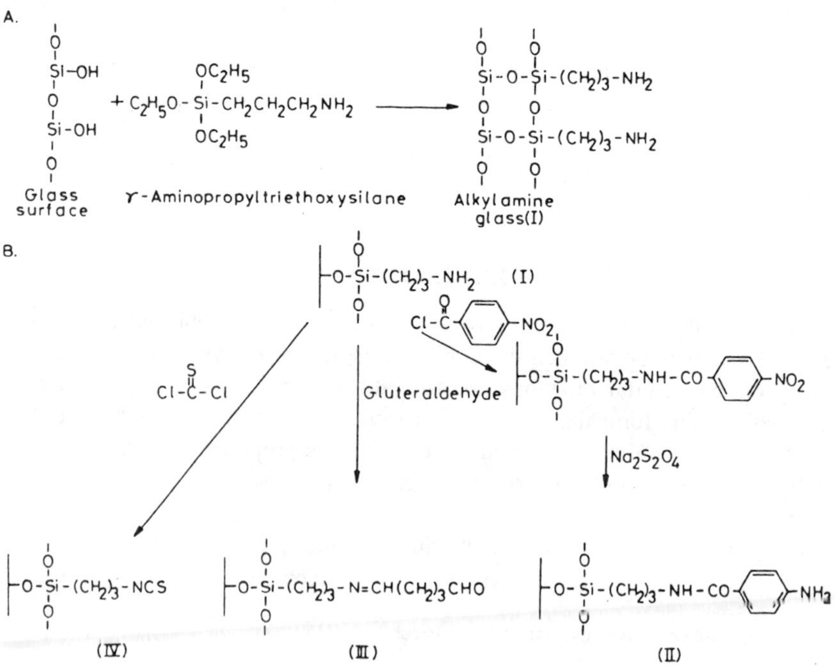

SCHEME 1.

Because of our long-standing interest in polyamino acids, we were particularly intrigued by the recent finding of Minamoto and Yugari [8] that polymethylglutamate can serve as a highly suitable carrier for enzymes such as urease and uricase. Enzyme binding to the carrier was made possible by converting some of its ester groups into the corresponding azides using standard techniques. The bound enzymes showed high activity and remarkable stability.

A new technique for enzyme and cell entrapment was recently developed by Freeman at the Department of Biotechnology of the Tel-Aviv University [9]. It involves preparing a linear polyacrylamide of desired chain length, transforming some of the amide side chains into the corresponding hydrazides, admixing the polymer derivative with the enzyme, bacteria, or yeast, and crosslinking with a dialdehyde such as glyoxal or glutaraldehyde. The method has been found particularly useful for the preparation of gels in which viable cells are entrapped.

The availability of monoclonal antibodies (mABs) which interact specifically with enzymes without affecting their catalytic activity has enabled the recent development of a new technique for the preparation of highly active immobilized enzymes. The method consists of two steps: covalent binding of the specific antibody to a polymer carrier, such as agarose or Eupergit C, to yield the corresponding polymer-mAb conjugate, followed by specific adsorption of the enzyme and formation of the highly active polymer-mAb·E complex. Using this method it was possible to prepare highly active immobilized carboxypeptidase A [10] and guinea pig liver transglutaminase [11].

A list of the various immobilized enzymes prepared so far may be found in References 2–5, 12.

STABILITY OF IMMOBILIZED ENZYMES

One of the most surprising observations made in the early stages of research on immobilized enzymes was that the stability of certain enzymes is markedly increased on their immobilization [3, 4, 13]. In the following we shall restrict our comments to one or two examples which illustrate the marked effect of the carrier on enzyme stability. In the case of glucose isomerase, for example, it was demonstrated that whereas the enzyme derived from *Streptomyces phaeochromogenes* loses practically all of its activity when incubated in solution for 40 h at 70°C and pH 8.0, if the enzyme is immobilized on DEAE-cellulose under similar conditions most of its activity is retained. Another example is β-D-glucosidase: in solution, the native enzyme loses more than 80% of its original activity when incubated for 10 h at 37°C and pH 5.0, whereas the immobilized enzyme, β-D-glucosidase-cellulose carbonate, loses no more than 20% of its activity even after 50 h under similar conditions.

ENZYME REACTORS

The availability of relatively stable immobilized enzyme derivatives in the form of enzyme beads, enzyme capsules, enzyme columns, and enzyme membranes has led to the construction of enzyme reactors of the type presented in

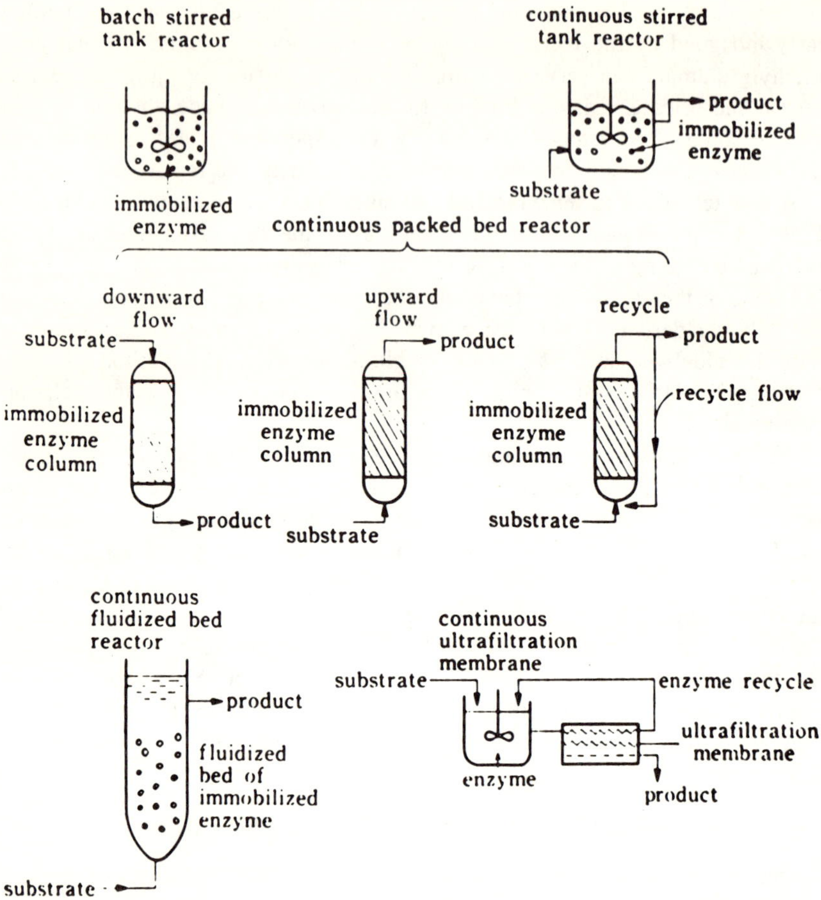

FIG. 1.

Figure 1. Both in industry and in the laboratory the most popular are the batch stirred tank reactors, the continuous packed bed reactors and the continuous fluidized bed reactors.

KINETIC BEHAVIOR

A quantitative evaluation of the kinetics of immobilized enzymes requires a thorough understanding of the factors determining the mode of action of bound enzymes. Conformational changes might occur as a result of the immobilization technique employed; steric effects caused by the matrix might prevent interaction with high molecular weight substrates; partition effects may lead to an increase or a decrease in substrate and/or product concentrations within the domain of the immobilized enzyme; and external and internal diffusional limitations might markedly alter the kinetic characteristics of the bound enzyme.

The last decade has seen considerable progress in the theoretical analysis of the kinetics and mode of action of enzymes bound on surfaces, embedded in porous spheres, and entrapped in membranes [2, 14, 15]. Analysis has clarified the following: (a) the effect on enzyme activity of the microenvironment produced by the carrier; (b) the effect on enzyme activity of the environment produced by the immobilized enzyme itself; (c) the restrictions imposed on an immobilized enzyme system as a result of external diffusion; (d) the restrictions imposed as a result of internal diffusion.

INDUSTRIAL APPLICATIONS

The availability of different immobilized enzymes and the understanding, in broad terms of their mode of action as heterogeneous catalysts has encouraged biochemists and organic chemists to test their possible use as specific catalysts in various organic reactions. A few remarkable successes have been reported. So far mainly immobilized hydrolases and isomerases are being used on a large scale. It seems likely, however, that once adequate techniques become available for economic cofactor recycling, the use of immobilized enzymes will be extended to other organic reactions, particularly those involving stereospecific synthesis of simple or complex organic molecules.

Penicillin acylase (penicillin amidase) from *Escherichia coli*, immobilized by adsorption or covalent binding to organic or inorganic carriers, is used industrially for the production of 6-aminopenicillanic acid (6APA), an important intermediate in the synthesis of semisynthetic penicillins from penicillin G, according to Scheme 2. The preparation of some of the immobilized penicillin

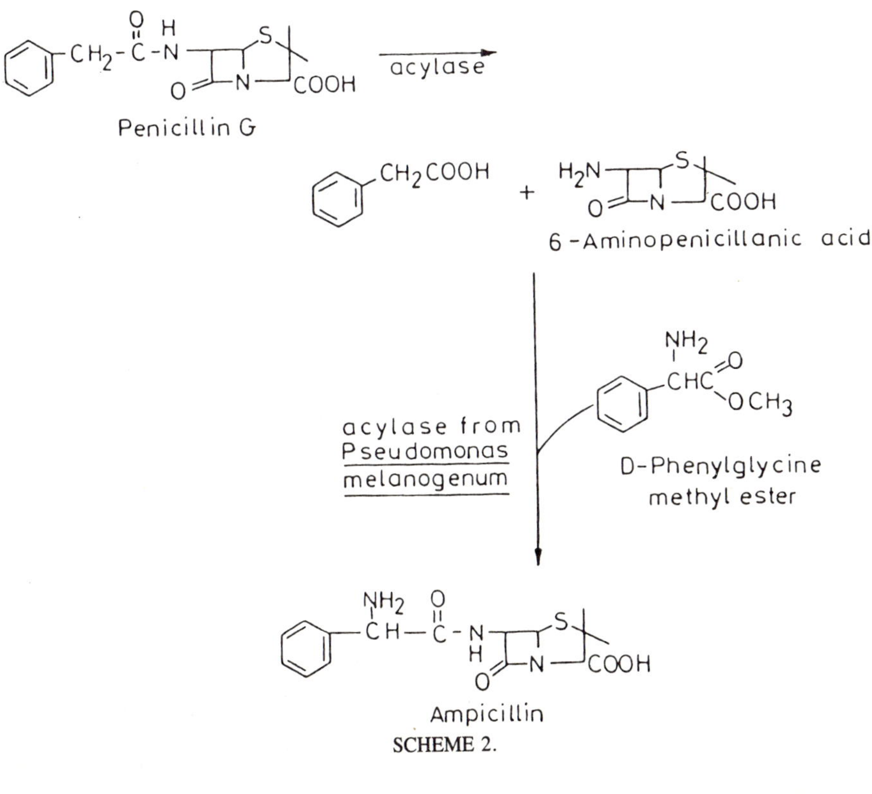

SCHEME 2.

amidases used by the Bayer Company in Germany and the Beecham Company in England is based on a procedure worked out in our laboratory at the Weizmann Institute of Science.

One should note here that immobilized acylase from *Pseudomonas melanogenum* can be used as a catalyst for the synthesis of ampicillin from 6APA and D-phenylglycine methyl ester [4] [see also Ref. 16]. The synthesis in this case undoubtedly occurs as a result of the acyl activation of the phenylglycine residue by the enzyme.

Hydrolytic enzymic reactions similar to those seen with the penicillins can also be carried out with the cephalosporins [17, 18]; the intermediate products obtained, 7-aminocephalosporanic acid (7ACA) and 7-aminodeacetoxycephalosporanic acid (7ADCA), can be used in the enzymic synthesis of new semisynthetic cephalosporins. The formulae describing the enzymic synthesis of cephalexin from 7ADCA and D-phenylglycine methyl ester, using the appropriate cephalexin synthesizing enzyme as catalyst, are given below.

Aminoacylase from *Aspergillus orizae* is known to hydrolyze acetyl-L-amino acids but not the corresponding acetyl-D-amino acids. Chibata and collaborators, working at the Tanabe Seiyaku Company in Japan, were thus able to use immobilized aminoacylase, obtained by binding the enzyme electrostatically to DEAE-Sephadex, as a stereospecific catalyst in the production of native optically active amino acids from the corresponding racemic acetyl-DL-amino acids prepared synthetically [3]. One-thousand-liter aminoacylase columns are used to produce several hundred kilograms of L-methionine, L-phenylalanine, L-tryptophan, and L-valine, within 24 h.

Another industrial process devised by the Japanese workers involves the production of L-aspartic acid from ammonium fumarate using columns packed with immobilized *E. coli* cells showing high aspartate activity [3]. This new technique, based on the utilization of immobilized cells, was found superior to conventional fermentative or enzymatic techniques.

Immobilized glucose isomerase is another enzyme derivative used in the United States, Japan, and Europe for the large-scale industrial production of

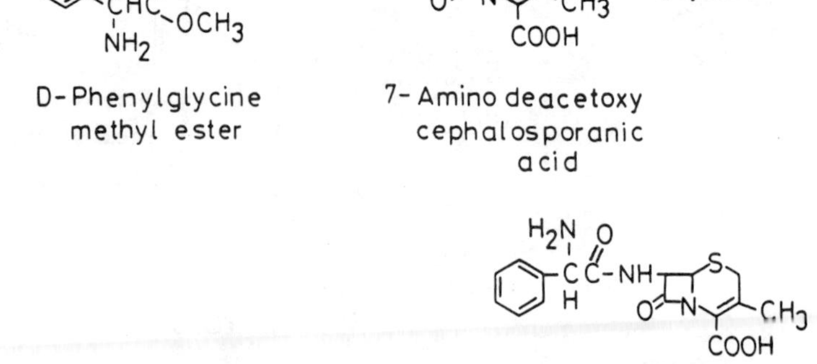

SCHEME 3.

high fructose syrups by partial isomerization of glucose derived from starch [19, 20]. The enzyme is produced by various strains of *Bacillus* and *Streptomyces,* and can be immobilized by adsorption or covalent binding to suitable carriers, or alternatively by immobilization of the microorganisms containing the enzyme. The glucose isomerase preparations mainly used in industry at present are either those in which the enzyme has been immobilized by ionic binding to DEAE-cellulose (a preparation supplied by Standard Brands, USA), or in which microbial cells are immobilized by crosslinking with glutaraldehyde (a preparation supplied by Novo Industri, Denmark) (Fig. 2).

Millions of kilograms of high fructose syrup are produced annually in the United States, Europe, and Japan by the above enzymatic process. Immobilized glucose isomerase is thus to date the most widely used of all the immobilized enzymes. The industrial success of the process undoubtedly derives from the following facts: glucose derived from starch is relatively inexpensive; fructose is sweeter than glucose; and the high-fructose syrup contains approximately equimolar amounts of glucose and fructose, and is thus similar to sucrose in nutritional value.

COFACTOR RECYCLING

None of the immobilized enzymes used in the industrial processes described above require cofactors for activity. However, a significant increase in the

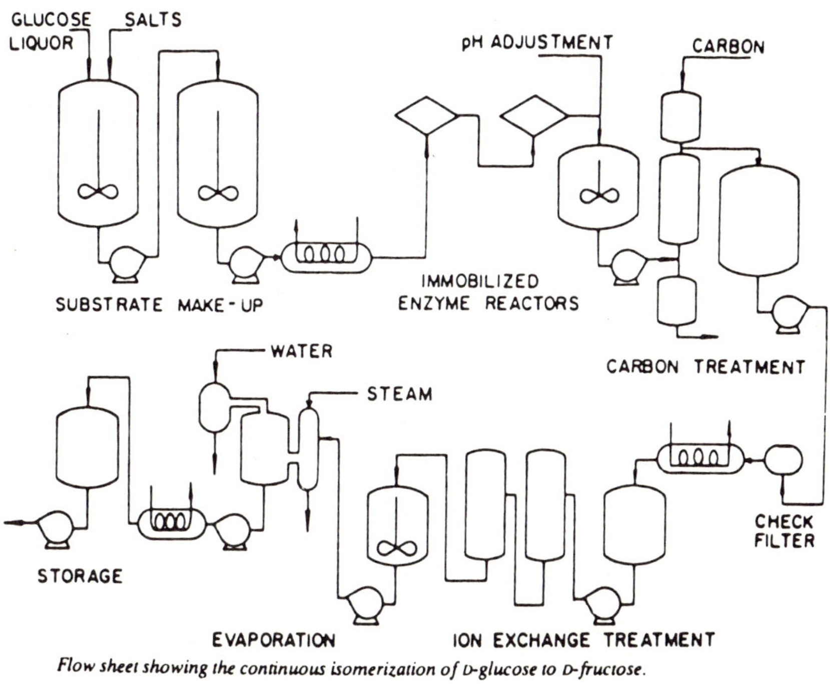

Flow sheet showing the continuous isomerization of D-glucose to D-fructose.

FIG. 2.

number of applied enzyme-catalyzed processes will require the development of efficient techniques for cofactor utilization. Of the 2000 or so different enzymes that have been assigned numbers by the International Union of Biochemistry, more than one third require one of the five adenine coenzymes (NAD, NADP, ATP, FAD, and CoA) for catalytic activity. All of these are rather expensive, and efforts are thus being made to develop efficient techniques for their recycling. This can be achieved by means of enzymatic, chemical, or electrochemical methods [21, 22]. A few examples of the possible recycling of NAD-polymer conjugates in characteristic oxidoreduction reactions are given below.

Wandrey et al. [23, 24] used an NAD-polyethylene glycol conjugate (NAD-PEG) as a nondialyzable water-soluble coenzyme derivative in their cofactor-recycling experiments in which L-leucine was produced enzymatically from α-ketoisocaproic acid and ammonia. A membrane-enclosed reactor containing L-leucine dehydrogenase, NAD-PEG, and formate dehydrogenase, in addition to the dialyzable substrate and product, was employed. Reduction of the NAD-derivative was carried out by formate dehydrogenase in the presence of formate. Under the experimental conditions employed each molecule of NAD-PEG could be recycled about 1800 times before being consumed. Other soluble and insoluble cofactor-polymer conjugates have been synthesized by Mosbach and his collaborators [25, 26]. Some of these—particularly the insoluble ones—have been employed successfully in the isolation and purification, by affinity chromatography, of enzymes requiring cofactors. However, their efficiency seems to be rather low in the enzymic reactions in which cofactor recycling is required.

A chemical redox coenzyme-recycling process can be illustrated by the production of NADH in the yeast alcohol dehydrogenase oxidation of ethanol to acetaldehyde, and its chemical reoxidation using phenozine methosulfate as a redox mediator capable of transferring electrons from NADH to 2,6-dichlorophenol-indophenol [27, 28].

A number of laboratories are currently engaged in the preparation and possible recycling of coenzyme-polymer conjugates in enzymically catalyzed oxidoreduction and synthetic reactions; new developments in the field should therefore be expected in the near future.

USE IN ORGANIC SYNTHESIS

The interest of chemists in enzymes as homogeneous or heterogeneous catalysts has grown with the realization that biocatalysts can enhance a great variety of synthetic reactions. Of particular interest is the fact that many of the enzymes can chemically activate their corresponding substrates and induce high stereospecificity in the reactions being catalyzed. Some of the procedures used in the synthesis of organic compounds with the aid of immobilized enzymes are described in the literature [29, 30].

USE IN ANALYTICAL AND CLINICAL CHEMISTRY

The techniques developed in enzyme immobilization have facilitated the preparation of enzyme columns and enzyme tubes, which can be used repeatedly as specific heterogeneous catalysts in assays worked out to determine the amounts of the corresponding substrates by standard analytical procedures. They have already been successfully inserted into instruments designed to perform continuous automatic analyses in the clinic and the laboratory (for a general review see Refs. 15, 31). Particularly appealing in this connection is the design of enzyme thermistors by Mosbach and collaborators in Sweden [32]. Here the immobilized enzyme system is brought into contact with a thermistor, and the changes in temperature occurring as a result of the specific reaction are measured in a thermally isolated container. Enzyme-thermistor calorimeters have been developed for single measurement as well as for continuous recording of substrate concentration. The methods can be used generally, and the relative simplicity of the instrumentation ensures its wide acceptance.

Firefly luciferase, in the presence of luciferin, ATP-Mg and molecular oxygen, produces light. The specificity of the enzyme for ATP has served as a basis for the development of many quantitative assays for this energy-rich compound. The preparation of an immobilized firefly luciferase by covalent binding of the enzyme to glass [33] made it possible to develop a useful procedure for the routine assay of small amounts of ATP (adenosine triphosphate (up to $10^{-8}M$))with great sensitivity and precision. De Luca et al. [34–36] also prepared immobilized bacterial luciferase and oxidoreductases derived from *Beneekea harvey*. The enzymes were covalently bound to arylamine glass beads which were cemented to glass rods. The resulting immobilized enzyme rods were stable, reusable, specific for either NADH or NADPH, and could be used to monitor reactions producing NADH and NADPH. Picomole levels of malate dehydrogenase, lactate dehydrogenase, alcohol dehydrogenase, glucose-6-phosphate dehydrogenase, and hexokinase have been assayed using such enzyme rods.

Enzyme electrodes [15, 37, 38], a fairly recent development, represent a new group of specific detectors designed for the potentiometric or amperometric assay of substrates such as glucose, urea, amino acids, alcohol, and lactic acid. The electrode consists of a given electrochemical sensor in close contact with a thin permeable enzyme membrane capable of reacting specifically with the given substrate. The enzymes embedded within the electrode membrane yield oxygen, hydrogen ions, ammonium ions, carbon dioxide, or other small molecules, depending on the enzymic reaction which takes place, and these are readily detected by the sensor to yield a response whose magnitude is determined by the concentration of the substrate being studied.

A considerable amount of literature on enzyme electrodes is already available [see, for example, Ref. 15]. Figure 3 shows the structure of the widely used glucose oxidase electrode [39]. It is composed of a Clark electrode for oxygen assay, which is in close contact with a glucose–oxidase-permeable membrane.

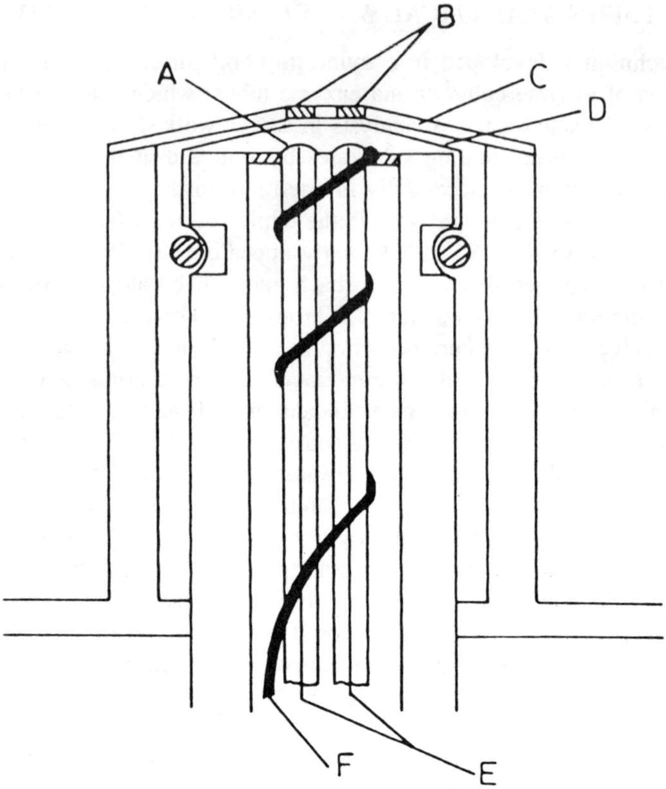

FIG. 3. Glucose oxidase electrode. *A*, electrolyte; *B*, gel-immobilized glucose oxidase; *C*, nylon net impregnated with silicone rubber; *D*, oxygen permeable membrane; *E*, P_t cathodes; *F*, Ag anode.

When the electrode comes into contact with a glucose-containing solution, glucose and oxygen diffuse into the immobilized glucose oxidase. The rate of diffusion of oxygen through the oxygen-permeable membrane into the electrode is gradually reduced as a result of the enzymic oxidation of glucose. A practically linear relationship exists between the concentration of oxygen reduced and the concentration of glucose oxidized when the glucose concentration is well below the K_m for the immobilized enzyme and the oxygen is in non-rate-limiting excess. The Clark electrode thus indirectly records the concentration of glucose in the solution in which it is immersed.

PREDICTED DEVELOPMENTS

As the work of genetic engineers and biotechnologists leads to new methods for the large-scale production of required known enzymes, and even to the biosynthesis of novel enzymes whose catalytic characteristics are not yet known, it will be possible to extend their use in all of the above fields. Basic research is still needed, however in order to strengthen the theoretical foundations on which future applied work will be based. The factors determining enzyme stability must be elucidated and general procedures for enzyme stabilization developed. Further efforts must be made to investigate the structure and mode of action of enzymes embedded in biological membranes, as well as their interaction with adjacent compounds, since an understanding of the action of immobilized enzyme systems *in vivo* will greatly facilitate the design of such systems *in vitro*.

AFFINITY CHROMATOGRAPHY

Affinity chromatography, a method of purification based on biological recognition, has become an important technique for the isolation and purification of biologically active molecules. The basis of the technique is simple. A given biopolymer molecule usually has a native or an artificial molecule that it specifically recognizes. If one of these recognized "partners" is immobilized on a polymeric carrier the "partner"-polymer conjugate can be used to isolate the corresponding biopolymer molecule simply by passing a medium containing the latter through the immobilized counterpart. Molecules not possessing an appreciable affinity for the immobilized ligand will pass unretarded through the column, whereas molecules capable of binding to the ligand will be adsorbed. The desired biomolecules can then be eluted in the presence of an excess of the soluble ligand, or by changing the external conditions such as pH, ionic strength, solvents, and temperature.

Since its introduction in 1968 [40], affinity chromatography has been used widely to isolate and purify antibodies and antigens, enzymes and enzyme inhibitors, hormone-binding proteins and receptors [41].

A new offshoot of the above technique, immunoaffinity chromatography is developing rapidly. This special type of chromatography involves the use of immobilized antibodies, particularly immobilized monoclonal antibodies, for the purification of the corresponding antigens, such as haptens, proteins, enzymes, receptors, and other binding molecules.

The polysaccharide agarose, following its activation by CNBr, is one of the most widely used polymer carriers for ligand binding. The agarose matrix has a loose, porous network which permits easy entry and exit of molecules. It is hydrophilic enough to avoid nonspecific binding of proteins, and it is stable in aqueous solutions of varying pH and ionic strength and in mixtures of organic solvents.

The preparation of an affinity chromatography column is a two-step process. The first step involves a process of activation in which reactive groups are introduced into the chemically inert polymeric resin. In the second step the ligand is covalently attached to the activated resin.

The study of the mechanism of activation of agarose by CNBr (Scheme 4) was facilitated by the development of analytical methods for the determination

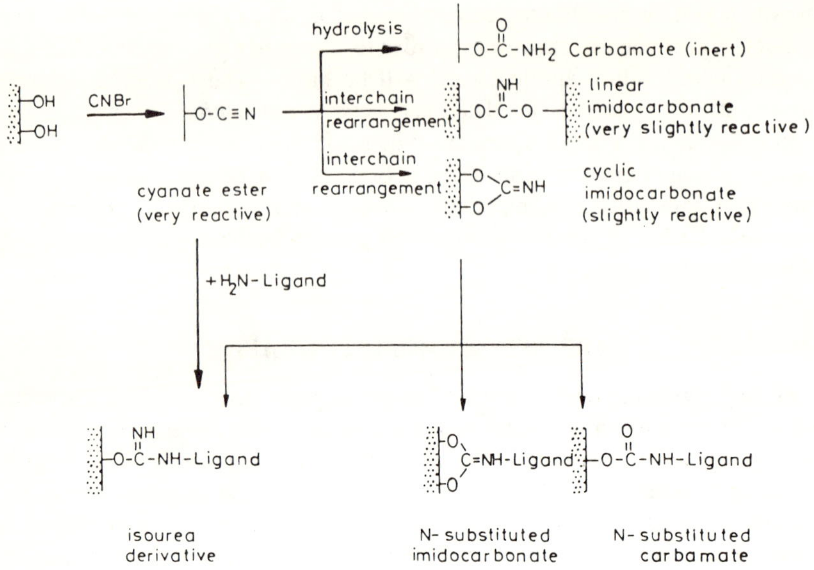

SCHEME 4.

of each of the active and inactive species produced [42]. Three main products were found to be present: carbamates, imidocarbonates, and cyanate esters. The last two are chemically active. Cyanate esters are stable below pH 4.0, whereas the imidocarbonates are most stable under basic conditions. The establishment of this mechanism for activation by CNBr led to the development of new and more efficient activation procedures [43–45]. The coupling of compounds containing amino groups to the CNBr-activated carriers results in isourea bonds, which are characterized by ion-exchange properties and are rather unstable under extreme conditions. Attempts were therefore made to develop new methods of activation and immobilization which would yield stable and uncharged columns. Of particular interest among the newly introduced reagents are *p*-nitrophenyl chloroformate, *N*-hydroxysuccinimide chloroformate, and trichlorophenyl chloroformate [46], all of which form reactive carbonates and, upon reaction with amines, produce stable and uncharged carbamates (urethans) (Scheme 5).

Once a biologically active molecule (e.g., enzyme, antibody, receptor) has been adsorbed onto a corresponding specific ligand-polymer conjugate, it has to be eluted with a substrate analog or salts. Binding to an affinity column may in some cases be so strong that the extreme conditions necessary for its elution destroy the activity both of the column and of the product. Under normal conditions, whereby a compound can be eluted with substrate analogs or salts, the column may be reused as often as 50–100 times; in cases where guanidine or chaotropic salts are applied, on the other hand, the column can be used only once.

Some of the most important classes of compounds purified by means of affinity chromatography are listed in Table I. For a complete list of the

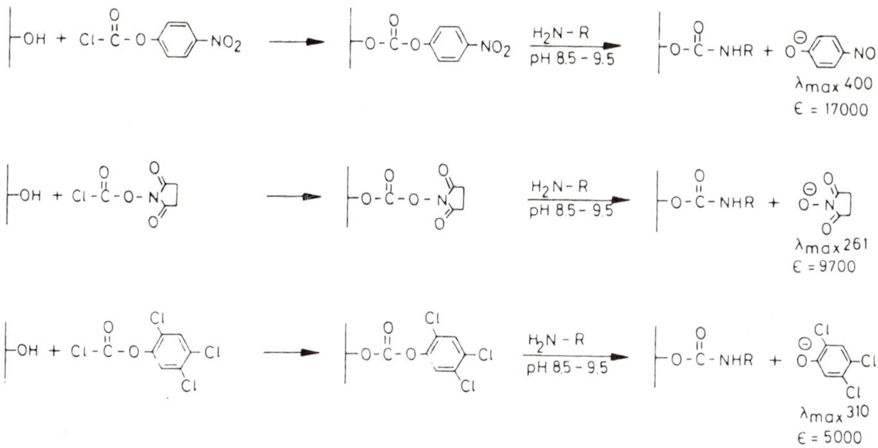

SCHEME 5.

TABLE I
Biomolecules Purified by Affinity Chromatography

1. Antibodies and antigens	8. Receptors
2. Enzymes and inhibitors	9. Glycoproteins
3. Regulatory enzymes	10. Lectins
4. Dehydrogenases	11. Genes
5. Transaminases	12. Cells
6. Hormone-binding proteins	13. Viruses, phages
7. Vitamin-binding proteins	14. Bacteria
	15. Others (miscellaneous)

enzymes, antigens, antibodies, lectins, and receptors isolated and purified by affinity chromatography see Ref. 41, Tables II–IV.

IMMUNOAFFINITY CHROMATOGRAPHY

Affinity chromatography has been used for many years for the purification of antibodies by employing antigens bound to carriers. In recent years there has been a considerable increase in the use of immobilized antibodies for the purification of proteins (including enzymes). Under ideal conditions, antibody columns allow specific proteins to be separated from crude mixtures by a one-step procedure. Antibodies of both the conventional and the monoclonal type have been used. Since a given protein may give rise to the production of a large number of mABs, low-affinity antibodies should be selected for immobilization. The lower affinity allows milder conditions for elution and thus less likelihood of irreversible denaturation. Monoclonal antibodies are being increasingly used for purification in all fields of biochemistry. If the present trend continues and mild conditions for elution become the norm, there is likely to be a shift from affinity chromatography to immunoaffinity chromatography as a standard procedure.

ENZYMES

The enzymes represent the largest category of proteins purified by affinity chromatography. Because of their specificity the variety of available ligands is potentially very wide, and includes competitive enzyme inhibitors, coenzymes, and—to a lesser extent—other substrates and cofactors. In cases where a general ligand is available, that is, one that is common to a relatively large group of enzymes, each of the enzymes sharing the ligand can be purified by the common ligand-polymer conjugate [41]. Since about 30% of the known enzymes require for their activity one of the following coenzymes: NAD, NADP, ATP, and CoA, immobilization of a coenzyme on a carrier might facilitate the purification of a wide range of proteins. All of the above cofactors have been immobilized and used in the preparation of affinity chromatography columns. (A list of enzymes purified by this procedure is to be found in Ref. 41, Table II.)

LECTINS

Lectins are sugar-binding, cell-agglutinating proteins of nonimmune origin and of wide distribution [47]. They are being used extensively as macromolecular carbohydrate-specific reagents. Because of their ability to combine specifically and reversibly with carbohydrates they can be purified on immobilized sugars [41]. Conversely, immobilization of the lectin itself facilitates the isolation of carbohydrate-containing compounds, including glycoproteins. Of the many lectins available, two are frequently used in affinity purification of carbohydrates: concanavalin A and *Ricinus communis* agglutinin. The general method for the use of lectins in affinity chromatography has been reviewed elsewhere [48].

RECEPTOR-BINDING PROTEINS

A receptor is a molecule or molecular ensemble which is present on a cell surface and which recognizes a specific chemical entity, binds to it, and initiates a series of biochemical events resulting in a characteristic physiological response. The interaction between a hormone, neurotransmitter, or drug and its receptor is selective and specific. Moreover, the interaction between the ligand and its receptor is characterized by high affinity, and extreme conditions are therefore necessary in order to elute the receptors from the corresponding columns. Some of these difficulties have been overcome by using monoclonal antibodies, which bind rather weakly to the receptor. Table IV of Ref. 41 contains examples of specific ligands and monoclonal antibodies used for the isolation of receptors and other binding proteins. Of particular interest is the fact that the affinity chromatography technique has facilitated the isolation of the receptors for insulin, acetylcholine, cAMP, and low- and high-density lipoproteins by means of the corresponding ligand-carrier conjugates.

POLYMERIC REAGENTS

Polymeric reagents can be thought of as polymers which are capable of reacting chemically with appropriate low or high molecular weight compounds. They thus represent a class of polymers which combine the physical properties of a polymer with the chemical properties of the chemically reacting groups. Polymeric reagents, if insoluble, can be readily removed from the reaction mixture by filtration or centrifugation, and if soluble, by ultrafiltration or selective precipitation. One may therefore apply a large excess of easily removable polymeric reagents in order to increase the reaction rates and improve the yields of final products. Polymeric reagents, like polymer-enzyme conjugates or polymer-ligand conjugates, can be packed in columns or used in batches.

POLYMERIC TRANSFER REAGENTS

Polymeric transfer reagents of the type Ⓟ-(α-A)$_n$ are capable of transferring group A from the polymer to a soluble compound B, yielding practically pure AB after the polymer is filtered off and the solvent removed. The reaction can be depicted as shown in Scheme 6.

The use of a polymeric acylating agent for the acylation of amines and the use of a polymeric amine for the deacylation of esters may serve as examples for the above reaction scheme. In the broadest sense, polymeric acids and bases are also transfer agents, having a proton as the functional group. The use of polymeric active esters of *N*-blocked amino acids for the acylation of free amine peptides has been reported by Fridkin et al. [49]. Extension of this method led to the elaboration of a new technique for peptide synthesis [50, 51]. An advantage of this method is the high degree of accessibility of the attached species to the reacting terminal amine of the free amino acid of peptide. A second advantage is the ease with which the intermediate growing peptides remaining in solution can be recovered and purified. Synthesis of protected Leu-enkephalin, Boc-Tyr(OBzl)-Gly-Gly-Phe-Leu-OBzl, was accomplished with a yield of 92%. Each coupling step took one hour. The active esters are insensitive to water and alcohols and are therefore convenient to handle (Scheme 7).

Recently a nitrophenol type of polymeric reagent was prepared by Friedel-Crafts acylation of polystyrene with a substituted benzoyl chloride [52] (Scheme 8). The resulting carbonyl group serves both as an anchor to the polymer and as an activator.

A polysulfonyl chloride was prepared by chlorosulfonation of macroporous resin with chlorosulfonic acid [53]. This polymer is an excellent dehydrating

SCHEME 6.

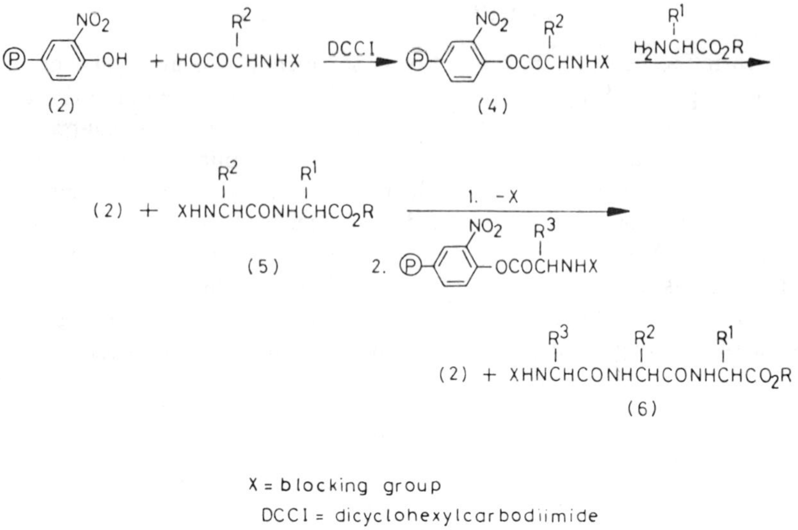

SCHEME 7.

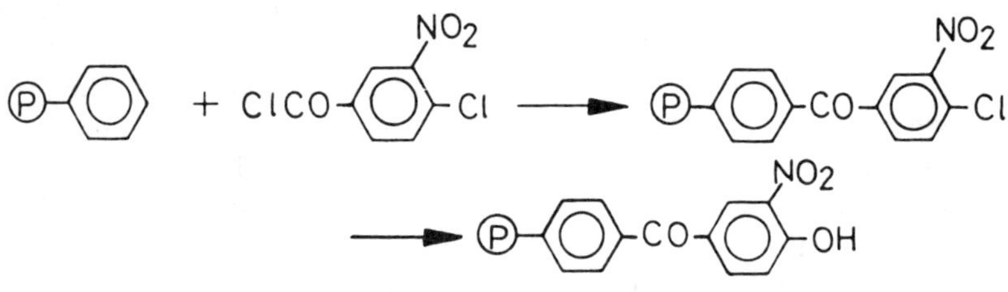

SCHEME 8.

agent in the conversion of carboxylic acids to carboxylic acid anhydrides. Interestingly, this polymer can be used to dehydrate nitrate anions, yielding a soluble reagent for nitration under neutral conditions. Presumably, the active species is N_2O_5.

Polysulfonyl chloride was also found to be an excellent reagent for the formation of l-hydroxybenztriazole (OBt) esters of *N*-blocked amino acids, as shown in Scheme 9.

Hydroxybenztriazole reacts with Ⓟ-SO_2Cl to give an intermediate benztriazole sulfate ester. In the presence of carboxylate salts, OBt esters are formed.

The way in which some other polymeric transfer reagents are used in organic synthesis is outlined in the following. The heterogeneous hydrogenation of organic halogen compounds by silica- and alumina-supported organotin hydrides was reported by Schumann and Pachaly [54]. The yields on reduction compare favorably with those obtained using polystyrene-supported organotin hydrides [55] and the chemical stability of the reagent is somewhat improved.

Colwell et al. [56] have described the synthesis of α-substituted and α,α'-disubstituted carboxylic acids via polymer-bound oxazoline derivatives.

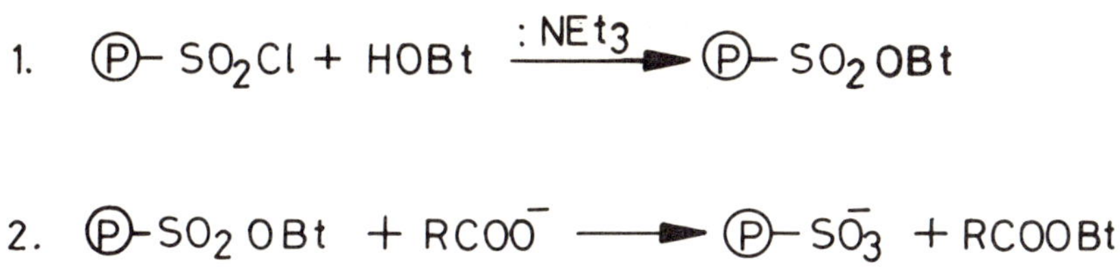

SCHEME 9.

A polymer-attached NADH mimic was synthesized by Mathis and co-workers [57]. The reagent proved to be superior to similar compounds used in solution chemistry for the reduction of substituted benzaldehydes.

Hayashi et al. [58] recently described a polymer-bound chiral ligand which complexes Ni(II) and promotes the asymmetric coupling of Grignard reagents with vinyl bromides. Chemical yields of 73–96% and optical yields of 34–53% were achieved.

An ion-exchange resin bearing the thioacetate anion has been described by Cainelli and co-workers [59]. The thioacetylated products can be converted into a variety of sulfur-bearing compounds.

An interesting and potentially useful method for functionalization of chloromethyl polystyrene was devised by Hodge and Waterhouse [60]. The chloromethylated polystyrene is first converted to its polytriphenylphosphonium salt by the action of triphenylphosphine. In the presence of base, the resulting phosphorane reacts with various aldehydes containing a range of attached groups, R, to give the corresponding substituted polymers P⁄⁄\R (Scheme 10).

THE "CASCADE" REACTION

The sequential use of two or more polymeric transfer agents on a single starting substrate represents a novel approach to organic synthesis. This "cascade" reaction is represented in (Scheme 11).

Polymer A reacts with substrate S to give S^1 and transformed polymer A′. Product S^1 serves as the substrate for polymer B in the same manner. In the penultimate step, the substrate S^2 may react with polymeric transfer agent Ⓟ-D to yield soluble product S^3, or substrate S^2 may react with the polymer carrier Ⓟ-C to yield the bound product Ⓟ-C-S^2. The use of a cascade in the former

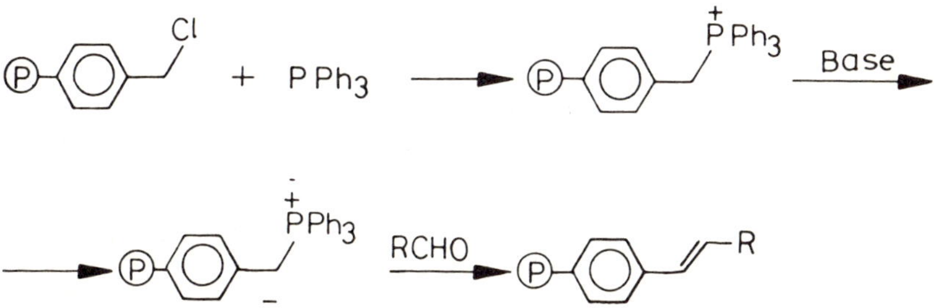

SCHEME 10.

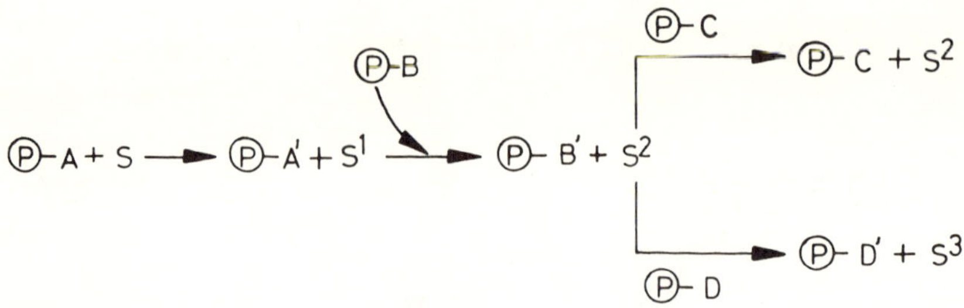

SCHEME 11.

case, when the final product is in solution, has been demonstrated in the synthesis of pyrazoles via the sequential application of three polymer transfer agents on a starting ketone [61].

For a cascade to succeed, substrate S must be entirely converted to S^1 prior to the reaction of S^1 with polymer B. There are, however, examples in which S^1 undergoes secondary reactions during its lifetime. In these cases a cascade approach fails, and another approach is necessary.

THE "WOLF AND LAMB" REACTION

When sequential synthesis is precluded by the short lifetimes of the soluble intermediates, a multipolymer approach may still be used provided that the substrate and the polymers are mixed together in the same reaction vessel. Such a reaction can be represented as in (Scheme 12).

Polymer A reacts with substrate S to give the reactive species S*. Species S* then travels to polymer B, where it is converted to product S′ (or becomes attached to a carrier polymer B). This type of reaction is of special interest when Ⓟ-A and Ⓟ-B are reagents which would react with one another in their nonpolymeric, soluble forms (i.e., Scheme 13).

This cohabitation of antagonistic reagents is the basis for the term "wolf and lamb" reaction. The concept was first developed by Rebek [62] with the aim of studying the nature of S*, and later by our group with the object of rendering the system synthetically useful. We have studied the "wolf and lamb" reaction of ketones in the presence of a strong polymeric base (polytrityllithium) and a polymeric acylating reagent. The sequence of reaction is shown in Scheme 14 [63].

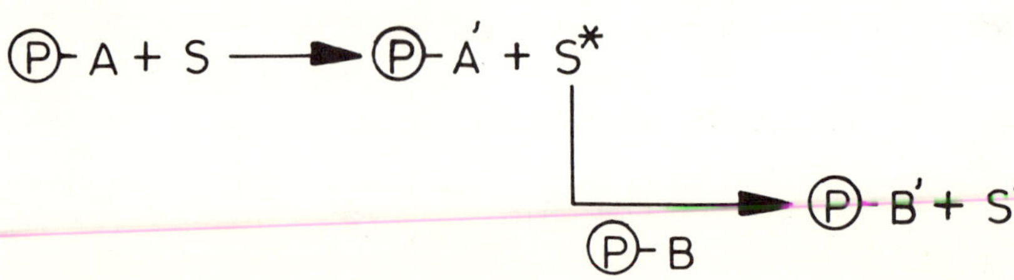

SCHEME 12.

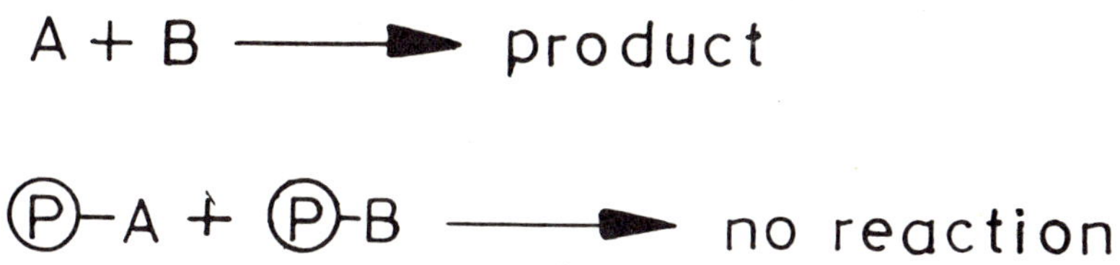

SCHEME 13.

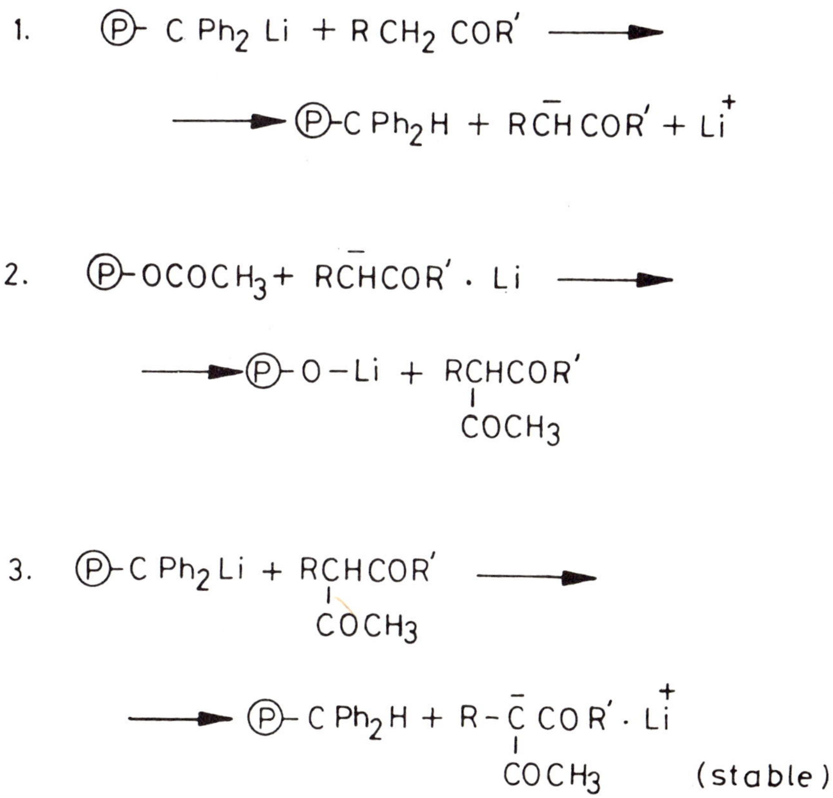

SCHEME 14.

The blood-red color of the polytrithyllithium disappears as the starting ketone is first deprotonated, then acylated, then deprotonated again. The beta-diketo anion product is stable and does not react further under these conditions.

THE "SHADCHAN" REACTION

A reaction similar to that of the "wolf and lamb" occurs when a polymeric transfer reagent is reacted with a substrate which is itself attached to a polymer. Here again the mutual inaccessibility of the polymer-bound materials prevents any reaction from taking place. For reaction to occur, a soluble reagent is em-

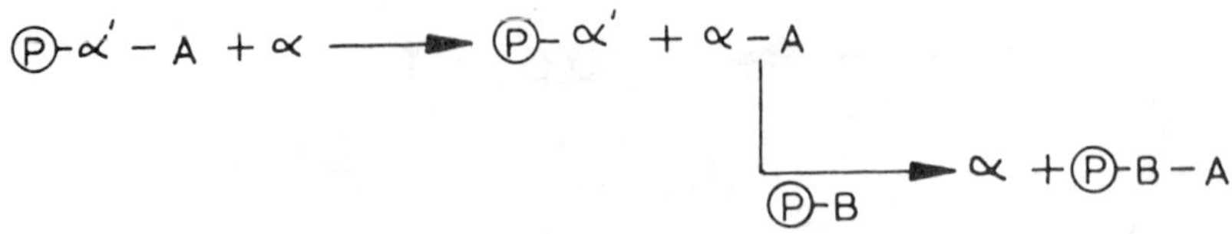

SCHEME 15.

ployed to carry the active species from the transfer polymer to the supporting polymer (Scheme 15).

We refer to the soluble reagent, α, as a "shadchan" or matchmaker, whose function is to deliver group A from the one polymer to the other. The "shadchan" α reacts with the polymeric transfer reagent Ⓟ-α'-A to give the intermediate α-A. The intermediate is then quenched by polymer B, yielding product Ⓟ-B-A and reconstituted α. Whereas in the "wolf and lamb" reaction the product remains in solution, the desired product in the "shadchan" reaction is the insoluble modified polymer Ⓟ-B-A [64]. The usefulness of the "shadchan" reaction in peptide synthesis was recently demonstrated [52].

In the "shadchan" method an amine-blocked amino acid is attached to an activated nitrophenol to form a highly active ester. The growing peptide chain is attached by a benzyl ester to a second polymer in a separate vessel. A dilute solution of a mediator molecule, such as imidazole, is then pumped into a closed loop between the two vessels. It picks up the acyl group from polymer I and transfers it to the free amine of the amino acid attached to polymer II. The free imidazole is then recycled through the loop and reused until the process is virtually complete. The reagent blocking the amino group is then removed quantitatively with trifluoroacetic acid and the process is repeated with the next amino acid in the sequence. By means of this technique it was possible to synthesize the blocked form of the pentapeptide Leu-enkephalin with a purity of 99%.

The mediator technique described above can be readily automated and the chemical reactions which occur can be easily monitored.

REFERENCES

[1] K. Mosbach, *Immobilized Enzymes. Methods in Enzymology,* Vol. 44, Academic, New York, 1976.

[2] L. Goldstein and G. Manecke, in *Applied Biochemistry and Bioengineering,* Vol. 1, L. B. Wingard, E. Katchalski-Katzir, and L. Goldstein, Eds. Academic, New York, 1976, pp. 23–126.

[3] I. Chibata, (Ed.), *Immobilized Enzymes,* Wiley, New York, 1978, pp. 1–284.

[4] I. V. Berezin and K. Martinek, (Eds.), *Introduction to Applied Enzymology,* Moscow University Press, Moscow, 1982, pp. 1–382.

[5] H. H. Weetall, in *Methods in Enzymology,* Vol. 44, K. Mosbach, Ed., Academic, New York, 1976, pp. 134–169.

[6] R. Axén, J. Porath, and S. Ernback, *Nature (London), 214,* 1302 (1967).

[7] J. Kohn and M. Wilchek, *Enzyme Microb. Technol. 4,* 161 (1982).

[8] Y. Minamoto and Y. Yugari, *Biotechnol. Bioeng., 22,* 1225 (1980).

[9] A. Freeman and Y. Aharonowitz, *Biotechnol. Bioeng., 23,* 2747 (1981).

[10] B. Solomon, R. Koppel, and E. Katchalski-Katzir, *Biotechnology 2(8)* 709 (1984); B. Solomon, R. Koppel, G. Pines, and E. Katchalski-Katzir, *Biotechnol. Bioeng., 8,* 1213 (1986).
[11] K. Ikura, K. Okumura, M. Yoshikawa, R. Sasaki, and H. Chiba, *J. Appl. Biochem., 6,* 222 (1984).
[12] K. Mosbach, (Ed.), *Immobilized Enzymes and Cells. Methods in Enzymology,* Academic, New York, in press.
[13] A. M. Klibanov, *Anal. Biochem., 93,* 1 (1979).
[14] L. Goldstein, *Methods in Enzymology,* Vol. 44, K. Mosbach, Ed., Academic, New York, 1976, pp. 397–443.
[15] P. W. Carr and L. D. Bowers, *Immobilized Enzymes in Analytical and Clinical Chemistry,* Wiley, New York, 1980, pp. 1–460.
[16] V. K. Svedas, A. L. Margolin, I. L. Borisov, and I. V. Berezin, *Enzyme Microb. Technol., 2,* 313 (1980).
[17] D. K. Rhee, S. B. Lee, J. S. Rhee, D. D. Y. Ryu, and J. Hospodka, *Biotechnol. Bioeng., 22,* 1237 (1980).
[18] W. G. Choi, S. B. Lee, and D. D. Y. Ryu, *Biotechnol. Bioeng., 23,* 361 (1981).
[19] R. V. MacAllister, in *Immobilized Enzymes for Food Processing,* W. H. Pitcher, Jr., Ed., CRC Press, Boca Raton, FL, 1980, pp. 81–112.
[20] R. L. Antrim, W. Colilla, and B. J. Schnyder, in *Applied Biochemistry and Bioengineering,* Vol. 2, L. B. Wingard, E. Katchalski-Katzir, and L. Goldstein, Eds., Academic, New York, 1979, pp. 97–155.
[21] S. S. Wang and Ch. -K. King, in *Advances in Biochemical Engineering,* Vol. 12, T. K. Ghose, A. Fiechter, and N. Blakebrough, Eds., Springer-Verlag, Berlin, 1979, pp. 119–146.
[22] S. W. May and S. R. Radgette, *Biotechnology, 1,* 677 (1983).
[23] C. Wandrey, R. Wichmann, and A. -S. Jandel, in *Enzyme Engineering,* Vol. 6, I. Chibata, S. Fukui, and L. B. Wingard, Eds., Plenum Press, New York, 1982, pp. 61–67.
[24] R. Wichmann, C. Wandrey, A. F. Buckmann, and M. R. Kula, *Biotechnol. Bioeng., 23,* 2789 (1981).
[25] K. Mosbach, P. -O. Larsson, and Ch. Lowe, in *Methods in Enzymology,* Vol. 44, K. Mosbach, Ed., Academic, New York, 1976, pp. 859–887.
[26] C. R. Lowe, in *Topics in Enzyme and Fermentation Biotechnology,* Vol. 5, A. Wiseman, Ed., 1981, pp. 13–146.
[27] C. H. Wong and G. M. Whitesides, *J. Am. Chem. Soc., 103,* 4890 (1981).
[28] S. S. Godbole, S. F. D'Sonza, and G. B. Nadkarni, *Enzyme Microb. Technol., 5,* 125 (1983).
[29] O. Andresen and P. B. Poulsen, in *Enzyme Technology,* R. M. Lafferty, Ed., Springer Verlag, Berlin, 1982, p. 179.
[30] J. Tramper, H. C. Van der Plas, and P. Linko, (Eds.), *Biocatalysts in Organic Synthesis,* Elsevier, New York, 1985.
[31] L. B. Wingard, E. Katchalski-Katzir, and L. Goldstein, (Eds.), *Applied Biochemistry and Bioengineering,* Vol. 3, Academic, New York, 1981, pp. 1–314.
[32] B. Danielsson and K. Mosbach, in *Methods in Enzymology,* Vol. 44, K. Mosbach, Ed., Academic, New York, 1976, pp. 667–676.
[33] Y. Lee, I. Lablonski, and M. DeLuca, *Anal. Biochem., 80,* 496 (1977).
[34] E. Jablonski and M. De Luca, *Proc. Natl. Acad. Sci. USA, 73,* 3848 (1976).
[35] C. Haggerty, E. Jablonski, L. Stav, and M. DeLuca, *Anal. Biochem., 88,* 162 (1978).
[36] G. Wienhausen and M. De Luca, *Anal. Biochem., 127,* 380 (1982).
[37] G. G. Guilbault, *Handbook of Enzymatic Methods of Analysis,* Dekker, New York, 1976, pp. 460–511.
[38] G. A. Rechnitz, *Science, 190,* 234 (1975).
[39] S. J. Updike and G. P. Hicks, *Nature, 214,* 986 (1967).
[40] P. Cuatrecasas, M. Wilchek, and C. B. Anfinsen, *Proc. Natl. Acad. Sci. USA, 61,* 636 (1968).
[41] M. Wilchek, T. Miron, and J. Kohn, *Meth. Enzymol., 104,* 3 (1984).

[42] J. Kohn and M. Wilchek, *Enzyme Microb. Technol.*, *4*, 161 (1982).
[43] J. Kohn and M. Wilchek, *Biochem. Biophys. Res. Commun.*, *107*, 878 (1982).
[44] J. Kohn and M. Wilchek, *FEBS Lett.*, *154*, 209 (1983).
[45] J. Kohn, R. Lenger, and M. Wilchek, *Appl. Biochem. Biotech.*, *8*, 227 (1983).
[46] M. Wilchek and T. Miron, *Biochem. Intern.*, *4*, 629 (1982).
[47] H. Lis and N. Sharon, *Ann. Rev. Biochem.*, *42*, 541 (1973).
[48] K. Yamamoto, N. Katsuda, and K. Kato, *Eur. J. Biochem.*, *92*, 499 (1978).
[49] M. Fridkin, A. Patchornik, and E. Katchalski, *Israel J. Chem.*, *3*, 69p (1965).
[50] M. Fridkin, A. Patchornik, and E. Katchalski, *J. Am. Chem. Soc.*, *88*, 3164 (1966).
[51] M. Fridkin, A. Patchornik, and E. Katchalski, *J. Am. Chem. Soc.*, *90*, 2953 (1968); R. Kalir, M. Fridkin, and A. Patchornik, *Eur. J. Biochem.*, *42*, 151 (1974).
[52] A. Patchornik and B. J. Cohen, in *Perspectives in Peptide Chemistry*, A. Eberle, R. Geiger, and T. Weiland, Eds., Karger, Basel, 1981, pp. 118–128.
[53] Y. Shai and A. Patchornik, unpublished results (1980).
[54] H. Schumann and B. Pachaly, *Ang. Chem. Int. Ed. (English)*, *20*, 1043 (1981).
[55] N. Weinshenker, G. M. Crosby, and J. Y. Wong, *J. Org. Chem.*, *40*, 1866 (1975).
[56] A. R. Colwell, L. R. Duckwall, R. Brooks, and S. P. McManus, *J. Org. Chem.*, *46*, 3097 (1981).
[57] R. Mathis, G. Dupas, A. Decormeille, and G. Queguiner, *Tetrahedron Lett.*, *22*, 59 (1981).
[58] T. Hayashi, N. Nagashima, and M. Kumada, *Tetrahedron Lett.*, *21*, 4623 (1980).
[59] G. Cainelli, M. Contento, and F. Manescalhi, *Synthesis*, *4*, 302 (1981).
[60] P. Hodge and J. Waterhouse, *Polymer*, *22*, 1153 (1981).
[61] B. J. Cohen, M. A. Kraus, and A. Patchornik, *J. Am. Chem. Soc.*, *99*, 4165 (1977).
[62] J. Rebek, Jr., *Tetrahedron*, *35*, 723 (1979).
[63] T. H. Maugh, II, *Science*, *217*, 719 (1982).
[64] Y. Shai and A. Patchornik, unpublished results (1982).

Ephraim Katchalski-Katzir was born in Kiev, Ukraine, May 16, 1916. He married Nina Gotlieb, Feb. 14, 1939, and received a M.Sc. from The Hebrew University of Jerusalem (1937), and a Ph.D. from The Hebrew University of Jerusalem (1941).

APPOINTMENTS HELD, RESEARCH, TEACHING, ADMINISTRATIVE

1941–1945	Assistant, Department for Theoretical and Macromolecular Chemistry Hebrew University, Jerusalem
1946–1948	Research Fellow, Polytechnic Institute, Brooklyn and Columbia University, New York
1949–1951	Acting Head, Department of Biophysics, The Weizmann Institute of Science, Rehovot
1951–1973	Head, Department of Biophysics, The Weizmann Institute of Science, Rehovot
1953–1961	Visiting Professor of Biophysics at the Hebrew University, Jerusalem
1957–1959	Guest Scientist at Harvard University, Cambridge, Mass.
1961–1965	Visiting Professor, Rockefeller University, New York Visiting Professor, University of Michigan, Ann Arbor
1964	Senior Foreign Scientist Fellowship, University of California, Los Angeles
1966–1968	Chief Scientist, Israel Defense Ministry
1971	Visiting Professor, Battelle Seattle Research Center, Seattle, Washington
	Elected 4th President of the State of Israel, April 11, 1973. Sworn into office May 24, 1973. Termination of office: May, 1978
1978–	Institute Professor, The Weizmann Institute of Science, Rehovot
1978–	Head, Center for Biotechnology, Tel-Aviv University, Ramat Aviv
1979	Regents Professor, University of California, San Diego (March-May)

1979 First incumbent of the Herman Mark Chair in Polymer Science, Polytechnic Institute of New York (September-October)

MEMBERSHIP IN SCIENTIFIC BODIES ABROAD

1966–	American Academy of Arts and Sciences (Foreign Honorary Member)
1976–	American Chemical Society (Centennial Foreign Fellow)
1976–	American Philosophical Society (Foreign Member)
1961–	American Society of Biological Chemists (Honorary Member)
19 –	Association of Harvard Chemists
1951–	American Association for the Advancement of Science
1961–	Biophysical Society
1966–	Ciba Foundation (Scientific Advisory Panel)
1963–	European Molecular Biology Organization (Founding Member)
1963–1968	European Molecular Biology Organization (Member of Council)
1961–	Federation of American Societies for Experimental Biology
1964–	Harvey Society (Honorary Member)
1983	International Society of Biorheology
1964–1973	International Union of Biochemistry (Council Member)
1968–	Leopoldina Academy of Science, Halle, Germany
1966–	National Academy of Sciences, USA (Foreign Associate)
1947–	New York Academy of Sciences (Honorary Life Member 1973)
1977–	Royal Society, London (Foreign Member)
1967–	World Academy of Art and Science (Fellow)

EDITORIAL AND ADVISORY BOARDS

1967–	Advances in Experimental Medicine and Biology, series Plenum Press (Editorial Committee)
1969–1974	Analytical Biochemistry (Editorial Board)
1978–1981	Analytical Biochemistry (Advisory Board)
1976–	Applied Biochemistry and Bioengineering, series Academic Press (Editor)
1960–1970	Archives Biochemistry Biophysics
1961–1965	Biophysical Journal (Editor)
1963–	Biopolymers (Editorial Board)
1983–	Bio/Technology (Nature Publishing Co.) (Advisory Board)
1968–1977	European Journal Biochemistry (Editor)
1978–	Journal of Applied Biochemistry (Hon. Editor)
1975–1980	Journal of Solid-Phase Biochemistry (Advisory Board)
1967–	Molecular Biology, Biochemistry and Biophysics, series Springer Verlag (Advisory Editor)
1973–1982	Molecular and Cellular Biochemistry (Advisory Board)
1972–	Progress in Surface and Membrane Science, series Academic Press (Advisory Board)

MEDALS AND AWARDS

1948	Tchernikhovski Prize (exemplary translation of scientific literature into Hebrew)
1950	Weizmann Prize (Municipality of Tel-Aviv)
1959	Israel Prize in Natural Sciences
1961	Rothschild Prize in Natural Sciences
1962	Incumbent of the Theodore R. Racoosin Chair of Biophysics at the Weizmann Institute of Science
1969	Linderstrøm Lang Gold Medal, Copenhagen

1972	Hans Krebs Medal awarded by the Federation of European Biochemical Societies
1976	Inauguration of Ephraim Katzir Chair of Biophysics, Bar Ilan University, Ramat Gan
1979	Alpha Omega Achievement Medal
1979	First Incumbent of the Herman Mark Chair in Polymer Science, Polytechnic Institute of New York
1980	Incumbent of the Leigh Family Chair in Biophysics and Biotechnology, Tel-Aviv University
1985	Japan Prize

HONORARY DOCTORATES

1973	Hebrew University, Jerusalem
1973	Brandeis University, Massachusetts
1975	University of Michigan, Ann Arbor
1975	Polytechnic Institute of New York (Hon. Professorship)
1975	Hebrew Union College, Jerusalem
1976	The Weizmann Institute of Science, Rehovot
1978	Northwestern University, Evanston
1978	Harvard University, Cambridge
1980	McGill University, Montreal
1980	Eidgenossische Technische Hochschule (E.T.H.), Zurich
1981	Thomas Jefferson University, Philadelphia
1981	University of Oxford
1983	University of Miami, Coral Gables, Florida
1983	Technion, Israel Institute of Technology, Haifa

CHEMICAL SYNTHESIS ON POLYMER SUPPORTS

BRUCE MERRIFIELD
The Rockefeller University, New York, New York 10021

SYNOPSIS

Classical chemical synthesis of organic molecules has advanced to a very sophisticated level and even complex molecules like peptides and oligonucleotides can be prepared by experienced chemists. However, it was shown some years ago that large biopolymers could be synthesized more easily and in higher yields if the reactions were carried out on a polymeric support. In certain cases the synthesis of smaller compounds can also be conducted to advantage on a polymer matrix.

I first used this general idea for the synthesis of peptides and called the process solid phase peptide synthesis. The plan was to begin with an insoluble solid particle—some organic resin, to attach a reactive group, and to anchor the first amino acid of the peptide to that group. Subsequent amino acid units would be added to the first in a stepwise manner until the entire sequence was assembled. Finally, the anchoring bond would be cleaved, liberating the peptide into solution where it could be isolated. The main feature of this procedure was that the resin-bound peptide could be filtered after each reaction and washed free of soluble reagents and by-products. This greatly accelerated the entire process and minimized losses of material. A convenient polymeric support was found to be a polystyrene bead crosslinked with 1% of divinylbenzene. This suspension polymer gel solvates readily in nonpolar solvents and swells to several times its dry volume. It is highly permeable and the chains of the matrix are very mobile. One chemical scheme for achieving the synthesis consists of derivatizing the resin by a chloromethylation reaction and then attaching the first protected amino acid to the support as a benzyl ester derivative. Removal of the tert-butyloxycarbonyl protecting group with trifluoroacetic acid and neutralization with a tertiary amine gives a free α-amino group ready for reaction with the second amino acid. This is conveniently done by activation with dicyclohexylcarbodiimide. A final treatment with the very strong anhydrous acid, HF. removes side chain protecting groups and cleaves the anchoring benzyl ester bond, thus liberating the free peptide into solution. Many chemical variations of this general procedure are possible. Because this is a repetitive process, with many steps, it is very important that the reactions proceed rapidly and go to completion, with a minimum of side reactions.

Since all of the reactions occur in a single vessel, the process lends itself to automation. We designed and built a simple machine for that purpose in 1965. It has since been modified and developed commercially by several companies. I should point out that a similar apparatus, employing similar operations has been very successfully adapted to oligonucleotide synthesis. Several variations in the basic design of peptide synthesizers have been introduced. These include continuous flow models, a moving belt design, and a recent development that leads to the simultaneous synthesis of multiple peptides. We will surely see improvements on this theme in the near future. There is now considerable interest in adapting the new field of robotics to solid phase peptide synthesis.

There have been some developments in new polymer supports. The most useful of these have been based on substituted polyacrylamides. By preparing these polymers within a rigid, porous inorganic matrix they can be used at low pressures in a continuous flow process. The polystyrene beads can also be used in this mode. It would be helpful if the expert polymer chemists would give some thought to our needs and develop new polymers with improved properties that may offer certain advantages in supported synthesis. One possibility might be to construct a zipper polymer that would provide the usual properties of insolubility and physical stability for synthesis but which could be opened to a soluble form for the final peptide removal step; for example by reduction of disulfide crosslinks or photolysis of ester crosslinks.

An area of chemistry that has not received enough attention is the elaboration and thorough testing of orthogonal schemes of protection. The base-labile Nα-fluorenylmethoxycarbonyl group in conjunction with acid-labile side chain and anchoring groups has been successful and is semiorthogonal. A fully orthogonal scheme utilizing a photolabile anchoring bond, acid-labile side chain groups and the thiol-sensitive dithiasuccinoyl Nα-protecting group has been found recently to be very effective. Multi-detachable resin linkages also offer some interesting chemical advantages that should be explored in more detail.

The main purpose of developing synthetic methods is to be able to produce products needed for other studies. Some of the more important immediate goals are to provide a ready supply of various naturally occurring peptides, including hormones, releasing factors, neuropeptides, and growth factors. Superagonists and antagonists of hormones for therapeutic use are of great interest, and recently there has been much excitement about the use of antibodies to synthetic peptides for diagnostic purposes and as probes for the detection and isolation of new gene products. The possibility of producing synthetic vaccines for a variety of infectious diseases has created much enthusiasm and hope, and this area will surely be pursued vigorously in the near future.

For the longer range, one can speculate on some potential goals: 1. the synthesis of peptides by catalysis with supported enzymes; 2. the application of computer graphics and energy calculations to develop the concepts of molecular recognition in small rigid synthetic peptides; 3. the synthesis of small and medium sized proteins in good yield and purity; 4. the synthesis of new, non-natural proteins designed to bind to specific adducts and, eventually, the design and synthesis of new proteins containing both a binding site and a catalytic site which will function as an enzyme; 5. the synthesis of a self replicating and evolving system analogous to the Eigen hypercycle; 6. the construction of a virus from a synthetic nucleic acid and a synthetic protein coat. I believe that polymer supports will continue to aid the chemical synthesis of biopolymers for some time to come and that there will be many discoveries and surprises ahead.

I may be the only participant here who was not a student or colleague of Professor Mark, but that certainly does not diminish my great admiration for him.

The main interest of our laboratories has been the chemical synthesis of peptides and proteins on polymeric supports. Since both of the components are polymers, I am not certain which one I should emphasize in this Symposium on Polymer Science in the Next Decade. My compromise will be to touch on each.

The classical techniques for making peptides use the methods of homogeneous solution chemistry. However, after a little experience with such methods I became convinced that new approaches were needed, and I eventually hit upon the idea of using a solid polymeric support to hold the peptide chain during its synthesis [1]. The plan was to functionalize the polymer, attach the first unit of the peptide chain covalently, add the next unit to make the dipeptide, then add subsequent units one at a time until the polymeric peptide chain was assembled, and finally to break the anchoring bond to liberate the product into solution. The advantages were that the support and the peptide chain would be insoluble and after each reaction the soluble reagents and byproducts could be separated by simple filtration and washing, thus providing rapid purification. The use of excess reagents could be used to force reactions to go nearly to completion in a short time. It was only necessary to find a suitable polymer and an effective set of reagents.

THE RESIN

The resin we found to be most useful was a suspension polymer of polystyrene containing 1% divinylbenzene as a crosslinker. The resulting gel is in the form of beads about 50 μm in diameter. The beads have several properties that make them especially suitable for this purpose (Table I). They are physically stable and easily filtered. They are easily derivatized using a variety of aromatic chemistry. They are completely insoluble in all solvents and reagents, but are highly solvated by organic solvents such as dichloromethane or dimethylformamide (DMF) and swell to several times their original volume. Furthermore, after a peptide chain is synthesized on them it also becomes solvated and promotes further expansion of the bead. When a 60-residue model peptide was prepared at high loading the beads were found to contain 80% peptide and 20% polystyrene by weight, and in dimethylformamide their volume increased 28-fold [2]. Maximum expansion of a 1% crosslinked polystyrene bead was estimated to be as much as 200-fold. We showed that the beads had not become filled with peptide and that neither the polystyrene chains nor the peptide chains had collapsed during this synthesis. We conclude that there is room inside such beads for much larger peptides to be constructed.

Both the peptide chains and the polymer chains have been shown by ^{1}H and ^{13}C NMR (nuclear magnetic resonance) to be highly mobile, with correlation times of 10^{-10} sec for the α-carbon of a peptide and 10^{-8} sec for the aromatic and backbone protons of the polystyrene being observed [3]. Based on experiments showing that over 99.5% of short peptide chains could be involved in intersite reactions we concluded that segmental chain motion of at least 200 Å can occur in the polystyrene chains within a bead [4]. As a result of these properties, rapid chemical reactions between resin-bound peptides and soluble reagents were expected and were found. The second-order rate constants for

TABLE I.
Properties of the Resin

1. Chemically and physically stable
2. Insoluble
3. Easy separation from soluble reagents
4. Easily derivatized
5. Uniform distribution
6. Highly solvated
7. High swelling
8. No crowding with chain growth
9. Rapid molecular motion
10. Chain flexibility
11. Rapid synthetic reactions
12. Very rapid mass transfer
13. High synthetic efficiency
14. Low association of peptide and resin
15. Reduced peptide–peptide aggregation

coupling amino acids to the peptide chain were between 1 and 10 L/mol · sec and the half times are therefore in the order of seconds [5]. Preliminary data indicate that mass transfer of reagents into the beads is more than 10 times as fast as the peptide-forming reactions and is not rate limiting [6]. Because of the high swelling and chain mobility the synthetic efficiency was also very high (>99.8%), and furthermore, it was not affected by peptide chain length up to 60 residues [7].

We believe it is an advantage to have the resin and the protected peptide of somewhat different polarity so that each of them will preferentially associate with solvent rather than with each other. We see no evidence for binding of peptide and resin in solvents such as CH_2Cl_2 or DMF. There are examples, however where certain peptides self-associate in β sheets and become partially unreactive with incoming activated amino acid. For example, the reaction of Boc-Tyr(ClZ)-OH anhydride with H-Ile-Asn-Gly-resin in CH_2Cl_2 was slow and incomplete, due to β structure, and the resin showed broad NMR bands. In contrast, the reaction with H-Ile-Ala-Glu(OBzl)-resin was faster and went to completion and the peptide and resin had rapid correlation times [3]. The effect of the β self-association was overcome by carrying out the reactions in DMF.

It could be pointed out that protected peptides that are known to be very insoluble in organic solvents, and therefore unreactive, are highly solvated in the same solvents when they are attached to lightly crosslinked polystyrene resin. A simple thermodynamic argument has been proposed to explain this observation [2].

SOLID-PHASE CHEMISTRY

The solid-phase synthesis of a peptide requires seven basic steps [1]: derivatization of the support, attachment of the first amino acid, removal of the α-amine protecting group, neutralization of the amine salt, coupling of the next protected amino acid, cleavage of the finished peptide from the support, and purification of the final product, Deprotection, neutralization, and coupling are repeated at every cycle of the synthesis and therefore must be rapid, complete, and free of side reactions if good products are expected. Several different combinations of chemical groups and reactions have been devised but cannot be covered here [8]. The most common scheme uses the acid-labile *tert*-butyloxycarbonyl (Boc) group for N^α protection and more acid-stable benzyl derivatives for side chain protection and for anchoring to the support. The benzyl derivatives can be removed by very strong acids such as HF (hydrogen fluoride). New procedures that avoid the differential acid approach make use of orthogonal schemes in which each type of protecting group is selectively cleaved by a qualitatively different kind of chemical reaction and is stable to the other reactions. A useful example has been reductive cleavage of the N^α group, acidolysis of side chain groups, and photolysis of the anchoring bond.

AUTOMATION

Because all of the synthetic reactions can be done in a single reaction vessel, no transfers of peptide or resin are necessary. This feature, coupled with the insolubility of the peptide-resin and its ease of filtration and washing allows the process to be automated. A machine capable of this [9] was built in our laboratory in 1965 and many commercial versions have since been produced. They are basically the same, but incorporate more sophisticated electronics, with easier programming and more fail-safe devices. There are also continuous flow systems in which the solvents and reagents are pumped through a packed column of the resin support [10]. These can be readily monitored spectrophotometrically for completion of reaction. Recent designs of great interest allow the simultaneous synthesis of many peptides. In one instance they are assembled on the tips of small pegs which are dipped into wells containing the reagents [11]. Up to 96 different peptides could be made at the same time. In another apparatus the resin beads are placed in a series of plastic mesh bags and processed in a similar way [12].

Several of the peptide synthesis machines have been readily adapted to oligonucleotide synthesis. The basic manipulations are very similar although the chemistry is different. The results have been remarkably good.

PROSPECTS FOR THE FUTURE

Some of the prospects for the future of peptide synthesis that I can envision and speculate about are outlined in Table II. There are many possibilities for new polymeric solid supports. Although polystyrene and the polyacrylamides are quite satisfactory, there nevertheless is a vast literature on polymeric materials and very few have been examined as supports for peptide synthesis. It might be possible to make insoluble polymers that could be solubilized after the synthesis has been completed, for example by a kind of zipper effect in which all the crosslinks could be opened so that a soluble linear polymer would result. Reduction of disulfides, or hydrolysis of labile esters, or photolysis of crosslinks are possibilities.

New peptide chemistry will continue to be developed, although progress in this area has noticeably slowed. The current chemistry is quite good and the trend has been to use the available methods for various applications of interest rather than to work on small improvements in the methods themselves.

The new synthetic approach where the greatest developments are being made is in the area of enzyme-catalyzed synthesis. This is an old idea championed by Fruton and recently expanded by several groups. The work of Johansen and colleagues has been impressive and holds great promise [13]. Liquid-phase methods using soluble supports have been used [14], as has an elegant liquid phase–solid phase combination approach [15]. The matchmaker method of Patchornic [16] is also a potentially useful technique. Most impressive of all are

TABLE II
Prospects for Future of Polymers and Peptide Synthesis

A. New supports	1. Zipper, disulfides, photolysis
	2. New monomers
	3. Polymers with special properties
B. New chemistry	1. Orthogonal protection
	2. Multidetachable resins
	3. Strong acid improvements
C. Other routes	1. Liquid phase; matchmaker
	2. Enzyme catalysis
	3. Genetic engineering
D. Applications	1. Antigens, antibodies
	a. Mapping epitopes
	b. Vaccines
	c. Antibody diagnostics
	d. Probes for gene products
	2. Molecular recognition
	a. Computer graphics
	b. Energy calculations
	c. Nonpeptide analogs
	3. New proteins for binding and catalysts
	a. Modify existing proteins
	b. De novo proteins
	4. Self-replicating system
	a. Hypercycle
	b. Virus

genetic engineering methods which in principle will be feasible for a wide range of naturally occurring peptides and proteins and many of their analogs. I cannot discuss this major development here.

The main purpose of developing synthetic methods is to be able to prepare compounds that are needed for other studies. Some of the more important immediate goals are to provide a ready supply of various naturally occurring peptides, including hormones, releasing factors, growth factors, neuropeptides, toxins, antibiotics, and enzymes. This, of course, has been done and virtually all of the known peptides have been synthesized by one or more methods. As new peptides are discovered and sequenced they can usually be synthesized in a relatively short time. Superagonists and antagonists of hormones for therapeutic use are also of great interest and have been made in very large numbers. Some of these analogs are very important. An interesting example is vasopressin. Following du Vigneaud's first synthesis, many analogs were made in an effort to understand the relation of structure to function. Zaoral et al. [17] succeeded in synthesizing by classical methods an analog that was effective in separating the undesired pressor effect of the natural hormone from the desired antidiuretic effect, namely 1-deamino-[8-D-arginine] vasopressin. Manning et al. [18] then undertook a long detailed study involving several hundred peptides using solid-phase methods, and eventually evolved a molecule with an antidiuretic/pressor ratio of 15,000:1, which is highly effective in the treatment of diabetes insipidus. This work was followed by a search for peptide inhibitors of both activities that have no detectable agonist properties even at high concentrations, which would be useful for treatment of patients with excessive water retention and resultant hyponatremia. The best of these was [1-(β-mercapto-β, β-cyclopentamethylene)-propionic acid, 2-D-phenylalanine, 4-valine]arginine-vasopressin [19].

In our laboratory we have been searching for an analog of glucagon that is a strict antagonist, with strong binding to the liver membrane receptors but with no ability to activate adenyl cyclase. Most of the analogs with reduced or negligible cyclase activity were also unable to bind to the membrane and therefore did not function as effective inhibitors. Recently, the synthetic analog [Asp^3, Glu^9]glucagon was found to be a competitive inhibitor with no detectable cyclase activity [20]. Efforts are now being made to enhance the binding constant and to use the inhibitor to study the role of glucagon in diabetes mellitus.

Both epidermal growth factor (53 residues) [21] and transforming growth factor α (50 residues) [22] have recently been synthesized in our laboratories. These peptides stimulate cellular proliferation, inhibit gastric acid secretion, promote wound healing, retard hair growth, and play a role in embryonic development. They are highly crosslinked, compact molecules that have been difficult to synthesize in the past. The new syntheses were successful because the best conditions that have been developed over many years in our laboratory were applied and all of the many known precautions were taken. These included the Pam-resin support [23], the new cyclohexyl ester-protecting group [24], pure reagents, quantitative ninhydrin monitoring [25], and the new low/high HF cleavage method [26]. The yields were high and the purified products were shown by HPLC (high performance liquid chromatography) to be homogeneous and indistinguishable from the purified natural products and had full biological activity.

A few solid-phase syntheses of peptides have been scaled up for commercial purposes. Salmon calcitonin is an example [27]. This 32-residue hormone has been prepared in single batches of 50–100 g and both yield and purity were good. The synthetic hormone is highly effective for treatment of Paget's disease and for other conditions of hypercalcemia. It seems probable that much larger-scale syntheses could be carried out if the need were to arise.

MAPPING EPITOPES

The area of greatest current interest probably is the elucidation of the immunogenic and antigenic determinants of proteins and the development of synthetic vaccines against viral and other infectious diseases. For this purpose many peptides must be synthesized. Lerner [28] synthesized a number of small peptides covering nearly the entire length of a virus coat protein and showed that all of them elicited antibodies if they were attached to a carrier protein. As many as 75% of these antibodies were found to react with the parent protein, showing that most of the antigenic sites along the whole native protein were accessible. The converse situation has also been examined by Geysen et al. [11], who made all 208 overlapping hexapeptides of the foot and mouth virus coat protein. These were tested by an enzyme-linked immunosorbent assay to determine which would bind to an antibody of the coat protein. It was found in this case that only one or two regions of the protein were antigenic, suggesting that only a few continuous sites of the protein were immunogenic and that the major sites were probably discontinuous. Immunization of animals with short conjugated synthetic peptides has been shown to lead to active immunity against a challenge with the infectious agent, and there is hope that effective synthetic vaccines can be developed. They may have the advantages that they are free of the danger of injection of a virulent organism, and that they can be directed to regions of the protein that are nonvariant and essential to the organism and therefore not susceptible to rapid genetic change that makes the organism resistant to the antibody. Unfortunately, titers to small synthetic peptides are often not as high as to the whole organism or to a coat protein.

Antibodies to synthetic peptides are already being used as diagnostics to detect various diseases. Synthetic peptides corresponding to regions of various genes that have been sequenced are frequently used to induce antibodies to the gene product protein. These in turn have been useful in the detection and isolation of the new proteins.

MOLECULAR RECOGNITION

The field of molecular recognition promises to be one of the very active areas of research in the years immediately ahead. There is much interest in being able to understand the precise fit of enzymes and substrates, antigens and antibodies, and hormones and receptors. There is hope that model organic compounds can be designed and synthesized which will recognize one another and accurately conform in their three-dimensional structure so that very strong noncovalent binding can occur between them. Energy calculations, x-ray structures and computer graphics are being brought to bear on this problem and rapid results are to be expected. It will be possible to design proteins with high binding constants for the concentration and isolation of wanted ligands or for the removal of other unwanted ligands. One approach is to modify existing protein structures but the ultimate goal is to design *de novo* proteins with these special properties. The next step will be the introduction of a catalytic site in the proper relation with the binding site to produce a new enzyme. This goal is ambitious, but I think it is feasible.

Gutte' group has succeeded in synthesizing a 34-residue peptide containing two antiparallel β strands and a helical region that was designed to bind to a nucleotide [29]. Binding was observed and there was some hydrolysis of RNA. Moser and co-workers have also designed a 24-residue peptide [30] that will bind DDT, with $K = 10^5$. The latter peptide was crystalline and is being studied by x-ray analysis. The recent studies of the Erickson and Richardson groups [31] are also very interesting. On the basis of a large amount of x-ray data, they have designed a molecule that is expected to fold into a β-barrel structure and thereby to form a binding pocket at one end. A novel synthetic approach was used in which two identical 32-residue chains were synthesized simultaneously on a bifunctional resin support. Thus the final β-barrel was composed of two chains held together at one end by an aromatic ring containing two short extender arms.

Finally, it may be fair to speculate that the methods of synthetic organic chemistry will eventually be successful in the design and synthesis of self-replicating systems. This might include a simple virus containing a nucleic acid and a protein which would use the machinery of the host cell to reproduce, or the hypercycle of Eigen [32] which would function independently and catalyze its own evolution into a more and more efficient system. These would be monumental feats, but perhaps would represent no greater increments in achievement than some we have seen in the past.

REFERENCES

[1] R. B. Merrifield, *J. Am. Chem. Soc.*, *85*, 2149 (1963).
[2] V. K. Sarin, S. B. H. Kent, and R. B. Merrifield, *J. Am. Chem. Soc.*, *102*, 5463 (1980).
[3] D. H. Live and S. B. H. Kent, in *Peptides: Structure and Function*, V. Hruby and D. Rich, Eds., Pierce Chem. Co., Rockford, IL, 1983, pp. 65–68.
[4] K. K. Bhargava, V. K. Sarin, N. L. Trang, A. Cerami, and R. B. Merrifield, *J. Am. Chem. Soc.*, *105*, 3247 (1983).
[5] R. B. Merrifield, Reported at the International Symposium on Peptide Synthesis, Madison, WI, May 30, 1973.
[6] B. Hetnarski and R. B. Merrifield, unpublished results.
[7] V. K. Sarin, S. B. H. Kent, A. R. Mitchell, and R. B. Merrifield, *J. Am. Chem. Soc.*, *106*, 7845 (1984).
[8] G. Barany and R. B. Merrifield, in *The Peptides*, Vol. 2, E. Gross and J. Meienhofer, Eds., Academic, New York, 1979, pp. 1–284.
[9] R. B. Merrifield and J. M. Stewart, *Nature*, *207*, 522 (1965).
[10] R. P. W. Scott, K. K. Chan, P. Kucera, and S. Zolty, *J. Chromatogr. Sci.*, *9*, 577 (1971).
[11] H. M. Geysen, R. H. Meloen, and S. J. Barteling, *Proc. Natl. Acad. Sci. USA*, *81*, 3998 (1984).
[12] R. A. Houghton, *Proc. Natl. Acad. Sci., USA*, *82*, 5131 (1985).
[13] F. Widmer, K. Breddam, and J. T. Johansen, in *Peptides 1980*, K. Burnfeldt, Ed., Scriptor, Copenhagen, 1981, pp. 46–55.
[14] M. Mutter, H. Hagenmaier, and E. Bayer, *Angew Chem., Int. Eng. Ed.*, *10*, 811 (1971).
[15] H. Frank and H. Hagenmaier, *Experientia*, *31* 131 (1975).
[16] Y. Shai, K. A. Jacobson, and A. Patchornik, *J. Am. Chem. Soc.*, *107*, 4249 (1985).
[17] M. Zaoral, J. Kolc, F. Sorm, *Coll. Czech. Chem. Commun.*, *32*, 1250 (1967).
[18] M. Manning, L. Balaspiri, M. Acosta, and W. H. Sawyer, *J. Med. Chem.*, *16*, 975 (1973).
[19] M. Manning, W. A. Klis, A. Olma, J. Seto, and W. H. Sawyer, *J. Med. Chem.*, *25*, 414 (1982).
[20] D. Andreu and R. B. Merrifield, Proc. Ninth Amer. Pep. Symp., Toronto, 1985.
[21] W. F. Heath and R. B. Merrifield, Proc. Ninth Amer. Pep. Symp., Toronto, 1985.
[22] J. P. Tam, H. Marquardt, D. F. Rosberger, T. W. Wong, and G. J. Todaro, *Nature*, *309*, 376 (1984).
[23] A. R. Mitchell, B. W. Erickson, M. N. Ryabtsev, R. S. Hodges, and R. B. Merrifield, *J. Am. Chem. Soc.*, *98*, 7357 (1976).
[24] J. P. Tam, T. W. Wong, M. W. Riemen, F. S. Tjoeng, and R. B. Merrifield, *Tetrahedron Lett.*, 4033 (1979).
[25] V. K. Sarin, S. B. H. Kent, J. P. Tam, and R. B. Merrifield, *Anal. Biochem.*, *117*, 147 (1981).
[26] J. P. Tam, W. F. Heath, and R. B. Merrifield, *J. Am. Chem. Soc.*, *105*, 6442 (1983).

[27] J. Hughes, U.S. Patent 3926-938, December 16, 1975.
[28] R. A. Lerner, *Nature, 299,* 592 (1982).
[29] B. Gutte, M. Däumigen, and E. Wittschieber, *Nature, 281,* 650 (1979).
[30] R. Moser, R. M. Thomas, and B. Gutte, *FEBS Lett., 157,* 247 (1983).
[31] S. B. Daniels, C. G. Unson, J. S. Richardson, D. C. Richardson, and B. W. Erickson, *Fed. Proc., 44,* 694 (1985).
[32] M. Eigen, *Naturwissenschaften, 58,* 465 (1971).

Robert Bruce Merrifield was born in Fort Worth, Texas, July 15, 1921. He married Elizabeth Furlong in 1949 and has six children.

EDUCATION

Montebello High School, California, 1939
B.A., University of California, Los Angeles
Major in Chemistry, 1943
Ph.D., University of California, Los Angeles
Major in biochemistry, 1949
Research director, Professor M. S. Dunn
Thesis: "Microbiological Studies on Pyrimidines."
D. Sc., honoris causa, University of Colorado (1969),
Ph.D., honoris causa, Uppsala University, (1970),
Yale University (1971), Newark College of Engineering (1972),
Medical College of Ohio (1972), Colgate University (1977),
Boston College (1984)

EMPLOYMENT

1943–1944	Philip R. Park Research Foundation, Chemist
1944–1947	U.C.L.A. Chemistry Dept., Teaching Assistant
1948–1949	U.C.L.A. Medical School, Research Assistant
1949–1953	Rockefeller Institute for Medical Research, Assistant
1953–1957	Rockefeller Institute for Medical Research, Associate
1957–1958	Rockefeller Institute for Medical Research, Assistant Professor
1958–1966	Rockefeller Institute, Associate Professor
1966–	Rockefeller University, Professor
1984–	Rockefeller University, John D. Rockefeller Jr. Professor

SOCIETIES

Sigma Xi, Phi Lambda Upsilon, Alpha Chi Sigma, American Chemical Society, American Society of Biological Chemists, American Institute of Chemists, National Academy of Sciences

PROFESSIONAL ACTIVITIES

Associate Editor, *The International Journal of Peptide and Protein Research*
Editorial Board of *Analytical Biochemistry*
Nobel Guest Professor, Uppsala, Sweden, 1968
Lasker Award for Basic Medical Research, 1969

Gairdner Award, 1970
Intra-Science Award, 1970
American Chemical Society Award for Creative Work in Synthetic Organic Chemistry, 1972
Nichols Medal, 1973
The Instrument Specialities Co. Award of the University of Nebraska, 1977
Alan E. Pierce Award, 1979
Nobel Prize in Chemistry, 1984

MODERN STEREOCHEMISTRY

KARL SCHLÖGL
Institute of Organic Chemistry, University of Vienna, A-1090 Vienna, Austria

SYNOPSIS

Sterochemistry is a dominating factor in most fields of chemistry—especially in biochemistry, organic and polymer chemistry. Its more important aspects include asymmetric syntheses, optical resolution, enantiomeric purity, chiroptical properties (mainly circular dichroism) as well as relative and absolute conformation and configuration (chirality).

Progress in these fields may be illustrated by a few examples from our more recent work at Vienna University, where especially axial and planarchiral arenes together with metallocenes have been extensively investigated. Among metallocenes arene tricarbonylchromium complexes are of particular interest, since they can easily be prepared, separated into enantiomers (vide infra), are easily accessible to x-ray structural analysis and chiroptical methods and can finally—after suitable transformations—be cleaved again under very mild conditions.

Medium pressure chromatography on triacetylcellulose in ethanol or ether—especially by a recycling technique—is an extremely useful tool for the efficient optical resolution of various chiral aromatic compounds. For some of the enantiomers thus obtained (including optically labile ones—since the chromatographic separation can also be performed at lower temperatures) torsional and inversion barriers were determined by racemization and n.m.r. studies. Absolute conformations and chiralities could be established by x-ray diffraction and by chiroptical (CD) comparison. Thus a combination of all modern methods and techniques available seems to be a condition sine qua non for modern stereochemical studies.

"Alles Leben ist Chemie"—"All Life is Chemistry"—this was the catching title of an Austrian TV series several years ago. It was presented and commentated by Professor Herman Mark in such a fascinating manner that the audience was deeply impressed by his expertise and his charm. Since then, Herman Mark has been known among a relatively large part of the Austrian population.

Of course, no one who had known Professor Mark before, especially his friends and colleagues and those who had ever attended one of his lectures was surprised at this presentation. And, a large part of his long and successful life was chemistry, and particularly polymer chemistry.

It is my great privilege and a great honor to have been invited to this festive occasion of celebrating the 90th birthday of our most famous living colleague, and I have the pleasure to convey greetings and the best wishes of three Austrian institutions: (a) First of all of the Austrian Chemical Society, of which Professor Mark has been an honorary member for many years; (b) of the Austrian Academy of Sciences, which bestowed its honorary membership on Herman Mark in 1970; (c) and finally of the University of Vienna, which honored him in 1980 by bestowing the honorary degree of Dr.rer.nat. honoris causa.

All life is chemistry—this could be more precisely expressed as "all life is stereochemistry" and some recent aspects of stereochemical research will be the topic of my lecture at this remarkable occasion.

At the University of Vienna we have been interested for many years in stereochemical aspects of organic chemistry and especially of organometallic chemistry. In this field two main aspects are related to the work of Professor Mark: (a) Chiral polymers and (b) structural investigations by x-ray crystal analysis.

One of the most important features in stereochemical research is the problem of separating enantiomers, or optical resolution.

For this purpose liquid chromatography on chiral stationary phases seems to be one of the most promising approaches. Chiral polymers, such as chiral polyamides (W. H. Pirkle), polytriphenylmethacrylate (H. Yuki, K. Hatada and Y. Okamoto), or chiral polychloral (O. Vogl) have been employed so far. We have concentrated our efforts (like some others: G. Hesse and R. Hagel, G. Blaschke, A. Mannschreck) on the system microcrystalline triacetylcellulose (TAC)/ethanol or ether at medium pressure. This technique, especially after extending it into a recycling mode proved to be extremely versatile and efficient for the semipreparative optical resolution of various centro- and particularly axial and planar-chiral compounds. A further advantage in our hands is the possible chromatography at lower temperatures (down to −15°C), thereby permitting the resolution of optically labile compounds (*vide infra*). The equipment used is schematically depicted in Figure 1. Three examples illustrating the advantages of the recycling technique are shown in Figure 2. (a) Resolution of an ferrocene derivative, (b) resolution of 70 mg of a chiral spirobiindane, and (c) a simple example for establishing the enantiomeric purity (in asymmetric reductions of a ketone to the alcohol shown).

The second main aspect of Herman Mark's work—x-ray crystal structure analysis—is obviously also a very important feature of modern stereochemistry. The chemistry and stereochemistry of metallocenes (aromatic transition metal complexes) has been one of the major research topics in our laboratory for the last thirty years. This field is particularly accessible to x-ray crystal structure analysis because of the heavy metals (Fe, Ru, Mn, or Cr) incorporated in these chiral structures.

Titanium-induced reductive coupling of ferrocenylcarbinols, Fc—CH(R)OH (with R = e.g., methyl, phenyl, *t*-butyl, or even ferrocenyl), gives 1,2-diferro-

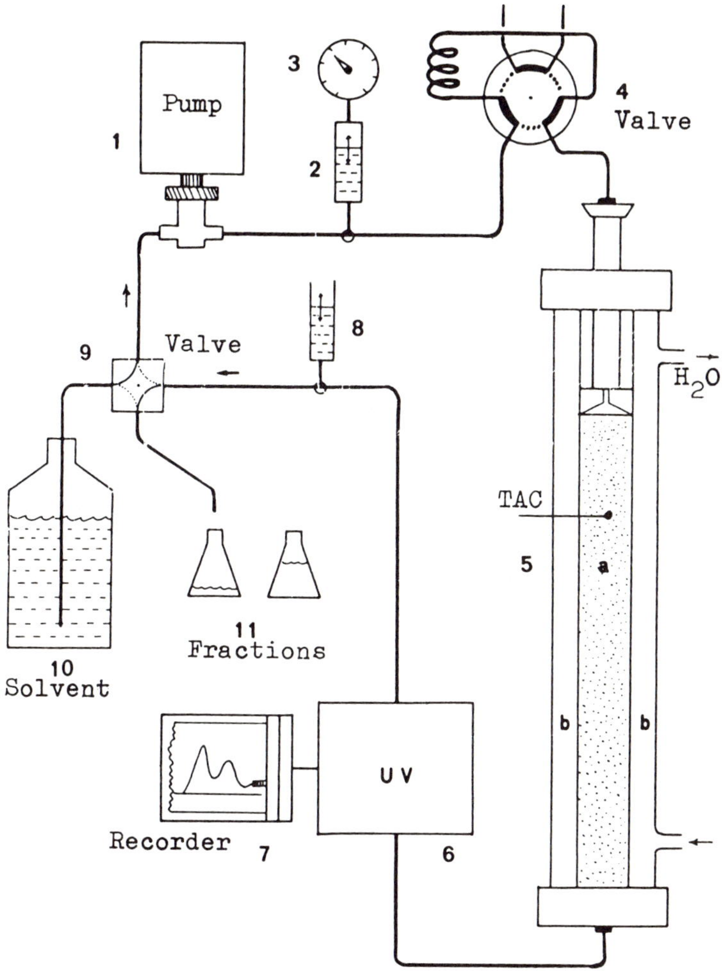

FIG. 1. Recycling chromatography on triacetylcellulose.

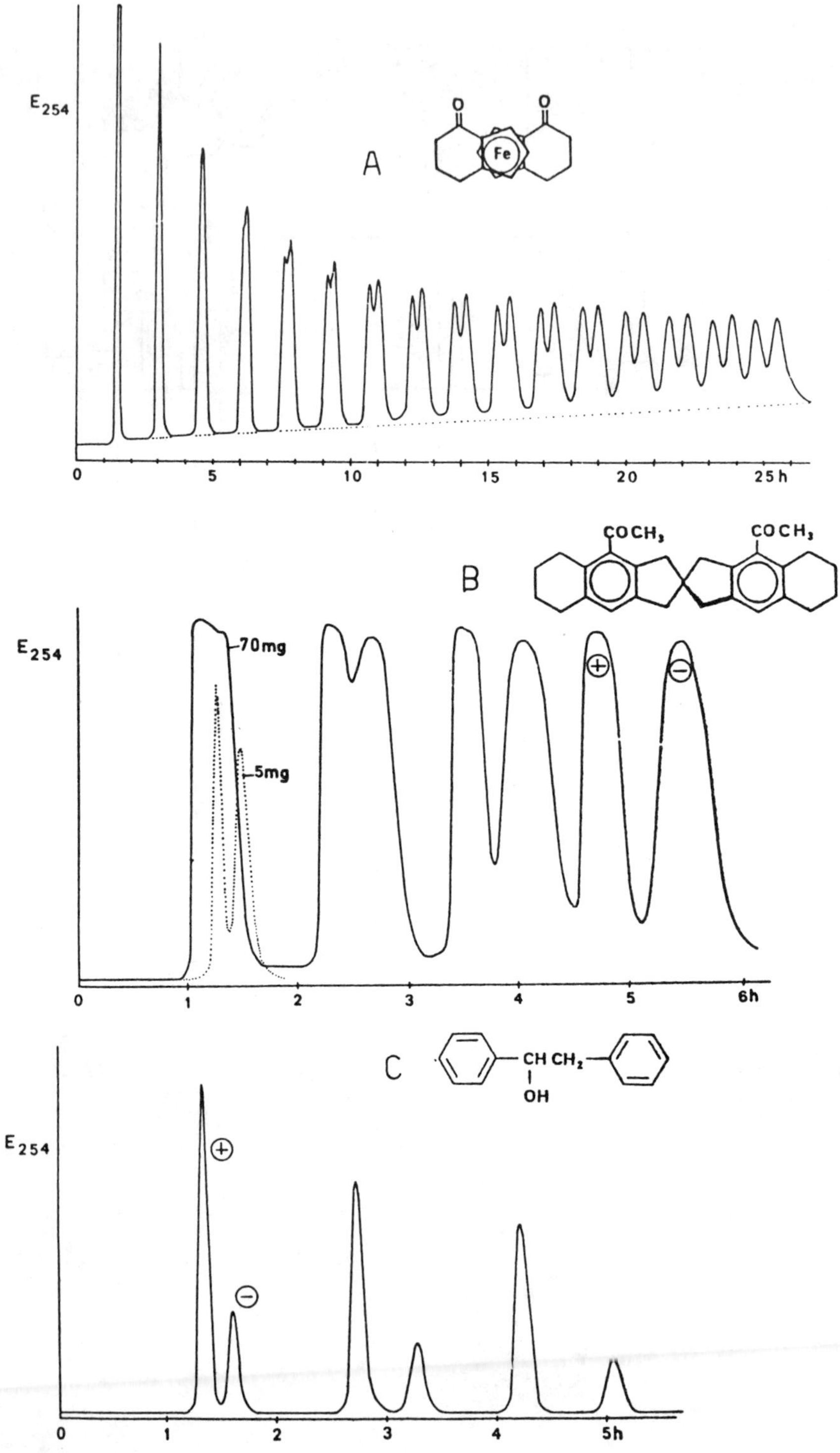

FIG. 2. Optical resolutions on triacetylcellulose in ethanol.

cenyl ethanes in good to excellent yields. For diferrocenyl-diphenyl ethanes (R = phenyl) the two diastereomers were separated and their configuration (meso or racemic, respectively) assigned by x-ray analysis. Of even more interest is 1,1′, 2,2′-tetraferrocenyl ethane, $(Fc)_2CH—CH(Fc)_2$. X-ray structure analysis revealed the presence of two enantiomers with C_2 symmetry resembling a four-bladed propeller (Fig. 3). Due to its low solubility, we had no chance to resolve this chiral compound by chromatography on TAC, and we therefore prepared a topologically related propeller, namely tetrakis(2,6-dimethyl-4-methoxy-phenyl)ethane by the above-mentioned coupling method. The *ortho*-methyl groups twist the structure into a four-bladed propeller, which could be (at least partly) resolved into enantiomers on TAC. The enantiomeric relationship was established by circular dichroism spectroscopy (CD). One of the enantiomers had a specific rotation of −500°. The relatively low inversion barrier ($\Delta G^\ddagger \sim 93$ kJ/mol), as established by racemization experiments, required chromatography at lower temperatures.

The so-called anulenes are bridged cyclopolyenes with a conjugated 10-, 14-, or higher π-electron system in the perimeter thereby representing an interesting type of Hückel-aromatic systems. If appropriately substituted, they become (planar)chiral and are thus fascinating models for stereochemical research. In collaboration with Professor Vogel and his group in Cologne (FRG) we have applied both TAC chromatography (for the separation of enantiomers) and x-ray crystal structure analysis (Bijvoet technique, for establishing absolute chiralities). 2,7-Methano[10]aza-anulene is a simple example of this type of chirality, which owes its chirality to the presence of nitrogen within the ring. It

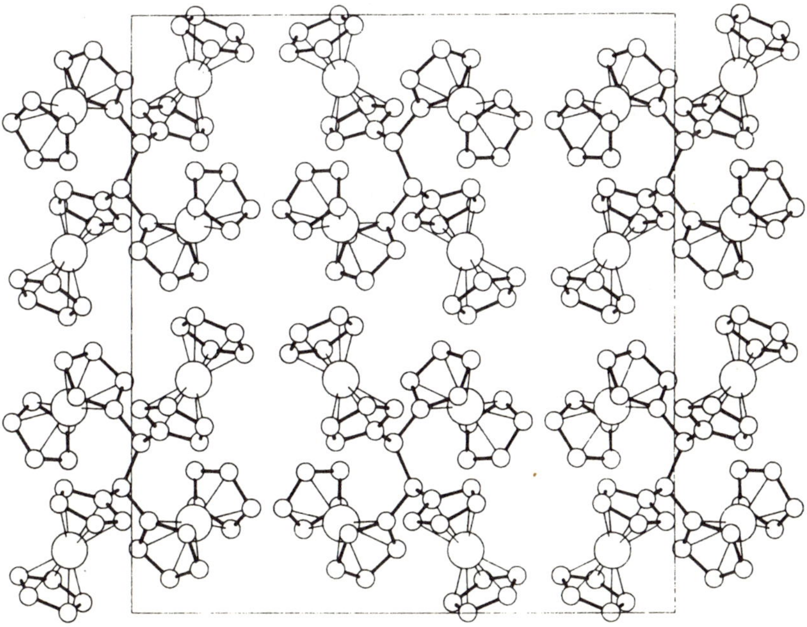

FIG. 3. X-ray crystal structure of tetraferrocenyl ethane.

was accessible only in small amounts: This together with its low basicity prevented a classical resolution. Chromatography on TAC, however, gave both optically pure enantiomers. This type of resolution was also successful for several other [10]- and [14] anulenes and aza-anulenes. Some typical examples of such separations are schematically depicted in Figure 4.

The dibromo-[14] anulenes with NH- or O- bridges (having C_2 symmetry) were of particular interest with regard to their optical stability: Racemization experiments revealed an inversion barrier of 134 kJ/mol for the bisoxido-[14]anulene, whereas the bisimino derivative is optically stable until 250°C (for 2 h), and consequently its inversion barrier is higher than 170 kJ/mol.

Application of the Bijvoet technique to the dextrorotatory enantiomers of 2,7-dibromo-1,6-methano[10]anulene and 2,9-dibromo bisimino [14] anulene (Fig. 4) established their absolute chiralities as $(+)(R)_p$ and $(+)(S)_p$, respectively. By thorough comparison of the CD spectra (which, e.g., are almost superimposable for the bisoxido and bisimino[14]anulenes) and several chemical correlations (e.g., interconversion of Br into $COOCH_3$) the absolute configura-

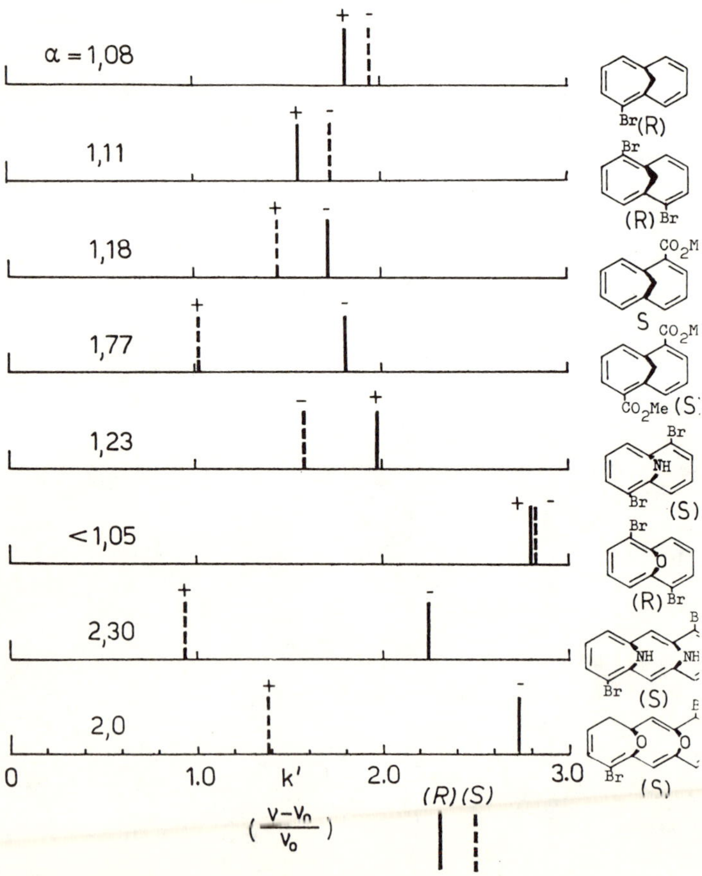

FIG. 4. Optical resolutions of several chiral bridged [10] and [14]anulenes.

tions (chiralities) of all anulenes investigated so far could be unambiguously established.

In the last part of this brief survey on our stereochemical work some recent results in the field of arene tricarbonylchromium complexes shall be presented:

Within the class of metallocenes these complexes (sometimes referred to as benchrotrenes) occupy a special position and proved to be very useful for stereochemical investigations for several reasons: They can be easily prepared from arenes and $Cr(CO)_6$ (or related chromium derivatives), are colored, which helps in chromatographic separation and isolation, and they are usually nicely crystalline which makes them excellent candidates for x-ray crystal structure determination (especially for establishing their absolute chiralities by the Bijvoet technique). Most important, by attaching the $Cr(CO)_3$ moiety, the symmetry of the parent arene is decreased, for example from C_s to C_1, thereby leading to chiral structures. And finally, $Cr(CO)_3$ can easily be removed under mild conditions (e.g., photochemically at low temperatures) thereby restoring an arene which might have been stereoselectively modified at the benchrotrene stage. This principle—using $Cr(CO)_3$ as a very useful stereoselective protecting group, is schematically shown in Figure 5.

For the optical resolution of benchrotrenes, classical methods (e.g., via diastereomeric derivatives) have been not very successful, whereas the above-described chromatography on TAC (particularly the recycling technique) has been proved to be the method of choice, since it permits for instance separation of labile enantiomers at lower temperatures.

Biphenyls (the stereochemistry of which has been extensively studied by Prof. K. Mislow and his school) are good examples to demonstrate the versatility of the chromium tricarbonyl approach. As schematically depicted in Figure 6, complexation by $Cr(CO)_3$ introduces a second element of chirality (namely the "metallocene chirality" in addition to the "biphenyl" = axial

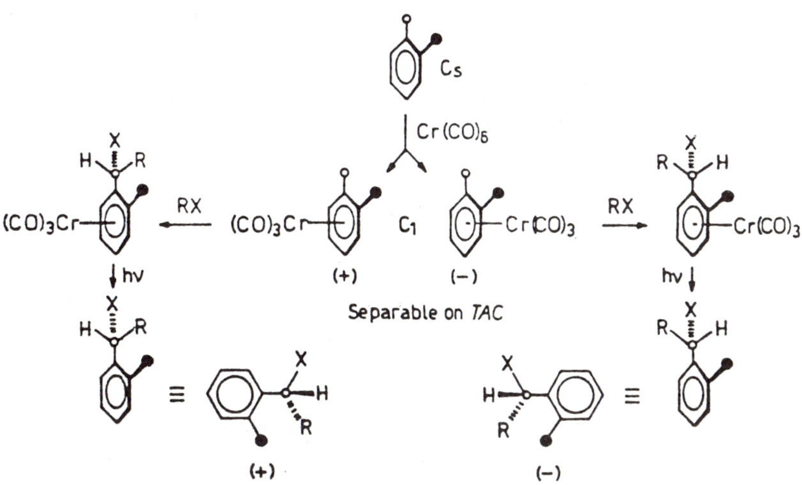

FIG. 5. Tricarbonyl chromium as stereoselective protecting group.

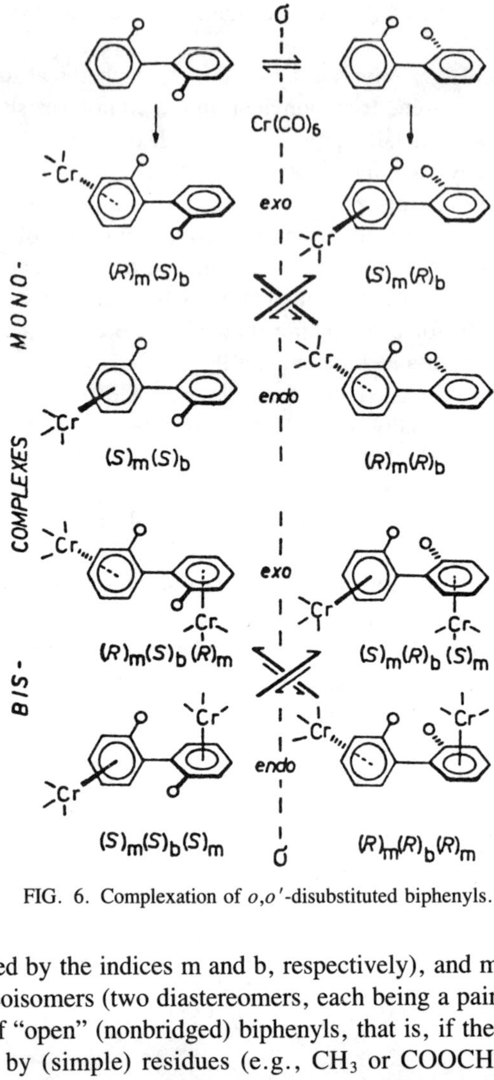

FIG. 6. Complexation of *o,o*′-disubstituted biphenyls.

chirality, denoted by the indices m and b, respectively), and may therefore give rise to four stereoisomers (two diastereomers, each being a pair of enantiomers).

In the case of "open" (nonbridged) biphenyls, that is, if the *ortho*-position(s) are substituted by (simple) residues (e.g., CH_3 or $COOCH_3$), x-ray crystallography revealed that only one of the two possible diastereoisomers is formed; namely this one which we denote as "exo" and in which the large groups point into the four quadrants of the projection formula (see Fig. 7 for a typical example).

After some tentative conclusions on the absolute chiralities based on chiroptical comparison with simple benchrotrenes of established configuration, these results could be confirmed by a combination of methods as outlined in the Figure 8.

The $Cr(CO)_3$-complex of dimethyldiphenicacid dimethylester was partially resolved on TAC after 12 cycles. The (+) fraction was decomplexed, and the ligand obtained thereby was completely separated into its enantiomers on TAC, thereby revealing an enantiomeric excess of 65%. The biphenyl chirality of the

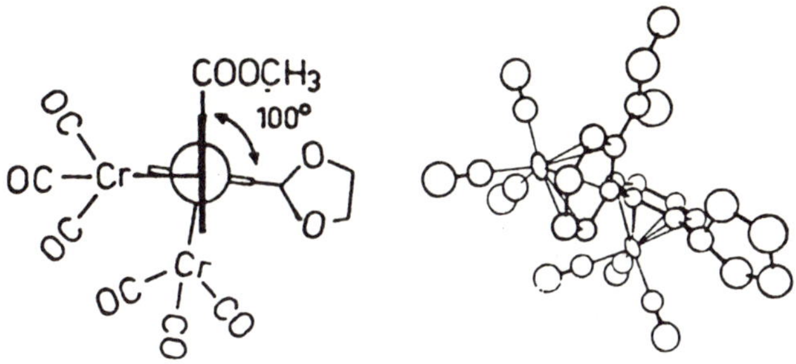

FIG. 7. X-ray crystal structure of a biphenyl bis-$Cr(CO)_3$-complex.

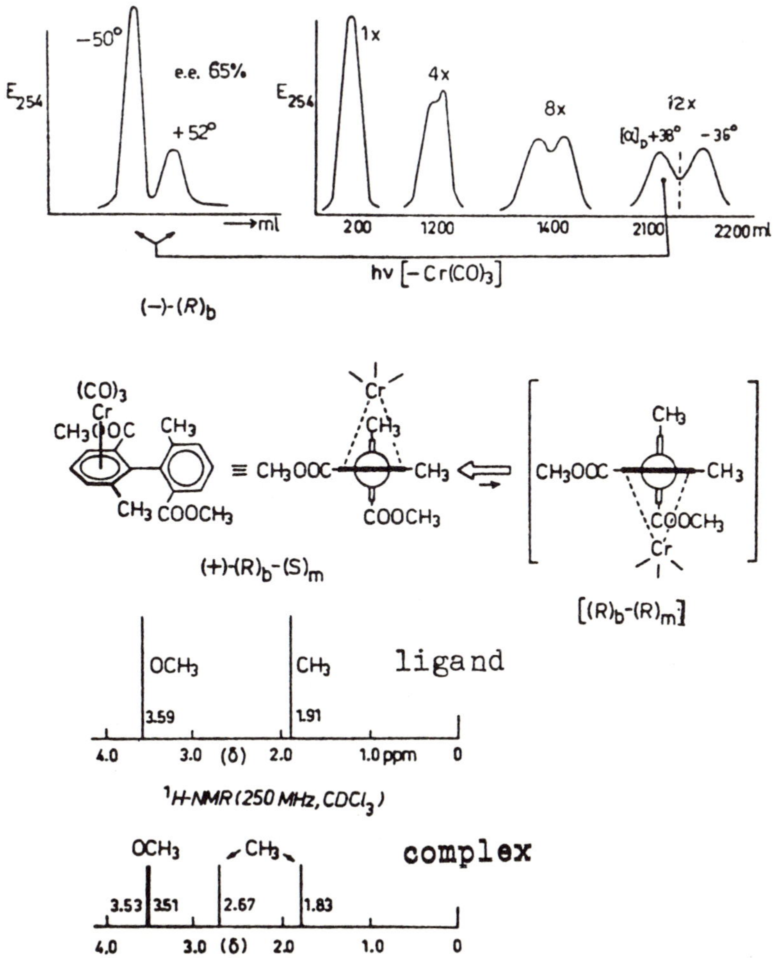

FIG. 8. Absolute chirality of the mono-$Cr(CO)_3$ complex of dimethyl diphenicacid dimethylester by combined TAC chromatography and NMR spectroscopy.

ligand was known to be $(-)\,(R)_b$. The metallocene chirality, on the other hand, was established by nuclear magnetic resonance (NMR) spectroscopy both of the ligand and the complex. Since in the complex one of the methyl signals is shifted downfield, it is evident that the $Cr(CO)_3$-moiety is on the same side as methyl, acting as an internal shift reagent for the latter. These results confirm the chirality $(+)\,(R)_b(S)_m$ and of several other mono and bis complexes which had been correlated with that shown.

Several new and promising aspects, such as with respect to torsional barriers, were found in $Cr(CO)_3$- and related complexes of bridged biphenyls. Again, the general situation is depicted in Figure 9.

Here, too, two diastereomers—exo and endo—are feasible which can be interconverted by torsion around the central biphenyl bonds, thereby changing $(R)_b$ into $(S)_b$ and vice versa, whereas the metallocene chirality is stable and no racemization takes place as has been proved by experiment. The methylated ligands (Fig. 9) give an almost equimolar mixture of both stereoisomers, whereas in the unmethylated ones the exo isomers predominate by more than

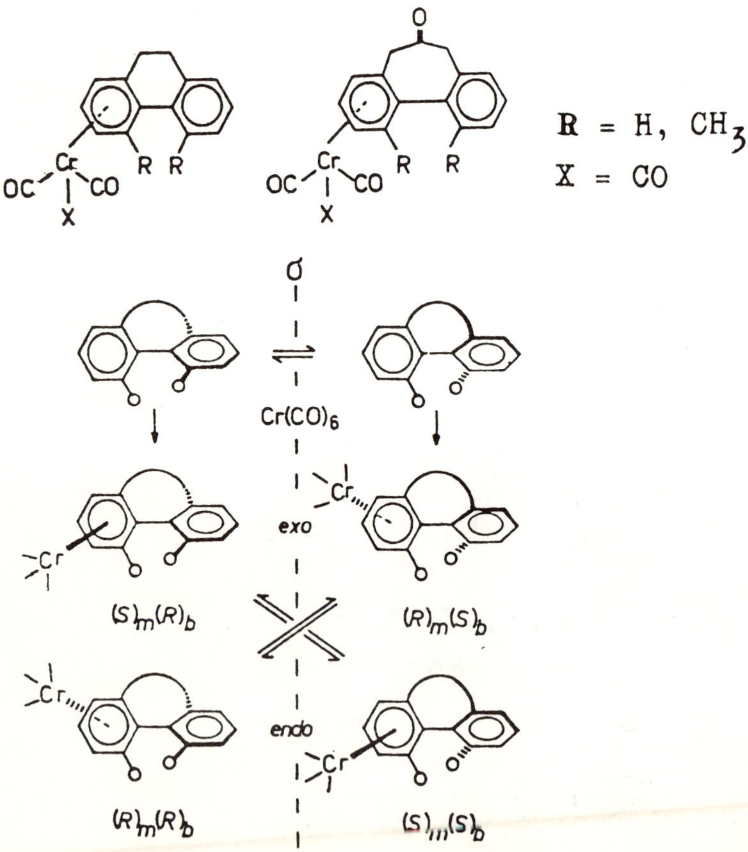

FIG. 9. Complexation of *o,o′*-bridged biphenyls.

90%. These results can be explained by steric interactions between the bridges on one side and $Cr(CO)_3$ or methyl on the other. The configurational assignments as well as the exo/endo ratios are based on temperature-dependent NMR spectra; the rotational barriers could also be deduced from racemization experiments. In the case of dimethyldihydrophenanthrene, complexation decreases the barrier by approximately 10 kJ/mol.

The stereoisomeric complexes can be separated on TAC: e.g., for the dimethylated bridged ketone all four isomers—(+)-, (−)-exo, (+)-, and (−)-endo—were obtained in one run!

One of the bridged biphenyl complexes, namely the bis complex of a lactone (a biphenyl bridged in *o* positions with $—CH_2—O—CO—$) was the first example for the preparation of an optically labile biphenyl by the $Cr(CO)_3$ approach. Photochemical decomplexation at −50°C afforded the optically active ligand; according to its low temperature CD spectrum (where it is stable for at least one hour at −50°C), it has a half time of racemization of approximately 10 min at −20°C. This in very good agreement with a recently established rotational barrier for the lactone of 65 kJ/mol.

In order to study a possible influence on the rotational barrier(s) and the exo/endo ratios, one of the CO groups in the bridged tricarbonyl chromium complexes was exchanged against CS or $P(Ph)_3$. These modified complexes could also be separated by chromatography on TAC (into enantiomers) and subsequently on silica gel (at lower temperatures) into the exo/endo stereoisomers. Whereas the $\Delta G^{\ddagger}$ values differ only slightly from those of the $Cr(CO)_3$ complexes, there are significant differences in the exo/endo ratios of the $P(Ph)_3$ complexes (obviously due to the bulkiness of the latter). These ratios are for dimethyldihydrophenanthrene: $Cr(CO)_3$ 50:50; $Cr(CO)_2P(Ph)_3$ 98:2.

Compounds with two aryl rings attached to a central unit such as CH_2 or CO are the simplest representatives of polyaryls which may undergo correlated rotation. This means that rotation of one ring causes the rotation of others within the same molecule. In order to sudy this phenomenon for $Cr(CO)_3$ complexes, we have extended our work to the complexation of diaryl methanes, ethanes, and ketones, where aryl is 2-methyl-6-ethyl- or -6-t-butyl-phenyl (also substituted in position 4 by OCH_3 or CH_3). The general situation is depicted in Figure 10.

For $R^1 \neq R^2$, two pairs of diastereomers are feasible, in which two of these are interconvertible. This equilibrium is determined by the height of the barriers of interconversion and is responsible for whether $Cr(CO)_3$ can be used as an auxiliary group for optical resolution. In such unsymmetrically substituted diaryls, correlated rotation leads to two chiral isomers with C_2 symmetry, whereas uncorrelated movement gives only one chiral isomer.

NMR kinetics, also supported by forcefield calculations, revealed that for the ethyl-methyl compound, monocomplexation decreases the barriers in the methane and ethane by about 4 kJ/mol, whereas in the ketone this barrier is increased by as much as 30 kJ. The barriers of the bis complexes are insignificantly (2–3 kJ) lower than those of the mono complexes.

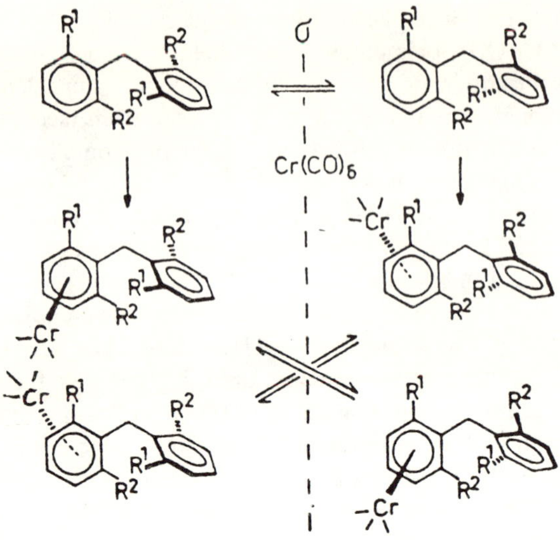

FIG. 10. Complexation of diarylmethanes.

In the case of bis(2,4-dimethyl-6-*t*-butyl-phenyl)ketone, monocomplexation yields a separable mixture of (stable) isomers in the ratio of 4:1. Their configurational assignment follows from their ^{1}H NMR spectra: Whereas the ligand shows a rather simple spectrum due to the low barrier (80 kJ/mol), one isomer exhibits a significant splitting of the *t*-butyl signal and has therefore the *t*-butyl group on the same side as $Cr(CO)_3$ and is thus thermodynamically less stable than the other isomer which is formed in 80% excess. Both isomers can be decomplexed by irradiation to one single compound—namely the original ligand.

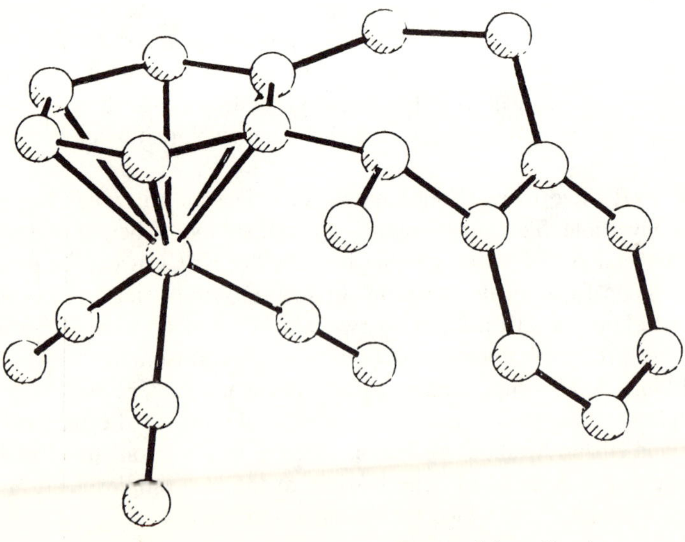

FIG. 11. X-ray crystal structure of dibenzocycloheptadienol.

These results represent an interesting example of stable torsional isomers = diastereomers.

From our work on the use of $Cr(CO)_3$ as a stereochemical "holding group," only one example is presented in Figure 11. The monocomplex of dibenzocycloheptadienol has a bowl-shaped (rigid) structure with the $Cr(CO)_3$ moiety "within" the bowl.

Karl Schlögl was born on October 5, 1924.

PRESENT POSITION

Professor of Organic Chemistry
Institute of Organic Chemistry
University of Vienna (Austria)
A-1090 Vienna, Austria
Director of Institute

EDUCATION

University of Vienna, Austria; Ph.D. 1950
British Council Scholar, University of Manchester (GB) 1954–1955

POSITIONS

Instructor, University of Vienna 1949–1958
Dozent (Assistant Professor) 1959
Full Professor, University of Vienna, 1970–
Dean, Faculty of Science, University of Vienna, 1977–1979
Director, Institute of Organic Chemistry, University of Vienna, 1974–
Visiting Professor, Japan, 1983

PROFESSIONAL ACTIVITIES

Austrian Chemical Society, President 1985–1987
Monatshefte für Chemie, Managing Editor
XII. International Conference on Organometallic Chemistry (Vienna, Sept. 1985) Chairman

AWARDS

Austrian Academy of Sciences, Fellow, 1982
Annual Vienna University Award, 1983

NEW CATALYSTS FOR POLYMER SYNTHESIS

GÜNTHER WILKE
Max-Planck-Institut für Kohlenforschung, Kaiser-Wilhelm-Platz 1, D-4330 Mülheim-Ruhr, Germany

It is my great privilege on behalf of the Max-Planck Society to congratulate Professor Herman Mark to his 90th birthday. A close relationship has always existed between Professor Mark and both the Kaiser-Wilhelm-Gesellschaft, now the Max-Planck-Gesellschaft, and the Max-Planck-Institut für Kohlenforschung. Therefore, it is a special honor and pleasure for me to be asked to contribute to this splendid birthday celebration. I would also like to take this opportunity to thank Professor Mark for all the help and interest he has extended to our institute over the years.

Having received my Ph.D. with Karl Freudenberg in Heidelberg, I would like to begin by reminding you of a fundamental publication by K. H. Meyer and H. Mark with the title "Über den Bau des krystallisierten Anteils der Cellulose" which appeared in 1928 [1]. This paper describes for the first time the elucidation of the structure of a high polymer by x-ray crystallography and was a real breakthrough in the field. The chemically based results of Karl Freudenberg [2] were confirmed in a most elegant manner (Figs. 1–3).

I would like now to show you, using an example from our own research, how the state of the art in the structural analysis by x-ray and neutron diffraction has developed. In 1961 we described the synthesis of the prototype of the so-called "pure" or "homoleptic" η^3-allyl transition metal complexes-*bis*-η^3-allylnickel, $(\eta^3\text{-}C_3H_5)_2$ [3] (Fig. 4).

This complex is of considerable interest for the theoreticians. At least four different groups have carried out *ab initio* MO calculations assuming that the allyl groups are planar ligands, that is, the 5 H atoms as well as the 3 C atoms occupy the same plane [4]. A very recent structural investigation by Krüger and Goddard in our institute applying both x-ray and neutron diffraction methods at very low temperatures proved, however, that the allyl groups are not planar. The syn- and meso-H-atoms are bent toward the metal whereas the anti-H atoms are pushed away [5] (Fig. 5).

At about the same time F. Mark in our neighboring institute, the Max-Planck Institute for Radiation Chemistry, carried out single-zeta and double-zeta calculations on $(\eta^3\text{-}C_3H_5)_2Ni$ [6] without any knowledge of the data obtained by Krüger and Goddard, and he found an energy minimum for the molecule if the H atoms are located at positions very close to those found in the structural analyses. The result is a stabilization of the molecule by about 50 Kcal compared with a planar arrangement (Fig. 6). This highly successful correlation be-

Berichte der Deutschen Chemischen Gesellschaft.

1928, Nr. 4 — Abteilung B (Abhandlungen) — 11. April.

95. Kurt H. Meyer und H. Mark: Über den Bau des krystallisierten Anteils der Cellulose.

[Aus d. Hauptlaborat. d. I.-G. Farbenindustrie A.-G. in Ludwigshafen a. Rh.]

(Eingegangen am 13. Februar 1928.)

FIG. 1. Title of the basic publication.

tween the theoretical and experimental results is particularly striking because of the size of the molecule involved.

The x-ray and neutron diffraction data give even more information on this molecule since we have been able to determine the electron deformation density and hence have detailed insight into the bonding situation within the molecule (Fig. 7).

Let me now turn to the use of these η^3-allyl complexes in catalysis. In many cases these "pure" η^3-allyl transition metal compounds [7] show high catalytic activity. Ethylene, for example, can be polymerized by $(\eta^3\text{-}C_3H_5)_3Cr$ to high molecular weight linear polyethylene [8]. This was the first example of the polymerization of ethylene by a single halide-free transition metal compound. Indeed, large-scale production using this catalyst system was contemplated by a Japanese company (Fig. 8). On the basis of $\eta^3\text{-}C_3H_5NiX$ (X = Cl, Br), we developed a process whereby propylene can be dimerized to 2, 3-dimethylbutene [9] which after hydrogenation gives 2, 3-dimethylbutane, a high octane number hydrocarbon and a valuable additive for gasoline. The $\eta^3\text{-}C_3H_5NiX$ must be activated by a Lewis acid such as $EtAlCl_2$, and modified by a phosphane ligand (Fig. 9). The data in Figure 10 give some impression of the activity of this type of catalyst [10]. The same catalyst which dimerizes propylene will also dimerize ethylene readily even at around −70°C. The ligand controlling the direction of reaction is (+—)$_2$P —< (Fig. 11).

If the isopropyl group in this ligand is replaced by a *t*-butyl group then we obtain a ligand (tris-*t*-butylphosphane) which modifies the catalyst drastically: instead of the dimerization of ethylene, polymerization is observed to give high molecular weight linear polyethylene. One methyl group more or less in the ligand can have such a marked influence upon the catalytic process [11] (Fig. 12).

Unfortunately, the catalysts described here can be applied only to unsaturated hydrocarbons. If the olefins are substituted by functional groups such as ether or other groups then the catalyst becomes deactivated as the result of cleavage of the complex anion formed by the Lewis acid with the halide atom of the η^3-

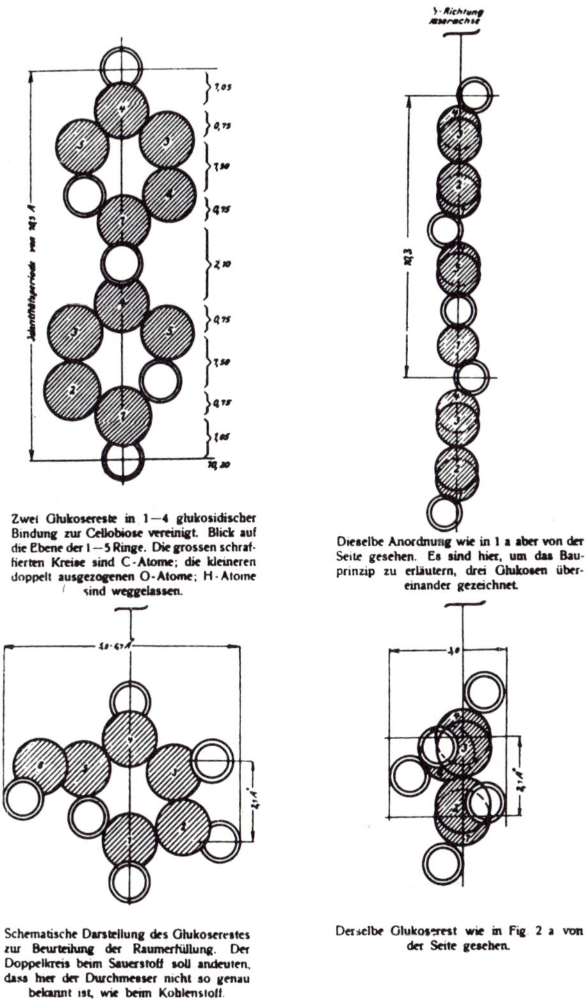

Kurt H. Meyer und H. Mark, B. **61**, 593 [1928].

FIG. 2. Schematic representation of the structure of cellulose.

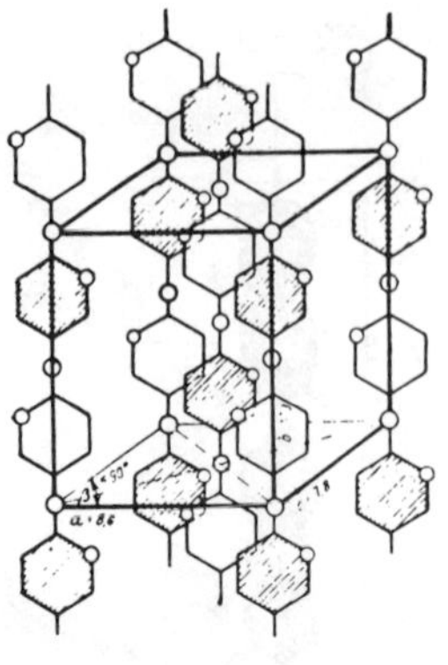

Perspektivische Darstellung der Anordnung der Cellobiosefäden im Elementarkörper. Die schraffierten bezw. nicht schraffierten Sechsecke (Hexosereste) gehen auseinander durch digonale Schraubung um die b-Achse hervor.

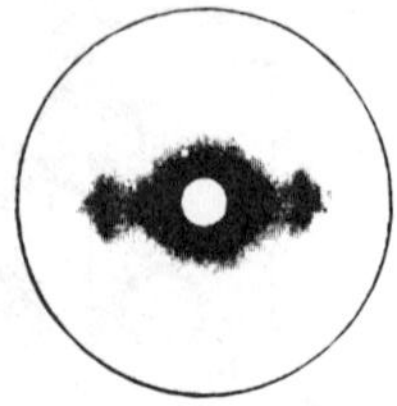

Diagramm des Cellulosetrinitrats. Nitrierung erfolgte unter Spannung bis gegen 13.9%; Cu K-Strahlung

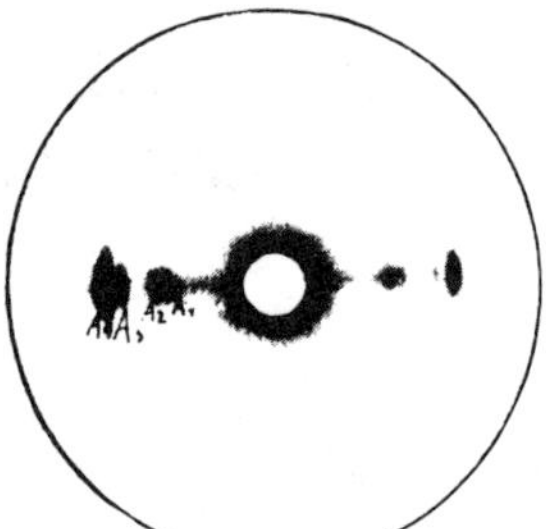

Faserdiagramm der nativen Cellulose mit Cu K-Strahlung (Ramie)

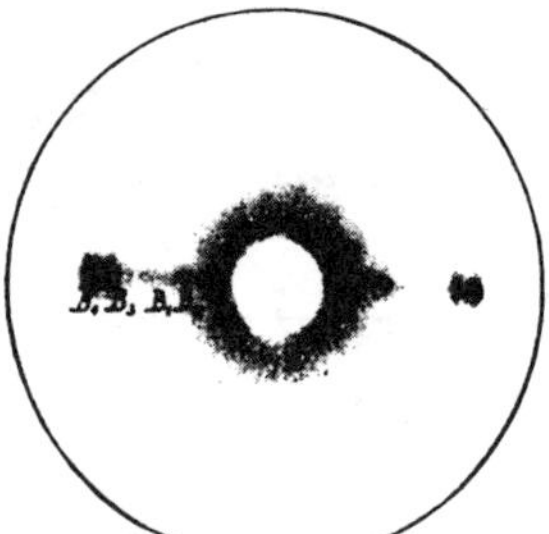

Faserdiagramm einer unter Spannung merzerisierten und nachher wieder ausgewaschenen Ramie; Cu K-Strahlung

FIG. 3. Perspective of cellulose chains and Debye-Scherrer diagrams.

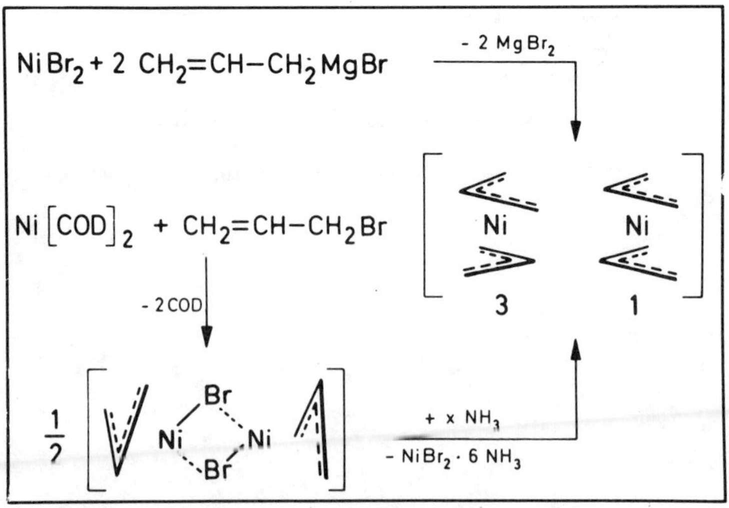

FIG. 4. Synthesis of bis-η^3-allylnickel.

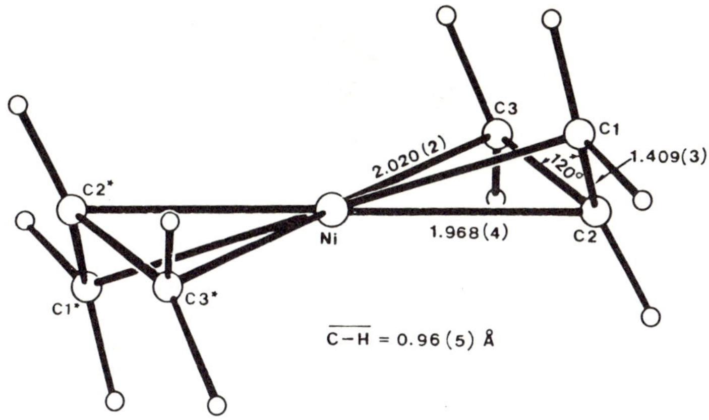

FIG. 5. X-ray structure of bis-η^3-allynickel.

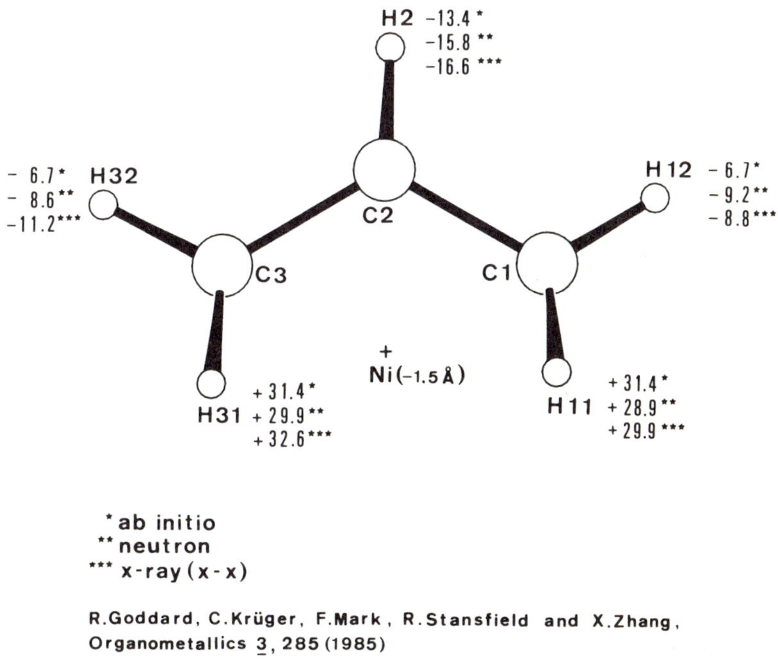

FIG. 6. Structural data by ab initio calculations, neutron- and x-ray diffraction.

C_3H_5-NiX moiety. Nevertheless, it did prove possible to develop a catalyst with which acrylic ester may be dimerized to monounsaturated adipic acid ester. This was accomplished by introducing a more stable complex anion, the BF_4-anion [12] (Fig. 13).

In addition to the monounsaturated adipic acid ester, minor amounts of a branched ester are formed. These esters are of some interest as monomers or

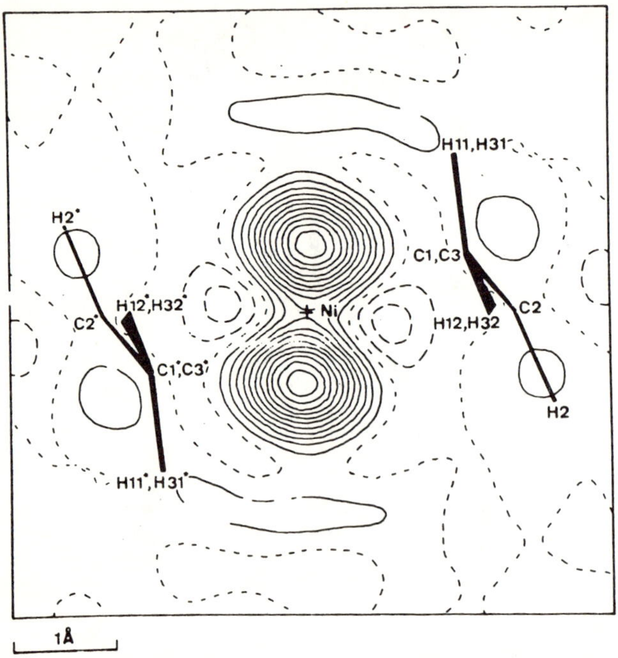

FIG. 7. Deformation electron density.

comonomers for the synthesis of unsaturated polyesters. The catalysts described here can also be modified by optically active phosphane ligands such as dimenthylisopropylphosphane. We have studied the codimerization of norbornene with ethylene in some detail [13]. The product, exo-vinylnorbornane, is a chiral compound and by decreasing the temperature, the enantiomeric excess (e.e.) can be increased to 70% at −70°C (Fig. 14). This reaction offers some futuristic aspects and it should be possible to increase the e.e. up to 95–100%. The resulting enantiomer containing three chiral centers would be a most interesting monomer or comonomer for the production of polymers (Fig. 15). The homopolymerization would lead to an optically active polymer which itself might exhibit interesting new properties.

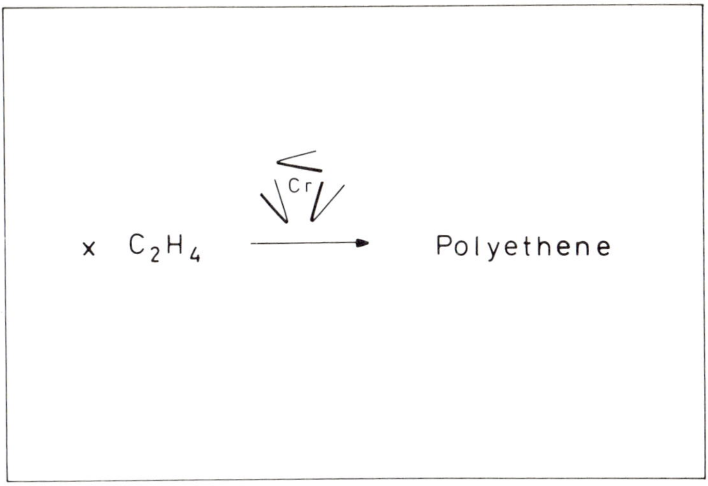

FIG. 8. $(\eta^3\text{-}C_3H_5)_3Cr$, a catalyst for the polymerization of ethene.

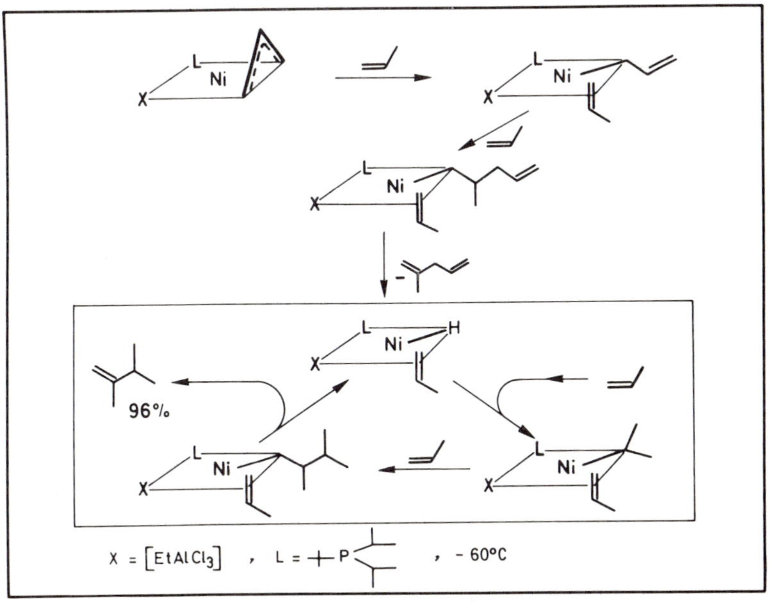

FIG. 9. Mechanism of the dimerization of propene.

Finally, I would like to mention a very recent development by Fink and Möhring in our institute [14]. If 1-butene is polymerized by a typical Ziegler catalyst one obtains a product which can be described as polyethylene with ethyl branches on every second carbon atom of the chain. An entirely new type of polymerization was observed using a special nickel catalyst and in this case

Dimerization of Propene

Catalyst : η^3-C_3H_5NiBr/ $P(C_6H_{11})_3$ / $EtAlCl_2$

Solvent : C_6H_5Cl

Temp. : - 55°C

Rate : 35 g C_6/µmol catalyst · h

35 t C_6/ mol catalyst · h

0,6 t C_6/1 g Ni · h

Turn-over number · sec^{-1} : 230

FIG. 10. Catalyst activity.

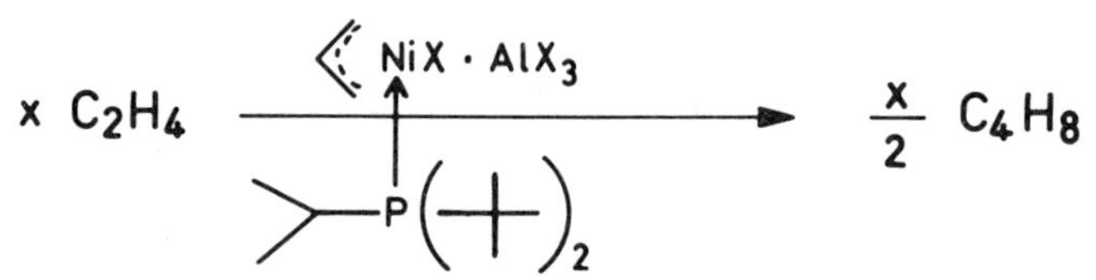

B. Bogdanović and H. G. Karmann , MPI für Kohlenforschung

G.Wilke, „Homogeneous Catalysis" Vol.3 , Plenum Press 1979

FIG. 11. Dimerization of ethene.

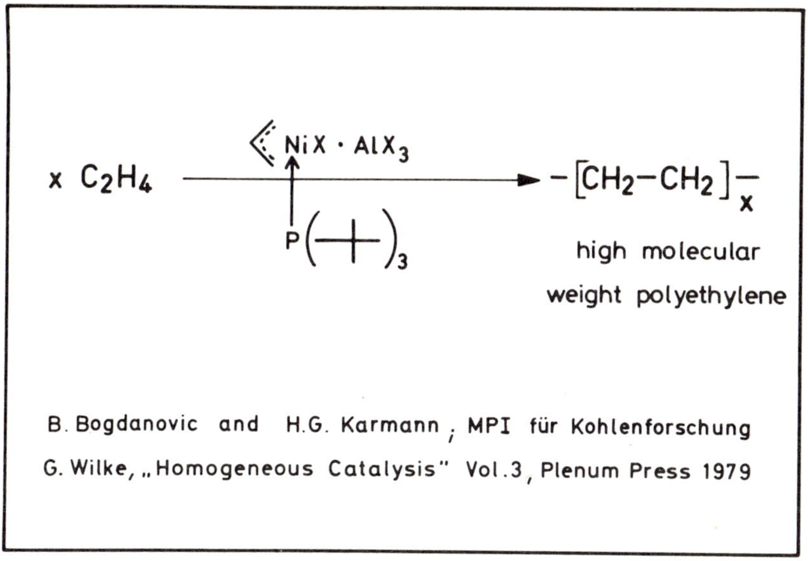

FIG. 12. Polymerization of ethene, influence of the ligand.

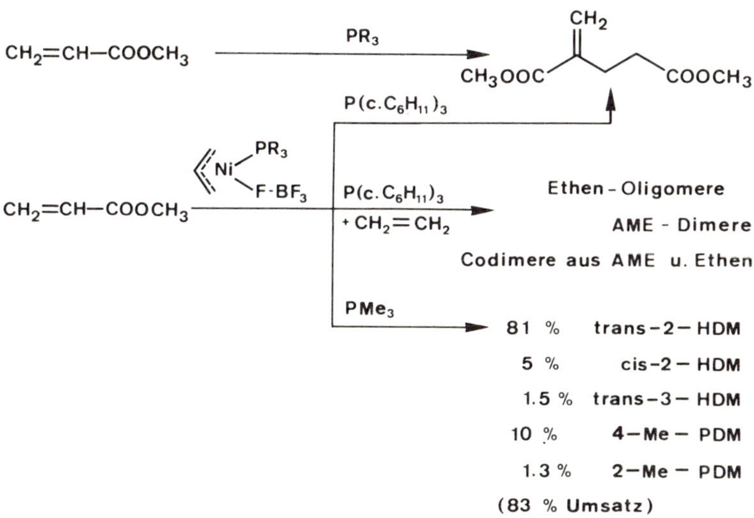

FIG. 13. Dimerization of acrylic acid ester.

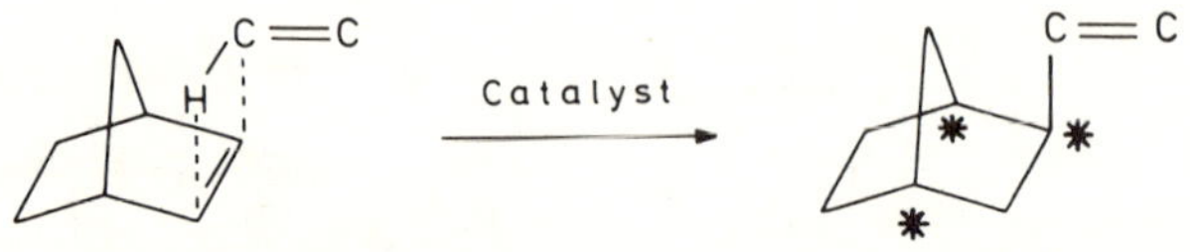

FIG. 14. Codimerization of norbornene and ethene by asymmetric catalysis.

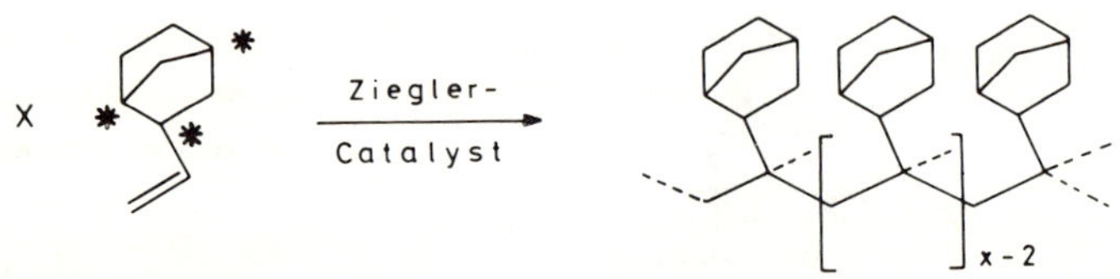

FIG. 15. Polymer from exo-vinylnorbornene.

1-butene yields in a polymer in which there is a methyl branch on every third carbon atom of the chain (Fig. 16).

Fink and Möhring have investigated a whole series of α-olefins from C_4 to C_{20} and in all cases they obtained methyl-branched polymers only. Whereas 1-butene gives a polymer which is substituted on every third carbon atom of the

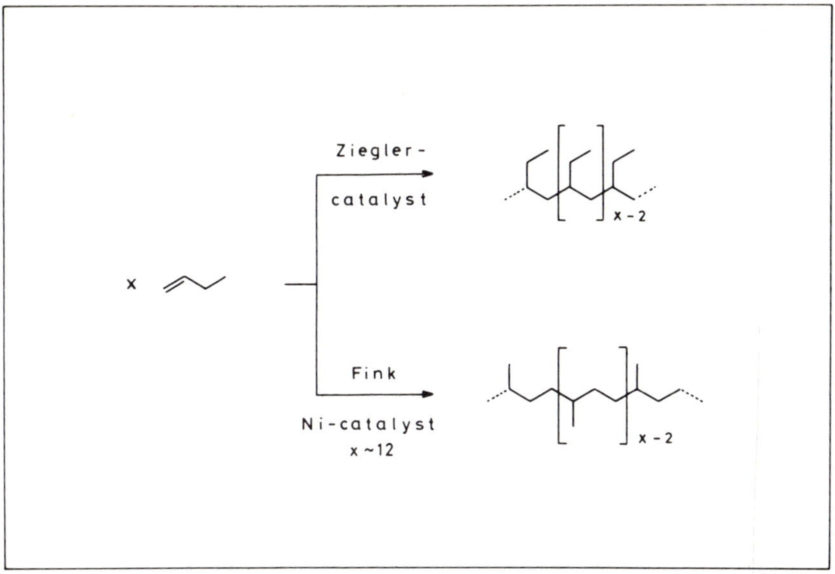

FIG. 16. Ziegler and Fink polymerization of 1-butene.

chain by a methyl group, the polymer formed by 1-eicosene is characterized by methyl substituents on every nineteenth carbon atom of the chain (Fig. 17). This remarkable type of polymerization can be mechanistically understood as a fast migration of the metal atom through the chain after each insertion step. The migration takes place by a sequence of β-hydrogenation/elimination and metal hydride addition steps and only when the metal reaches the end of the chain does the insertion of a further α-olefin take place (Fig. 18).

This is the first time that an α-olefin polymerization of this type has been observed. The structure of the polymers has been elucidated by detailed nucleus

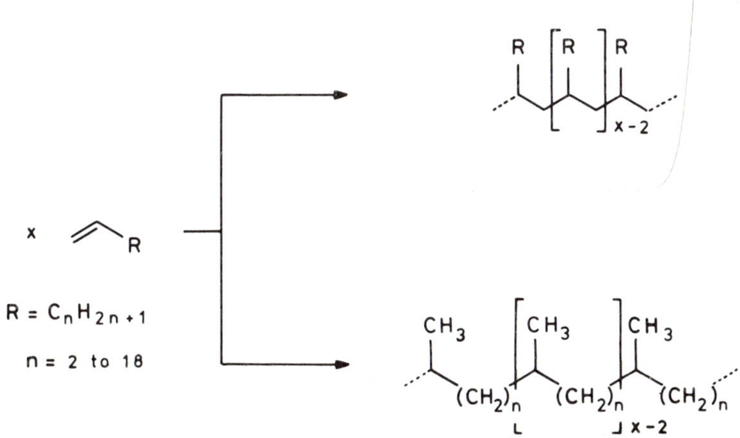

FIG. 17. Fink polymerization of α-olefines.

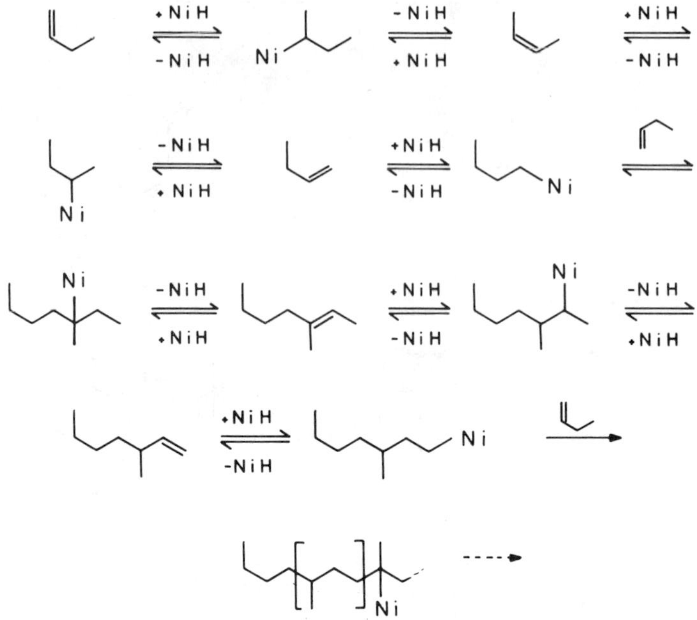

FIG. 18. Proposed mechanism of the Fink polymerization.

magnetic resonance (NMR) studies and the investigations are being continued with an eye to both, technical applications as well to the elucidation of the mechanism.

With this short retrospect as well as glimpse into future developments, I want to congratulate the "Altmeister" of polymer chemistry upon his ninetieth birthday.

REFERENCES

[1] K. H. Meyer and H. Mark, *Ber.,61,* 593 (1928).

[2] (a) K. Freudenberg, *Ber., 54,* 767 (1921); (b) K. Freudenberg and E. Braun, *Liebigs Ann. Chem.,460,* 288 (1928); (c) K. Freudenberg, *Liebigs Ann. Chem., 461,* 130 (1928).

[3] G. Wilke and B. Bogdanović, *Angew. Chem.,73,* 756 (1961).

[4] (a) A. Veillard, *Chem. Commun.,* 1022, 1427 (1969); (b) M. -M. Rohmer and A. Veillard, *J. Chem. Soc, Chem. Commun.* 250 (1973); (c) M. -M. Rohmer, J. Demuynck, and A. Veillard, *Theor. Chim. Acta, 36,* 93 (1974); (d) I. H. Hillier and R. M. Canadine, *Discuss. Faraday Soc., 47,* 27 (1969); (e) D. A.Brown and A. Owens, *Inorg. Chim. Acta, 5,* 675 (1971); (f) G. C. Hancock, N. M. Kostić, and R. F. Fenske, *Organometallics, 2,* 1089 (1983).

[5] R. Goddard, C. Krüger, R. Stansfield, and X. Zhang, *Organometallics, 3,* 285 (1985).

[6] F. Mark, *Organometallics, 3,* 285 (1985).

[7] G. Wilke, B. Bogdanović, P. Hardt, P. Heimbach, W. Keim, M. Kröner, W. Oberkirch, K. Tanaka, E. Steinrücke, D. Walter, and H. Zimmermann, *Angew. Chem., 78,* 157 (1966); *Angew. Chem. Internat. Edit., 5,* 151 (1966).

[8] (a) Studiengesellschaft Kohle m.b.H., Erf.: G. Wilke, U.S. Pat. 3379706 (Prior. 10.8.1963); (b) W. Oberkirch, dissertation, Bochum (1963).

[9] B. Bogdanović and G. Wilke, *Brennstoffchemie, 49,* 323 (1968).

[10] B. Bogdanović, B. Spliethoff, and G. Wilke, *Angew. Chem., 92,* 633 (1980); *Angew. Chem. Internat. Edit., 19,* 622 (1980).

[11] G. Wilke, *Pure Appl. Chem.*, *50,* 677 (1978).
[12] (a) G. Wilke, 5th IUPAC Symp. Org. Synth. Freiburg 1984, H. Streith, H. Prinzbach, and G. Schill, *Organic Synthesis an Interdisciplinary Challenge,* Blackwell, Oxford, 1985, pp. 1–15; (b) K. Sperling, dissertation, Bochum (1983).
[13] B. Bogdanović, B. Henc, A. Lösler, B. Meister, H. Pauling, and G. Wilke, *Angew. Chem.*, *85,* 1013 (1973); *Angew. Chem. Internat. Edit.*, *14,* 954 (1973).
[14] V. M. Möhring and G. Fink, *Angew. Chem.*, *97,* 982 (1985); *Angew. Chem. Internat. Edit.*, *24,* 1001 (1985).

Gunther Wilke was born in Heidelberg, February 23, 1925.

POSITIONS AND ACTIVITIES

Studied Chemistry in Karlsruhe and Heidelberg, 1946–1951.
Dr. rer. nat. with Karl Freudenberg, 1951.
Scientific assistant to Karl Ziegler at the Max-Planck-Institut für Kohlenforschung in Mülheim a. d. Ruhr, 1951.
Habilitation in Aachen, 1960.
Wissenschaftliches Mitglied of the Max-Planck-Instit für Kohlenforschung o. Professor at the Ruhr-University Bochum, 1963.
Offer of a chair at the ETH Zürich, 1966.
Second Director of the Max-Planck-Institut für Kohlen-forschung, 1967.
Dean of the Chemistry Department of the Ruhr-University Bochum, 1968–1969.
Appointed Director of the Max-Planck-Institut für Kohlenforschung and Managing Director of the Studiengesellschaft Kohle mbH, July 1, 1969.
Chairman of the Chemisch-Physikalisch-Technische Sektion and Member of the Senate of the Max-Planck-Gesellschaft, 1973–1976.
Chaire Francqui de l'Université de Louvain, Louvain-la-Neuve (Belgium), 1976.
Appointed Vice-President of the Max-Plank-Gesellschaft zur Förderung der Wissenschaften, 1978.
President of the Gesellschaft Deutscher Chemiker, 1980–1981.

HONORARY DEGREES

L'Université Louvain, Belgium, 1976.
University of Chicago, 1976.
Rheinisch-Westfälische Technische Hochschule Aachen, 1976.
Universidad de Oviedo, Spain, 1982.

AWARDS

The Ruhr Prize for Arts and Science of the town of Mülheium a.d. Ruhr, 1966.
Emil Fischer Medal of the Gesellschaft Deutscher Chemiker, 1970.
Chaire Francqui Medal, 1975.
Karl Ziegler Prize of the Gesellschaft Deutscher Chemiker, 1978.
Wilhelm Exner Medal of the Österreichischer Gewerbeverein, 1980.

HONORARY MEMBERSHIPS

Centennial Foreign Fellow of the American Chemical Society, 1976.
Honorary Member of the American Academy of Arts and Sciences, 1981.

ACADEMY MEMBERSHIPS

Rheinisch-Westfälische Akademie der Wissenschaften, 1974.
Deutsche Akademie der Naturforscher "Leopoldina", Halle/Saale, 1976.
Royal Dutch Academy of Sciences, 1977.
Akademie der Wissenschaften, Göttingen, 1979.

ADDRESS FOR PROFESSOR HERMAN F. MARK'S 90TH BIRTHDAY MAY 7, 1985

HANS MARK
Office of the Chancellor, The University of Texas System, Austin, Texas 78701

Chairman Vogl, Ambassador Klestil, Geheimrat, and ladies and gentlemen: It is a great pleasure for me to be here and to have the chance to talk about the Geheimrat—who is among many other things, my father. I hope you will forgive me if I speak in very personal terms today because this is a most auspicious occasion not only for polymer science and the Polytechnic Institute, but also for the family. It is as a member of the family, therefore, that I would like to speak to you.

One of my first conscious memories goes back 50 years to the Geheimrat's 40th birthday in 1935. I was five years old and we were living in Vienna at the time. I remember that we had a large party for him in the garden in the back of our house. There are two distinct things I remember about that splendid party: one is that it was an international party just as this meeting (and party) is international. This was the first exposure I had to the international nature of science and it was very impressive for this five year old to see people from India and Japan and the United States come to honor my father. The second thing I remember is that it was at this party that I first heard the word "polymer." I asked my mother what the word meant and she said that she did not know exactly what it meant but she told me that whatever it meant it was important. I have to confess that I still feel the way she did then about polymers. I don't understand much about them but I know they are important.

I also remember that it was in 1935 that the Geheimrat went on a mountain climbing expedition in the Caucasus. He is a great mountaineer and at that time he wanted to climb every mountain in Europe with a height above 3,000 meters. One of these is Mt. Elbruz in the Caucasus in the southern part of the Soviet Union. Now the Geheimrat had to find a good excuse to do this because the Russians were no more hospitable to plain tourism then than they are now. The excuse was soon invented. Harold Urey had already made his brilliant guess that the anomalies existing in the relationships between atomic weights and atomic numbers occurred because of the existence of isotopes. Urey and his collaborators had also succeeded in isolating the heavy isotope of hydrogen by taking advantage of the differential mobilities between "heavy" and "light" water molecules in the electrolysis of water. The Geheimrat and his collaborators speculated that "heavy" water would be slightly more prevalent at the bottom of deep glaciers than under normal circumstances on the surface of the earth. This also happens because of the differential mobility between the "heavy" and the "light" species of water molecules. Well, to make a long story short, they climbed Mt. Elbruz, they did bring back samples of water from the bottom of glaciers and sure enough, they determined in due course that these samples were indeed slightly richer in "heavy" water than a random sample of water. This whole incident was to have some amusing consequences for me later on.

The decade from 1935 to 1945 was an eventful and difficult one for the family because it marked our move from Europe to the United States. By the time we celebrated the Geheimrat's 50th birthday in Brooklyn, we had escaped from Europe and the terrible destruction of the Second World War was almost at an end. I was still in high school at the time but had already developed an interest in science. I remember spending lots of time at the polymer laboratories at the old Polytechnic Institute of Brooklyn at 85 Livingston Street in downtown Brooklyn. It is here that I first learned how to behave in a laboratory and how to develop the discipline necessary for experimental work. I am really pleased that one of my classmates from high school who shared that time with me, Dr. Burton Feinerman of Miami, Florida, is here with us tonight.

In 1947, I graduated from high school and went to the University of California at Berkeley to start my university education. The Geheimrat, of course, had many friends in California and one of them was one of our speakers here this morning, Professor Linus Pauling. The Geheimrat had some qualms about turning me loose in the wild west so he asked Professor Pauling, among others, to keep an eye on me. One of my fond memories is visiting the Pauling family in Pasadena in the late 1940s and staying with them over some of the school holidays. I also remember trying to understand some of the more esoteric principles of physical chemistry at the time, such as the Gibbs Phase Rule, and Linus Pauling telling me that I had a long way to go before I could come close to what my father knew about physical chemistry. I remember that Professor Pauling was not pleased with my progress and I am very much afraid that his pessimism about my scientific prowess has been justified by subsequent events.

The Geheimrat's 60th birthday in 1955 was important because that was when we bestowed the title "Geheimrat" on him. The idea originated with Professor Isidor Fankuchen who was one of the Geheimrat's colleagues at the old Brooklyn Poly. You all know that the title "Geheimrat" was used by the old Austrian Empire to honor and recognize distinguished academics. Geheimrat, translated literally means "Privy Counsellor" and, as you can imagine, the title was usually given to people who made much of their dignity and who would today be called stuffed shirts. Fankuchen thought it would be a huge joke to bestow the title on my father precisely because he was about as far away from being a stuffed shirt as one can get. Much to our surprise, the name stuck and everyone today—thirty years later—knows him as the Geheimrat.

By 1955, I was working for Edward Teller and I am most pleased that Edward is also here with us tonight. Now you may not know this, but Edward took a course in quantum mechanics from the Geheimrat at the University of Karlsruhe in 1927 and twenty-eight years later it fell to Edward to explain quantum mechanics to me. There is no doubt that Edward was a better student than I was, and I want to take this opportunity to thank Edward for the patience he has had with me over the years. In any event, I want you all to know that Edward is the link in the "generation gap" of the Mark family. There is one other story that I need to tell about my early association with Edward. In due course, I had to get a security clearance in order to work on some of the things we were doing at the time. Well, I filled out all the forms and turned them in. A couple of months later I was informed that there were difficulties and that I would have to have an interview. An interview was arranged with the FBI and they questioned me for several hours asking me about people who I knew were associates of my father's. Since I did not know most of them, I could not tell them very much so the interview did not go well. Finally, I decided to take the bull by the horns and I asked them what they really wanted to know. Well, after some hesitation, they finally came out with it: They wanted to know what the Geheimrat was doing in Russia in 1935 looking for deuterium and what did he tell the Russians about it. Well, I told my interrogator that he would not believe the story if I told him, but he indicated that he was willing to listen. So I told him what I recounted a few minutes ago. I don't know to this day whether he believed me or not, but I do know that I did get my clearance—and that I was therefore able to start the long and productive association that Edward and I have enjoyed now for thirty years.

In 1965, twenty years ago, we celebrated the Geheimrat's 70th birthday at the same hotel that we are in today. A couple of nights ago, at a small dinner party that our good friends, Sheldon and Helen Atlas hosted, I saw a photograph album that was put together for that occasion. Many of the people here today were at that party and were in the pictures. I have to confess that most of us looked better then than we look now—except for the Geheimrat who has changed least of all. I should mention here the two people in those pictures who are not with us tonight, my mother and my younger brother, the late Professor Peter Mark of Princeton University.

My mother was a remarkable woman who, by the time she died in 1970, had been married to the Geheimrat for nearly fifty years. She was a superb hostess and had a great sense of fun. And, in whatever she did, she helped all of us along. Many of you in this room tonight may remember the summer house that we had on Lake Peekskill about 50 miles north of New York. I remember that C. V. Raman, the famous Indian physicist, was a guest once and a few minutes before dinner my mother, Mimi, discovered that Raman was a vegetarian. We had a really delicious chicken dish that evening and I remember Mimi frantically asking my father what she should do to hide the chicken. Finally, we decided that dilution was the answer so we dumped more vegetables in the pot and served it. I don't think Raman noticed the chicken—at least he never told us if he did.

My brother, Peter, was one of those people who had the ability to see that something would be important long before others came to the same conclusion. He was one of the pioneers in surface physics and his most important contribution was to develop some of the first quantitative ground rules for the understanding of that very important area of physics. I remember that he told me once, shortly before he died in 1979, that what he wanted to do was to write a book on the quantum mechanics of surfaces. He never got to do this but I recently saw a book with that title and was not surprised to find many references to Peter's work in the index. I am pleased to acknowledge the presence tonight of Peter's wife, Delia, and his daughter, Catherine.

What kind of a man is the Geheimrat? What is the essence of the human being? This symposium is evidence enough that he is a great scientist. We heard four Nobel laureates talk about some of their work this morning and acknowledge their intellectual debts to him. What I would like to do is to pick up on something that Linus Pauling said which, I think, captures what this meeting is all about. He said that the last 75 years have been the most important in all the history of science. I agree with him. At the risk of quarreling a bit with Professor Pauling's time scale, I would argue that his statement applies to the 20th Century generally. Since the Geheimrat's life spans almost the entire century, he is, indeed, the quintessential 20th Century man. He was born in 1895, the year that Roentgen discovered x-rays and Bequerel discovered radioactivity. He was four years old when Max Planck wrote his seminal paper on quantum theory and ten when Albert Einstein wrote his monumental papers on relativity, the photoelectric effect, and on the determination of Avogadro's number using a fluctuation calculation. In short, he was already alive when the foundations of modern science were laid. He was eighteen when Nils Bohr unraveled the hydrogen spectrum in 1913 and by the decade of the 1920s he was not only there as a spectator but as a contributor. I well remember stories that the Geheimrat told us about his meetings with all of the great names who were his predecessors like Nernst, Haber, and Einstein and those who are his contemporaries, Schroedinger, Dirac, Pauling, and many others.

Over the years I have had the good fortune to work with the Geheimrat and to watch how he does things. You all heard today of his unique ability to combine profound theoretical knowledge with a practical outlook on life that have characterized his career. I saw this ability first hand in the work he helped NASA do on the prevention and control of fires in aircraft accidents. More recently he has been very active in helping NASA with polymeric materials used in the space shuttle program. Perhaps the most important of these are the materials that we want to use to replace the steel casings of the large solid fueled rockets used to boost the shuttle into Earth orbit. What we are planning to do is to use fiber-reinforced polymers for constructing these rocket cases. Recently, the Geheimrat served as the chairman of a committee to advise NASA on what to do about some of the problems we were having with the cases. About a year ago, I attended a meeting of this committee. There was no doubt that the Geheimrat was in charge and he talked most enthusiastically about the future and about how we would use synthetic materials to greatly enhance the efficiency of space vehicles. I reflected as I listened that here was someone confidently talking about space flight, its future, and what we were going to do, who was already eight years old when the Wright brothers made their first powered flight in the atmosphere in December 1903.

To me this incident captures the essence of his character. He is the quintessential man of our century, a century that has seen both great progress and much human suffering. The Geheimrat has seen, in full measure, the dark side of our times. He was a front line soldier in the Imperial Austrian Army during the First World War and, at great risk, he led the escape of our family from Europe just before the cataclysmic events of the Second World War began. He never looked back and he never became obsessed with the existence of evil in the world. He fought the evil of our time with the best weapon he could devise — the spread of human knowledge — and in doing that he has made life better for all of us. He does this because he has an unquenchable faith in the future and an indestructible optimism that the future will be better than the past. In that sense, he is not only the quintessential 20th Century man but also the quintessential American. The assumption that the future will be better than the past is one of the cornerstones on which this nation is built and this, I believe is why he has felt so much at home in his adopted country, the United States of America.

On behalf of all of us in the family, I would like to thank you all for being here today to help us honor the Geheimrat on his 90th birthday. For the grandchildren, Catherine, Rufus, and Jane, for the daughters-in-law, Delia and Marion, I would like, with much love, to add our own best wishes for a very happy birthday and for many more to come.

POLYMER PHOTOCHEMISTRY PERSPECTIVES AND PROMISES

G. SMETS
K. Universiteit Leuven, Laboratory for Macromolecular and Organic Chemistry, B-3030, Leuven, Belgium

SYNOPSIS

A better understanding of the physical and chemical processes occurring in photochemical reactions becomes increasingly a requisite for further development of polymer photochemistry. This requirement is even more severe when the phenomena to be analyzed occur in bulk polymers as is the case in most photochemistry applications. It is the main purpose of the present paper to stress the intimate relations existing between polymer photochemistry and polymer properties, and the need of fundamental research in this domain. Successively we will consider (i) the influence of the bulk properties of the solid polymer, (ii) the free volume distribution and distribution of reactive sites, (iii) cage effects and diffusion restrictions, (iv) local concentration and neighborhood effects.

INFLUENCE OF THE BULK PROPERTIES OF THE POLYMER

The primary photochemical processes are usually studied in dilute systems; they may be, however, completely superseded by the *bulk properties of the solid polymer,* as α and β transitions, segment mobility, and so on. This statement can be nicely illustrated by recent data concerning the time-dependent fluorescence intensity of DPAE, namely, *bis*(9,10(phenyl)anthrylmethyl)ether, so called diphant. De Schryver et al. [1] showed that a clear solvent viscosity dependence exists for excimer formation which requires an approach of both anthryl groups in an adequate position. Moreover, the activation energy of excimer formation E_{DM} increases from 14 kJ mol^{-1} in methylcyclohexane to 25 kJ mol^{-1} in hexadecane. Using single photon counting technique the lifetimes of the local excited state and excimer were measured. The monomer fluorescence intensity I_M follows a monoexponential decay, while that of the excimer fits with a difference of two exponentials with a rise corresponding to the decay of I_M. Monnerie and co-workers [2] examined intramolecular excimer formation of DPAE in bulk polymers as polybutadiene, polyisoprene, and so on.

At low temperatures, molecular motions are completely frozen, and the monomer lifetime τ_M converges to the same value of 6.8 ns (6.5 ns in methylcyclohexane). At higher temperatures τ_M and thus the rate constant of excimer formation varies with the host matrix. The activation energies of excimer formation E_{DM} (between 40 and 80°C) are 37.6 and 38.5 kJ mol^{-1} for polybutadiene and polyisoprene, respectively, which largely exceeds the values in solution. Moreover, the reciprocal of excimer formation varies with the temperature following the WLF equation, in other words, this behavior corresponds to the molecular motions of the chain surrounding the probe which must afford the free volume needed by the probe to achieve its conformational transition. The fluorescence behavior of bichromophoric probes of different size could thus permit evaluation of the lower volume limit from where excimer formation is controlled by segmental motions of the polymer matrix. In the same vein of thinking Smets et al. [3] studied the decoloration kinetics of photochromes of different molecular sizes in polystyrene and polyester films and concluded that the kinetic behavior below T_g is mainly due to the physical state of the glassy polymer and not to the chemistry of the photochrome reaction.

Remarkable also are the results of Mita et al. [4], using laser pulse excited triplet phosphorescence of benzophenone dispersed in polymethyl- and poly(isopropyl methacrylate) and poly(methyl acrylate). Below T_β, that is, below the onset of ester side-group rotation, and above T_g, the kinetics of phosphorescence decay follow single exponential law. Between these temperatures a nonexponential decay is observed, and the kinetics are interpreted on the basis of a diffusion-controlled rate coefficient, with a time-dependent transition term for the quenching of benzophenone triplet by the carbonylester side groups of the matrix. This bimolecular rate coefficient is expressed by the following equation:

$$k_q = 4\pi RDN\left(1 + \frac{R}{(\pi Dt)^{1/2}}\right) \tag{1}$$

where D is the sum of the diffusion coefficients for both reacting functions, R is the reaction radius, and t the time of reaction. Some values are given in Table I for a poly(methyl methacrylate) matrix.

These examples show clearly that the incorporation or addition of photosensitive probes in a polymer constitutes a source of information about the structure and molecular dynamics of solid polymers.

On the other hand, the polymer matrices in which photochemical processes have to take place influence considerably the primary photophysical processes, such as lifetime of the excited species. They have therefore to be chosen in such a way as to permit a high quantum yield, or assure an optimum photosensitization and photoinitiation. A very appreciable advantage of the fluorescence or phosphorescence methods resides in their exceptional sensitivity, which permits the study of the polymer without modification of its physical state.

TABLE I
Phosphorescence Decay Kinetics of BT^T in PMMA Matrix [4]

D cm^2 s^{-1}		E (kJ/mol) for D at		
$T = T_{\alpha'}$	$T = T_g$	$T_\beta < T < T_{\alpha'}$	$T_{\alpha'} < T < T_g$	$T_g < T$
6.7×10^{-14}	1.5×10^{-12}	29	40	131

$T_{\alpha'}$ (T_g-50 to 70°C) corresponding to *local* mode relaxation of the main chain.

INFLUENCE OF THE FREE VOLUME DISTRIBUTION

A second aspect of photochemical processes occurring in solid polymers concerns the *free volume distribution* and the correlated *distribution of reactive sites* which affect the kinetics of the reactions. Such distribution can be evaluated from the decrease of the quantum yields with the degree of conversion [5, 6]. This problem was tackled by Reise and Pitts in the case of the cross-linking of polymers by photodimerization of unsaturated side groups. With poly(vinyl cinnamate) the initial (unsensitized) 300-nm quantum yield is 0.26 and decreases to zero at a degree of conversion $\chi = 0.5$. On the contrary, copolymers with side diphenyl cyclopropene groups (triplet-sensitized film) start with $\phi = 1.7$ (380 nm); it decreases, however, rapidly and approaches zero at $\chi = 0.02$–0.03 (i.e., when very few chromophores have been converted.

Another example of the existence of sites of different mobilities and nonhomogeneity of the system was recently described in the bulk photopolymerization of bifunctional monomers [7] as hexanedioldiacrylate and *bis*(2-hydroxyethyl)bisphenol-A dimethacrylate. Nonhomogeneity of the systems was taken in evidence at conversions far beyond the gel point by the presence of trapped radicals and postpolymerization on postheating in the dark. In these systems double-bond conversion runs ahead of volume shrinkage; possibly pendent double bonds show in the beginning an enhanced reactivity due to local concentration. A high concentration of cross-linkable side groups is therefore only favorable if adequate spacers linking them to the main chain preserve a sufficient chain segment mobility. Undoubtedly, heterogeneity of the systems and accessibility of the functions should be taken into account for the formulation and optimization of industrial photoprocesses.

In this respect an interesting comparison was recently published concerning the photoisomerization behavior of azo-labeled polystyrenes in dilute solution and in the glassy state [8]. The azobenzene chromophores were incorporated at the chain ends (E-PS) 0.24%, the chain center (CPS) (0.12%), or as side groups (S-PS) (<1 mol% azo residue) and the *trans-cis* isomerization was followed by nanosecond-pulsed laser spectroscopy. In dilute solution a small difference is found for the sites of the chain (see Table I). In the glassy films the side and end labels isomerize much easier than the center ones though still more slowly than in dilute solution. It is assumed that only a certain fraction (α) of the azo chromophore can isomerize with the same rate as in dilute solution, while the rest cannot do so due to the lack of free volume necessary for such a reaction (see Table II).

This means that about 45% of the end and side azo labels can isomerize, and only 8% of the center ones can do so. The difference in mobility is much more pronounced than in dilute solutions.

In most cases the quantum yields of photochemical processes are much larger in solution than in polymer matrices. In *cis-trans* isomerizations it is the lack or restriction of segmental mobility around the isomerizable group, that is, the free volume distribution in the matrix which is rate deteminant. Photochrome systems belong to this category [9]. In some more elaborate processes the matrix may restrict the conformational changes required from a specific reaction intermediate, and consequently modify profoundly the overall course of the reactions. This can be illustrated by two recent examples of which the first relates to the photolysis of cyclooctyl nitrite [10]. In solution the main reaction product is the 4-nitroso compound (or its dimer) as expected from the Barton's reaction mechanism, besides 13% ketone. On the contrary, in polymeric media limited interconversion of ring conformation of the cyclooctyloxy radical intermediate not only decreases the total quantum yield of photolysis (0.3 instead of 0.6) but cyclooctanone becomes the major reaction product (68%).

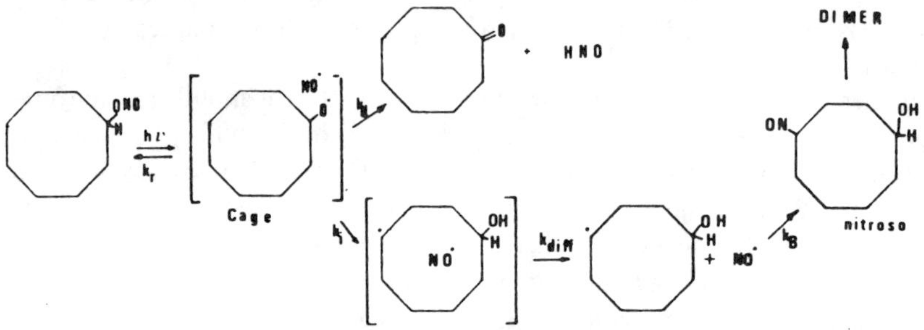

(2)

The second example is based on the photolysis of bicyclic aziridines where ring cleavage produces the anti-4-ylide that isomerizes into the syn-4-ylid [11]. This may either return to the initial compound, isomerize into enediimine, or be trapped in a cyclo adduct. Above T_g the ylide can achieve a planar structure and isomerize easily to enediimine; ring closure however requires a rotation of the bulky *p*-nitrophenyl group into a plane vertical to that of the ylide, rotation which is reduced in a high viscosity polymer matrix. The relative importance of the competitive ring closure is decreased and the quantum yield of enediimine attains 0.75 above T_g (instead of 0.3). In the glassy state the ylide cannot reach the planar structure, and sticks almost in its original position, and returns much more easily to the bicyclic aziridine.

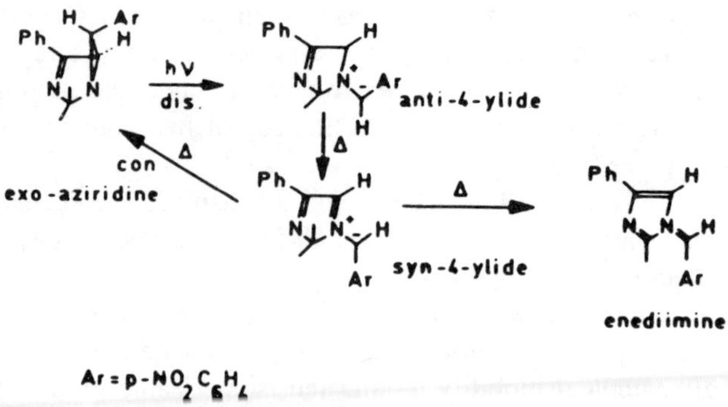

(3)

TABLE II
Trans-content at the Photostationary State of Azo-Labeled Polystyrene (λ_{irr} 353 nm)

	% azo (mol)	benzene	film	$(\alpha)^b$
C-PS[a]	0.12	0.26	0.94	0.08
E-PS	0.24	0.15	0.63	0.45
S-PS	1.0	0.33	0.70	0.44

[a]Azo group incorporated at the center (C), at the end (E), or as side group (S) of the polystyrene.
[b](α) is the fraction azo chromophore isomerizable in film.

CAGE EFFECTS AND DIFFUSION RESTRICTIONS

A third aspect of photoprocesses in bulk polymers consists in the existence of important *cage effects* and *diffusion restrictions* which limit severely the applicability of a reaction, particularly in dissociation phenomena. An extreme example is given by the photodegradation of PMMA containing 1-phenyl-1,2 propanedion-2-0 methacroyloxime [12] α-keto-0 acyloximes, which is known to photolyse easily ($\phi = 0.8$) with the production of free radicals. Such groups can be incorporated as side groups into PMMA through copolymerization of MMA with methacryloyl oxime. On irradiation with monochromatic light (λ_{irr} 367 nm) in benzene solution, strong degradation occurs as shown by gel permeation chromatography; the number of chain scissions N_s per molecule can be calculated from the decrease of the number average molecular weight. The degradation is rapid and practically independent of the temperature of irradiation. The reaction scheme is:

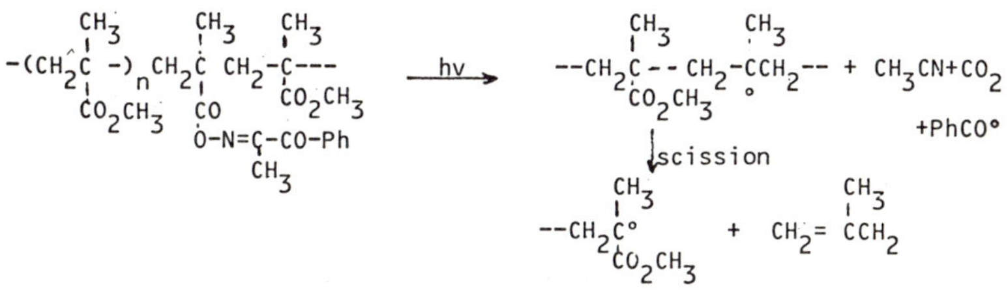

(4)

Some degradation data are summarized in Table III.

In the bulk, no chain scission can be observed neither at 25° nor at 60°C, obviously owing to radical recombination. This picture changes completely, however, if, instead of monochromatic light one uses deep UV laser (193 nm). Ejection of small fragments dissipates the excess energy and gives surface etching. This method is the basis of a deep UV-lithographic process for photoresists with high resolution (ablative photodecomposition). Thus, Chandross et al. showed that in these conditions a terpolymer methyl methacrylate-oximino-ester-methacrylonitrile (8-0.6-1.4) is 85 to 100 times more sensitive than

TABLE III
Photodegradation of co(MMA-methacroyloxime) Polymers
(λ_{irr}, 367 nm, t, 25°C)

Reaction medium	Oxime (mol %)	Irradiation Time (min)	Irradiation t (°C)	$M_n \cdot 10^{-3}$	N_s[a]
Benzene soln.	5.4	—	25	29.1	—
5 g/dl		15		10.5(10.2)[b]	1.77(1.85)
		60		4.76(4.1)	5.11(6.14)
	3	—	25	31.2	—
		15		13.3	1.34
		60		7.3	3.24
film + 30% diBu.phtal.[c]	3.5	—	60	20.9	—
		15		17.5	0.19
		30		6.7	2.12

[a] $N_s = (M_{n_0}/M_{n_t}) - 1$.
[b] Values between parentheses for experiments at 60°C.
[c] T_g 32°C.

PMMA [13]. Incorporation of indanone side groups instead of acyloximino groups seems especially attractive.

In most cases the difference between solution and bulk polymer are not so pronounced, for example, the photolysis of *p*-quinone-bisnitrones [14]. In PMMA the first dissociation step becomes reversible, the percent dark recovery depending on the nature of the group R. The very reactive nitrene recombines with the carbonyl group within the cage. On the contrary, in solution it diffuses away, forms dimer (azobenzene), or reacts with the solvent.

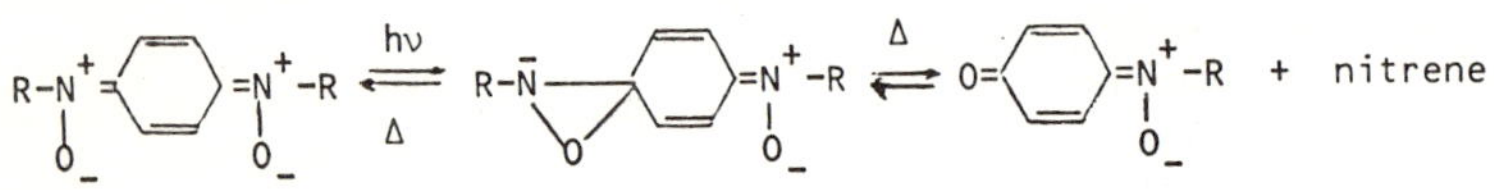

(5)

Homolytic dissociation-recombination reactions are evidently very sensitive to the reaction medium, to cage effects, and diffusion phenomena which will determine the availability of the free radicals in the system. Previous examples [14] were hexaphenyl-1.2′ bisimidazole, its bulky substituted ester, and 2.2′-substituted thioindigo white derivatives (R-TIW) of which the dissociation equilibrium is shown in Eq. (6):

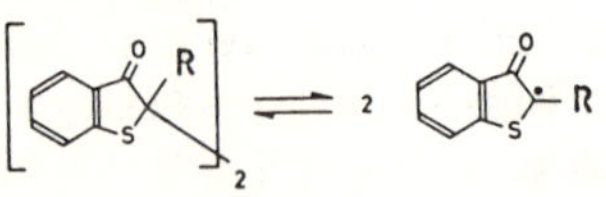

(6)

Interesting is the comparison between the dissociation-recombination parameters of these compounds in PMMA and polystyrene below their glass temperatures [15]. On the average, the diffusion rate constants in PSt are one-third or less of the values in PMMA while the residence life-times of the free radicals in the cage are usually higher in PSt than in PMMA. It is admitted that these noticeable differences are related to the local motions of the α and β transitions of the host polymers. Some data are given in Table IV; They were obtained on the basis of Eq. (1) (see above) assuming that below T_g cage recombination rate constants largely exceed those of diffusion.

LOCAL CONCENTRATION AND NEIGHBORHOOD EFFECTS

These effects are characteristic of photocrosslinking by a nonchain mechanism of polymers containing both electrodonor (D) and electron acceptor (A) groups implanted along the chain. The reaction scheme represented below for 4,methyl benzophenone (A) and *N,N*-dimethyltoluidine (D) is the model reaction for the crosslinking of vinylbenzophenone (VB) and dimethylamino styrene (DAS) polymers and copolymers.

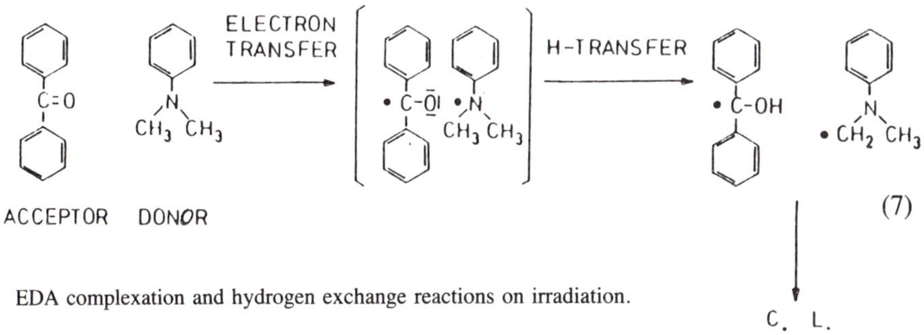

(7)

EDA complexation and hydrogen exchange reactions on irradiation.

Exciplex formation is the first step in the photochemical crosslinking; it is followed by a proton transfer within the photoexcited EDA complex or exciplex

TABLE IV
Diffusion Rate Constants k_D and Cage Residence Lifetime of 2,2′ Thioindigo White (TIW) Free Radicals in PMMA and in PSt

	PMMA (T_g 65–70°C)			PSt (T_g 86–87°C)		
TIW (4 wt%) subst. R is	t°/C	$k_d \cdot 10^2$ $mol^{-1}\ s^{-1}$	$r \cdot 10^{-2}$ s	t°/C	$k_d \cdot 10^2$ $mol^{-1}\ s^{-1}$	$r \cdot 10^{-2}$ s
Phenyl	22	4	96	23	1.4	280
	51	36	9.5	55	6	28
	71	122	4			
4-Nitrophenyl	22	2	25	21	0.5	17
	51	3	(3)	52	1.6	4
	72	35	1			
Methyl	21	8	5.5	21	1	8.5
	56	170	0.5	50	11	0.5
	74	370	—			
Ethyl	20	4	2	21	0.5	6.5
	51	64	0.5	50	5	2

giving rise to the formation of free radicals of which the combination results in crosslinking.

The dilute solution behavior of such VB-DAS copolymers of different composition has been described recently on the basis of absorption and emission spectrometries and of gel permeation chromatography data [12, 16]. The primordial role of intramolecular exciplex formation, even in the case of terpolymers of VB-DAS-methyl acrylate, was demonstrated experimentally. It obliges one to consider the concentration of crossed dyads (donor–acceptor direct neighborhood) as the basis of their complexation properties instead of the overall concentration in the solution.

Self-supporting films can easily be prepared on the basis of VB-DAS-butyl acrylate (BA)terpolymers. On irradiation of the films, free radicals are progressively produced, and their subsequent decay can be followed below and above T_g using EPR spectrometry. From these measurements kinetic parameters were again calculated on the basis of Eq. (1) for diffusion-controlled reactions; some values are given in Table V.

As expected the diffusion rate constant k_D and the diffusion coefficient D above T_g are much larger (about 5 times) than at room temperature. Typical stepwise radical decay was observed; at each temperature the radical concentration approaches a limiting value (Fig. 1).

In order to follow the rate of formation of radicals and their diffusion out of the cage, experiments were carried out by incorporating nitroxyl radicals (1.2×10^{-3} mol) (4-hydroxy-2.2.6.6. tetramethylpiperidinoxy) in the film: the total concentration of radicals was followed above and below T_g in function of the time of irradiation (Fig. 2).

At 79°C the free radical concentration decreases very rapidly owing to their reaction with the nitroxyl radicals after diffusion out of the cage. Below T_g at room temperature the radical decay is relatively slow and results mainly from cage recombination.

These results confirm previous observations. Indeed, the rate of crosslinking of these terpolymers measured gravimetrically on the basis of the weight fraction of insoluble product is practically insensitive to the presence or absence of air and is in large limits independent of the film composition [16].

CONCLUSION

The strong dependence of polymer photochemical processes from the physical properties of the basis polymer requires further chemical and physical inves-

TABLE V
Kinetic Parameters of Radical Decay in Terpolymers
(VB 47-DAS 14-BA 39) (T_g64°C)

	20°C	79°C
k_D ($mol^{-1}\ s^{-1}$)	0.03	0.13
D ($cm^2\ mol^{-1}\ s^{-1}$)	8×10^{-18}	4.1×10^{-17}
Kinetic cage radius (Å)	55	43
Lifetime (s)		14

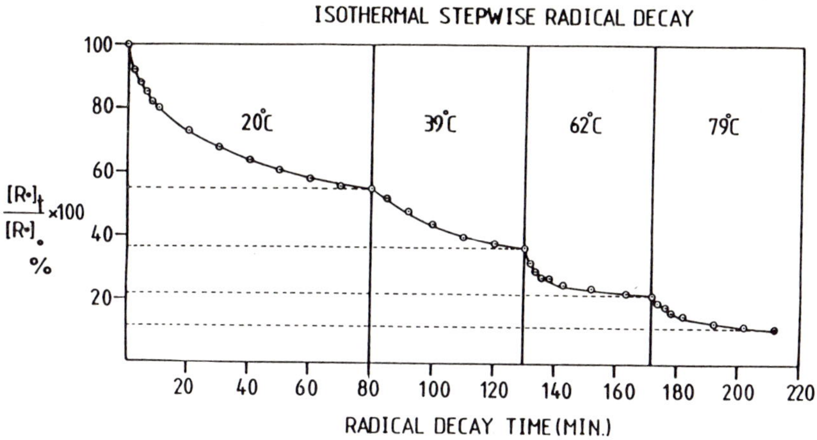

FIG. 1. Isothermal stepwise radical decay in terpolymer (BA39-VB47-DAS14).

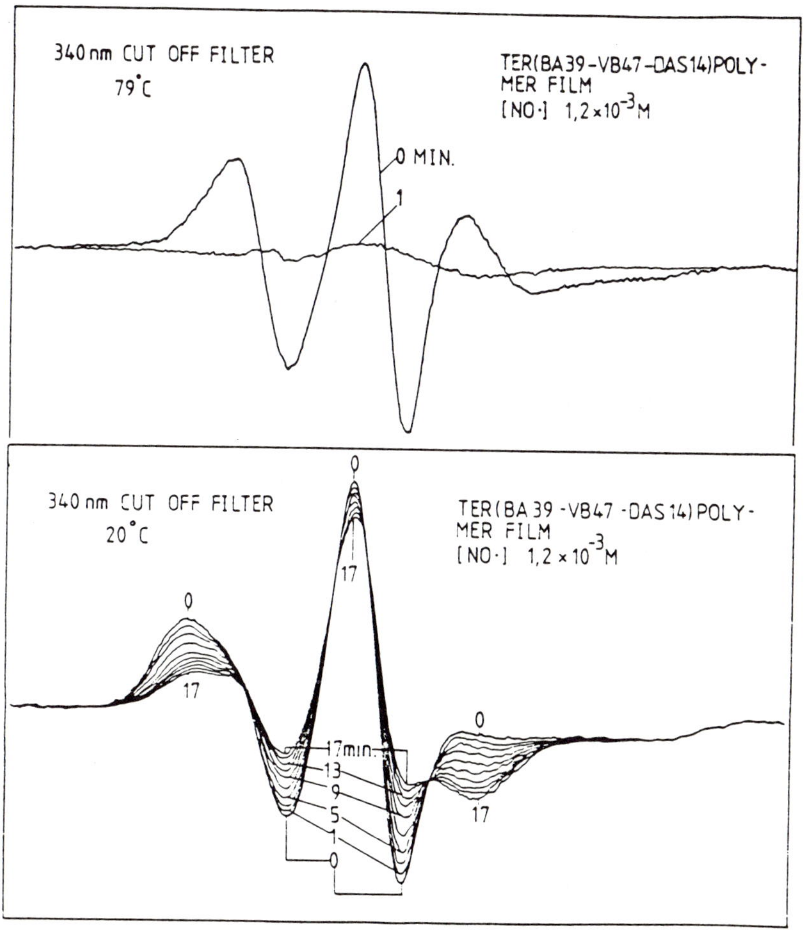

FIG. 2. Change of total radical concentration in terpolymer in the presence of nitroxyl free radicals (1.2×10^{-3} mol) at 20°C (after 1, 5, 9, 13, and 17 min) and at 79°C (1 min).

tigations in order to improve their application possibilities. On the other hand, the study of similar interactions in the case of micelles, of organized systems as bi- and multilayers and membranes is promising for the elucidation of their internal structure and internal mobility, and of the changes of their properties under the influence of light.

The author is indebted to the Ministry of Scientific Programmation, Belgium, for financial support to the laboratory.

REFERENCES

[1] F. C. De Schryver, K. Demeyer, J. Huybrechts, H. Bouas-Laurent, and A. Castellan, *J. Photochem., 20,* 341 (1982).
[2] E. Pajot-Augy, L. Bokobza, L. Monnerie, A. Castellan, and H. Bouas-Laurent, *Macromolecules, 17,* 1490 (1984).
[3] G. J. Smets, J. Thoen, and A. Aerts, *J. Polym. Sci. Symp., 51,* 119 (1975).
[4] K. Horie, K. Morishita, and I. Mita, *Macromolecules, 17,* 1746 (1984).
[5] P. L. Egerton, E. Pitts, and A. Reiser, *Macromolecules, 14,* 95 (1981).
[6] E. Pitts and A. Reiser, *J. Am. Chem. Soc., 105,* 5590 (1983).
[7] J. G. Kloosterboer, G. M. M. Van de Hei, and H. M. J. Boots, *Polymer (Comm.), 25,* 354 (1984).
[8] C. S. T. Sung, I. R. Gould, and N. J. Turro, *Macromolecules, 17,* 1447 (1984).
[9] G. Smets, *Adv. Polym. Sci., 50,* 18 (1983).
[10] C. Jakus and G. Smets, *Macromolecules, 15,* 435 (1982).
[11] C. Jakus and G. Smets, *Macromolecules, 17,* 1895 (1984).
[12] G. Smets, *Polymer J., 17,* 153 (1985).
[13] C. W. Wilkins, E. Reichmanis, and E. A. Chandross, *J. Electrochem. Soc., 127,* 2510, 2517 (1980).
[14] G. Smets and M. Schmitz-Smets, *Structural Order in Polymers,* F. Ciardelli and P. Giusti, Eds., Pergamon, New York, 1981, pp. 221–230.
[15] Y. Leest and G. Smets, *Abstracts 27th IUPAC Sympos. Macromol. Strasbourg, 1981, 1,* 475 (1981).
[16] G. Smets, Sabr Nal El Hamouly, and T. J. Oh, *Pure Appl. Chem., 56,* 439 (1984).

Georges J. Smets was born on August 11, 1915.

POSITION

Professor em. Kath. Universiteit Leuven and Department of Chemistry, Univ. Cath. Louvain Celestijnenlaan 200 F, B-3030 Leuven, Belgium

EDUCATION

Kath. Univ. Leuven, Belgium; Ph.D., 1940.
National Research Council, 1940–1942.
Advanced Fellow of the Belgian American Education Foundation, 1949.
Dr. rer. natur. h.c. Univ. Mainz FRG, 1979.
Dr. h.c. Univ. Aberdeen UK, 1973.
Polymer Chemistry Award of the A.C.S., 1971.
Centennial Foreign Fellow of the A.C.S., 1976.
Honorary Member Soc.; Polym. Sc. Japan (1982)
Real Soc. Espanola (1978).

POSITIONS

Research chemist Gevaert Photoproducten
Antwerp Belgium, 1943–1944
Doc. Univ. Leuven, 1942.
Prof. Univ. Leuven/Univ. Louvain, 1944–1984
Visiting professor at the universities of Tokyo-Kyoto, 1972.
Midland Macromolecular Institute, 1974.
Univ. of Massachusetts Amherst and College of Science Univ. Notre-Dame Indiana, 1981.
Acad. Sc. P.R.C., 1980.
Elected Member Bureau IUPAC (1969–1977).
President Elect IUPAC (1975–1977). President IUPAC (1977–1979).

PHOTOINITIATED REACTIONS OF ORGANIC POLYMERS

BENGT RÅNBY
Department of Polymer Technology, The Royal Institute of Technology, S-100 44 Stockholm, Sweden

INTRODUCTION

Photochemical reactions of polymers have been observed and even utilized since ancient times, for example, photocuring of asphalt tar by the Egyptians and photohardening of natural rubber by the Indians. The nature of the phenomena and the molecular mechanisms were unknown at the time. The reactions could of course not be understood before the concept of macromolecules was developed in the 1920s.

Without tracing the history of polymer photochemistry, at this symposium I like to refer to Gerald Oster's pioneering research around 1950 at the Polytechnic Institute of Brooklyn [1]. He studied *photopolymerization* and *photocrosslinking* and also demonstrated *image formation* in polymeric systems. One purpose of Oster's work was to find substitutes for the silver halide photographic process — systems which we are still looking for. But Oster also wanted to improve the properties of polyethylene, a new and promising polymeric material at that time.

PHOTORESIST TECHNOLOGY

The growing interest in polymer photochemistry has been triggered by the explosive development of microelectronics. This technology is largely made possible by the photochemical production of resists, called *photoresists* which carry the main functions of modern computers. Photoresist technology requires very sensitive photopolymer systems, reacting with high yields at increasingly faster rates [2]. The basic systems can be either positive or negative processes, referring to the "image" formation on the resist plate, namely, a silicon chip.

The classical *positive photoresist polymer system* is based on a Novolak (cresolformaldehyde) resin (N) to which a diazo-naphtho-quinone (I) is added as a sensitizer [3]. When a thin resin layer of (N) + (I) is exposed to near ultraviolet (UV) light, the photosensitive component (I) is rapidly photolyzed to a ketene which with water forms an acid and makes the resin (N) soluble in alkali (Fig. 1). This leaves the unexposed area under the mask as a positive "image" on the silicon chip.

The classical *negative photoresist polymer system* is based on *cis*-1,4-polyisoprene prepared by a Ziegler-Natta catalyst, cyclized to some extent by heating and/or chemical treatment and sensitized by adding a bisazide, which upon near UV irradiation initiates photochemical crosslinking (Fig. 2), which is a

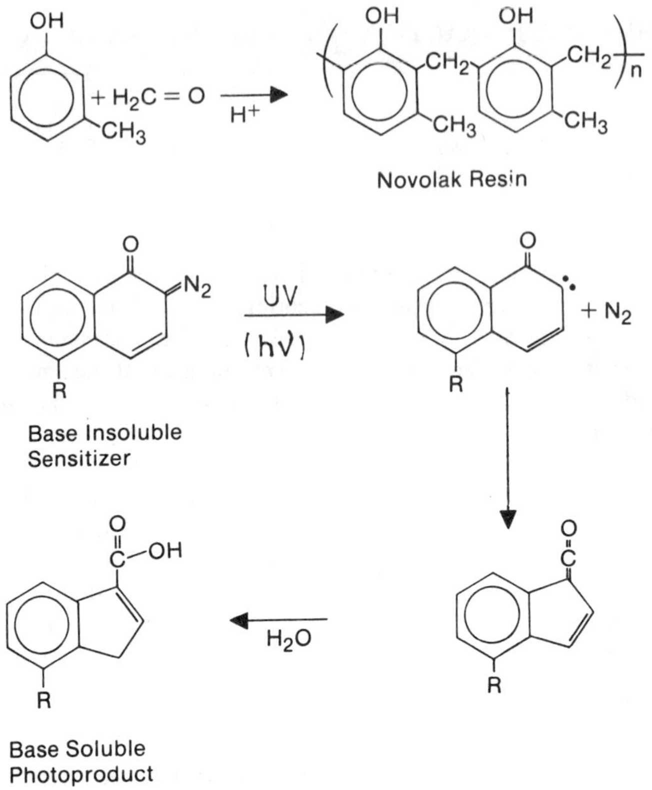

FIG. 1. The formation of Novolak resin from m-cresol and formaldehyde. The photochemical conversion of diazo-naphthoquinone (base-insoluble) to a carbene and further to a ketene which adds water and forms a base-soluble indene-carboxylic acid which solubilizes the Novolak resin.

rapid process [3]. The cyclized polyisoprene, which still contains double bonds, is very soluble in hydrocarbons and other nonpolar solvents while the photocrosslinked polymer shows only limited swelling and is retained. This is the basis for a negative "image" formation on the silicon chips.

Photoresist technology has now developed to modified systems with increased sensitivity and resolution using shorter UV wavelengths. *Mid UV resists* are processed at 313 nm commercially [4], using quartz optic for the lithography. *Deep UV resist* technology (below 300 nm) is under development, applying the strong mercury emission line at 254 nm or laser excimer radiation from krypton halides (220 to 250 nm) for higher intensity [5]. At the same time new photochemical systems are developed which allow higher photochemical yields, sharper "image" definition, and higher resolution in the "image" formed [6]. The photoresist technology is advanced and may find applications in other areas.

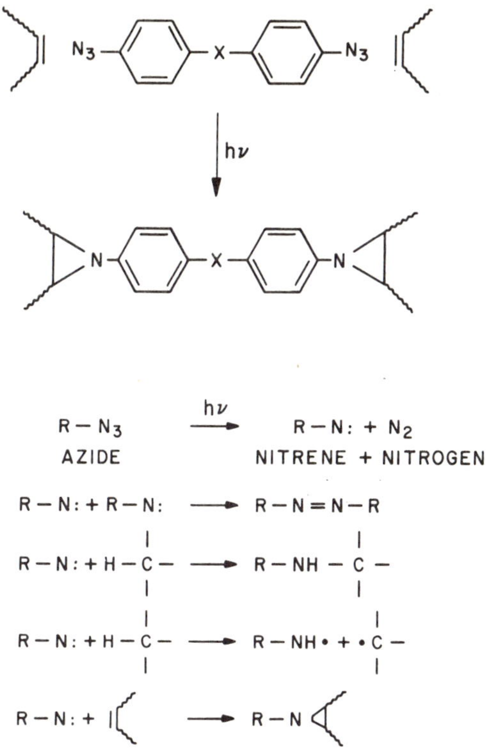

FIG. 2. Photochemical decomposition of a bisazide sensitizer which causes crosslinking of an unsaturated polymer (cyclized polyisoprene) as the main reaction.

PHOTOCURING OF ORGANIC COATINGS

Conventional surface coatings require large amounts of organic solvents (50 to 80%) for application. After application, the solvent has to be removed by evaporation, usually at elevated temperatures. To reduce air pollution and health hazard and to use energy more efficiently, photocuring of organic coatings and printing inks containing no solvent has been developed and applied industrially. The photoreactions used are, until now, mainly carried out by free radical mechanisms [7-9]. Initiating systems using ionic mechanisms have also become known and are presently under development for industrial use.

Our studies of UV curing of organic coatings deal with both *photofragmenting initiators,* such as peroxides, acetophenones, and azo compounds, and *hydrogen abstracting initiators,* such as benzophenones and other aromatic ketones like thioxanthones which react with a hydrogen donor [10]. Both types of initiators give very rapid reactions with acrylic monomers and oligomers as shown for a thioxanthone/triamine/acrylate system (Fig. 3), which cures with UV light to 95% (only 5% of the acrylate groups remain) in a few tenths of a second irradiation, both in a nitrogen atmosphere or in air. The amine radicals

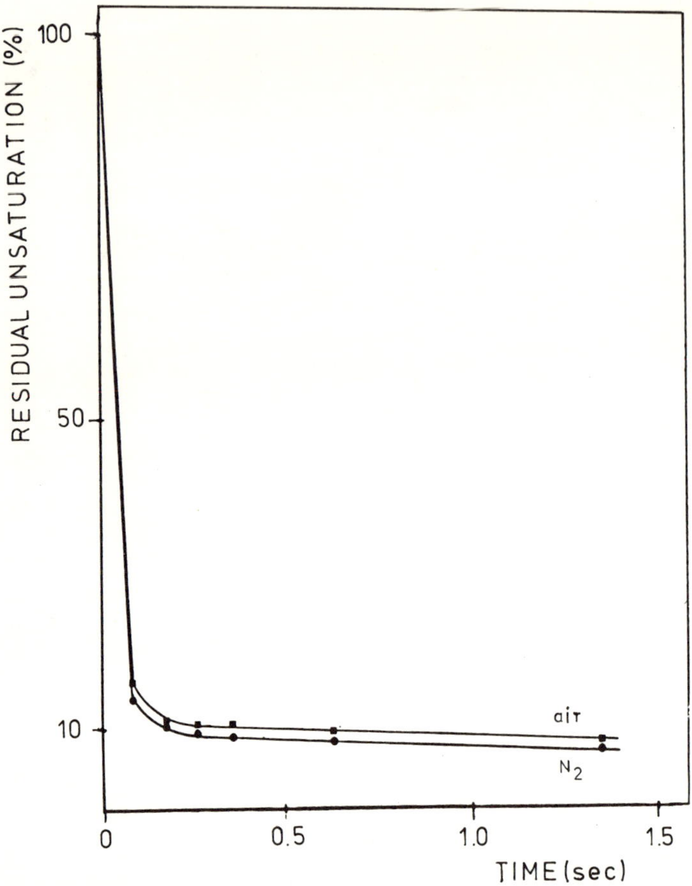

FIG. 3. Photocuring of an acrylated surface coating using UV irradiation. Residual unsaturation in a thin film is measured by infrared (IR) transmission spectra vs. irradiation time. UV irradiation in air and in nitrogen atmosphere at room temperature.

formed by hydrogen abstraction are the main initiators. The reaction scheme is rather complex; oxygen takes part in the curing reactions in several ways, for example, by deactivating the excited states of thioxanthone, formation of hydroperoxy and peroxy radicals, and groups which on decomposition give additional initiating radicals (Fig. 4). This means that oxygen is incorporated into the crosslinked polymer which decreases its thermo- and photostability.

We have found that the exposure of coatings in a weatherometer or outdoors to sunlight is largely a surface reaction which involves oxidation [11] starting at the exposed surface (Fig. 5). Two problems are considered: methods to prevent oxygen from reaction at the surface of the coating during curing and methods to stabilize the cured coating against further photodegradation.

To decrease oxygen reactions at the surface during photocuring, the irradiation is often made in a nitrogen atmosphere. Only a small amount of oxygen is

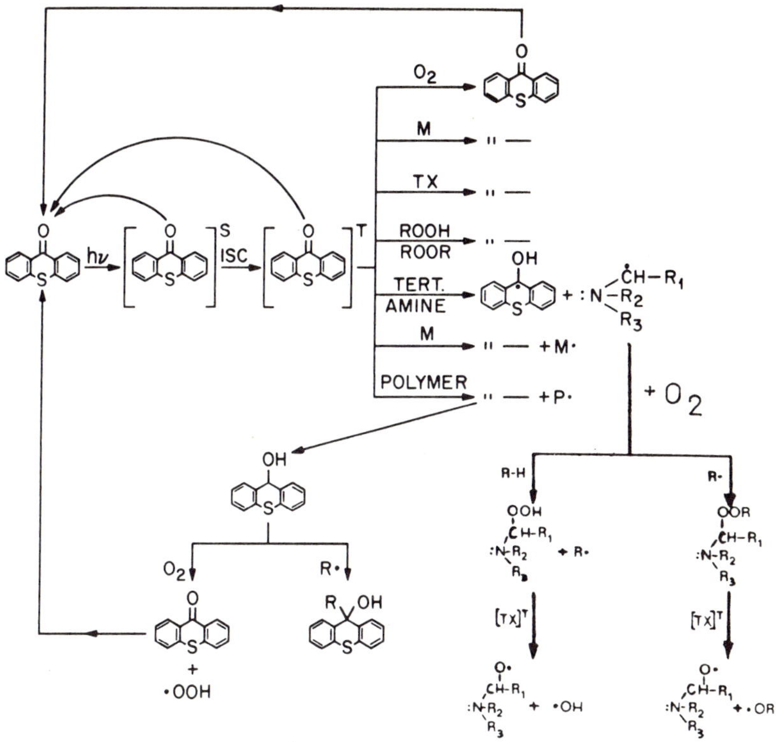

FIG. 4. Photochemically initiated thioxanthone/tert. amine/oxygen reactions giving amine, peroxy, hydroperoxy, and alkoxy radicals which may initiate polymerization of monomer (M) and/or give chain transfer to polymer (P).

dissolved in the coating. When curing is made in air, the highest concentration of oxygen is at the surface layer and may react there by addition to the radicals formed. To reduce the rate of oxygen reaction at the surface during curing and to increase the rate of curing, we have recently prepared and used a *surface-active initiator* (No. V in Fig. 6) which contains a fluoroalkyl group at one end and a photofragmenting initiator at the other end [12]. This initiator has a very limited solubility in the coating material and is dispersed in the lacquer by stirring. After application, the initiator V diffuses out and forms a surface layer on the coatings in a short time. During the UV irradiation the initiator V reacts at the surface by radical formation and gives a crosslinked surface layer. The rapid surface curing hardens the surface layer and may retard further diffusion of oxygen into the deeper layers. Using ESCA measurements we have found that the diffusion of the surfactive initiator V to the surface of the coating (thickness 30 μm) is rather complete 60 sec after application of the coating. The pendulum hardness (Fig. 7) develops more rapidly and reaches higher values with 0.14% initiator V and 2% conventional initiator than with 8% of the conventional initiator alone [13].

It is evident that the surface-active photoinitiator does give the effect anticipated and accelerates the curing of the surface layer of the applied coating.

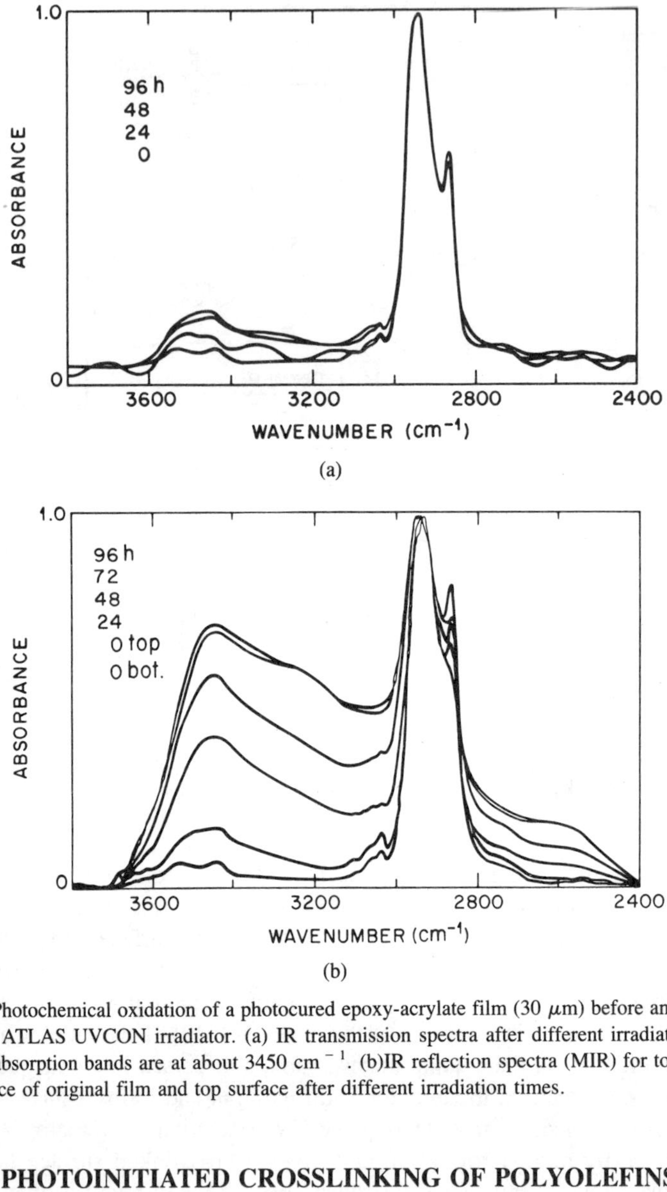

FIG. 5. Photochemical oxidation of a photocured epoxy-acrylate film (30 μm) before and after aging in an ATLAS UVCON irradiator. (a) IR transmission spectra after different irradiation times. The OH absorption bands are at about 3450 cm^{-1}. (b)IR reflection spectra (MIR) for top and bottom surface of original film and top surface after different irradiation times.

PHOTOINITIATED CROSSLINKING OF POLYOLEFINS

Commercial crosslinking processes are applied largely to polyethylene (PE) and initiated by thermal decomposition of peroxides, azo compounds or electron beam irradiation [14]. Photochemical crosslinking of polyethylene was reported some 30 years ago by Oster et al. [1] and later in patent specifications [15], but does not seem to be used industrially. With benzophenone or anthrone as initiators and UV lamps as radiation source (Figs. 8 and 9), Oster obtained moderate degrees of crosslinking (40-50% gel content), requiring long irradiation times (10 min to 2 hr). We have found [16] that with the proper combina-

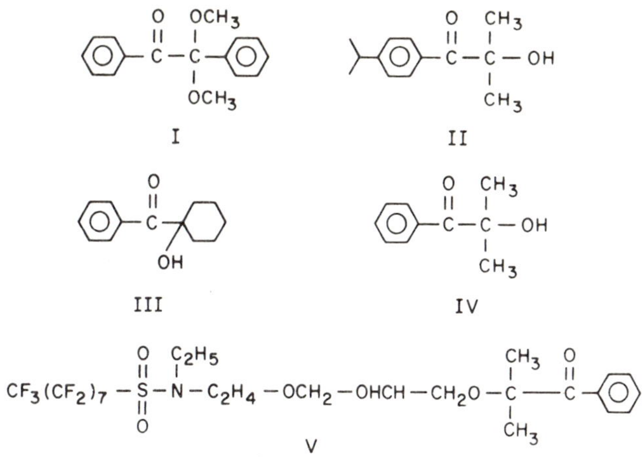

FIG. 6. Different types of photofragmenting UV initiators: I–IV are commercial and V is an experimental surface active initiator.

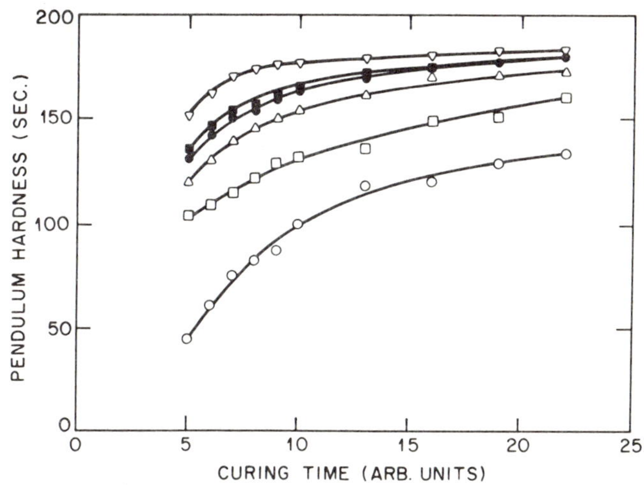

FIG. 7. Pendulum hardness of surface coating containing increasing amounts of initiator IV (Fig. 6); 1, 2, 6, and 8% and 2% IV + 0.14% V (top curve) as a function of UV curing time.

tion of a sensitizer and a multifunctional crosslinker added, 2 to 5 mm sheets of linear PE can be crosslinked to a high gel content (over 80%) by UV irradiation for short times (5 to 15 sec). The gel content can be analyzed by extraction with boiling xylene. An atmosphere of nitrogen is favorable to prevent surface oxidation of the substrate. As examples are shown, the rate of crosslinking of 2 mm sheets of three linear polyethylenes (HDPE) with 1% of sensitizer and 1% of a multifunctional crosslinker added which are UV irradiated for various lengths of time in nitrogen atmosphere (Fig. 10). For two of the HDPE samples, 80% gel content is reached in 10 sec. The degree of crosslinking is related to the temperature near the melting point (Fig. 11). Below the melting point

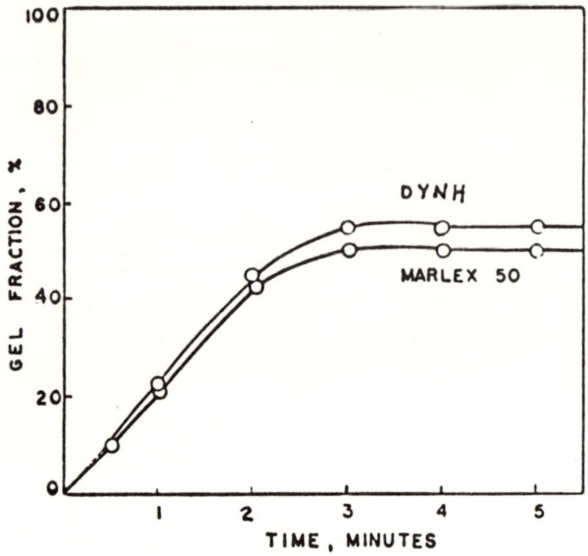

FIG. 8. UV-initiated crosslinking of thin polyethylene films containing 0.25% benzophenone as a function of irradiation time. Concentration of benzophenone in Fig. 8 after 3 min UV irradiation time. DYNH is a low-density and Marlex 50 a high-density polyethylene.

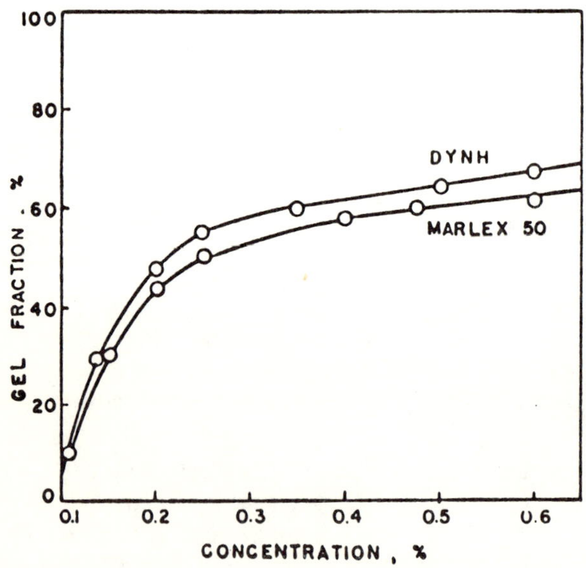

(which is about 135°C) the gel content is about 70% and above the melting point 85-90% for irradiation times of 15 sec. The presence of air has a detrimental oxidation effect on the surface properties of the PE substrate which may be discolored or even slightly degraded ("sticky") after the UV irradiated crosslinking. The effect of air during the process on the total gel content of the

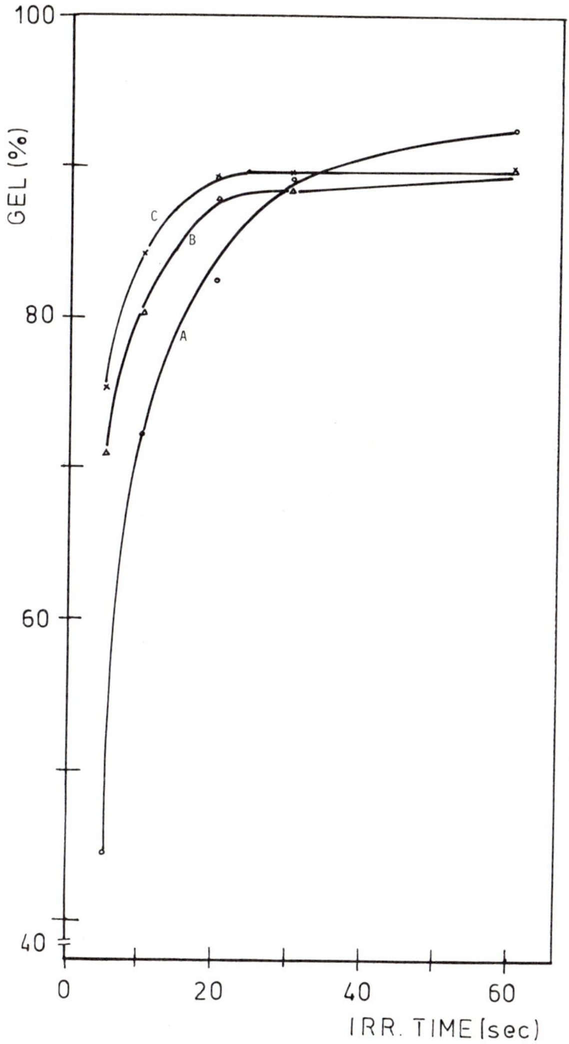

FIG. 10. UV-initiated crosslinking of three linear polyethylene sheets (2 mm thick) containing 1% sensitizer and 1% multifunctional crosslinker. Gel content as a function of irradiation time.

crosslinked PE sheet is rather small (a few % difference in gel content) as shown in Figure 12.

For an efficient UV crosslinking reaction with sensitizer added, the use of an added crosslinker is essential as shown in Figure 13. Without crosslinker, it takes 30-40 sec irradiation time to reach 80% gel content. With 0.5% crosslinker, 80% gel content is reached in 10 sec and with 2% crosslinker in about 5 sec.

An important problem is to be able to stabilize the photocrosslinked polyethylene against thermal oxidation, especially when the crosslinked polymer is used at elevated temperatures for extended periods of time. Most stabilizers

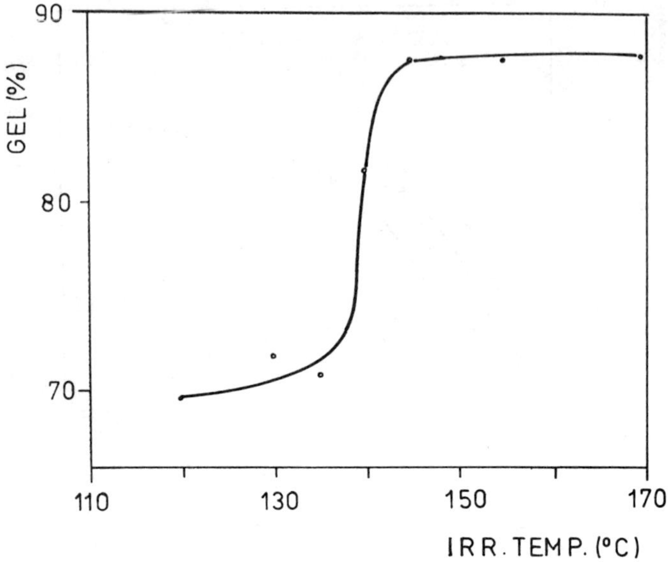

FIG. 11. Gel content of UV-crosslinked linear polyethylene sheet as a function of temperature below and above the melting point of the polymer.

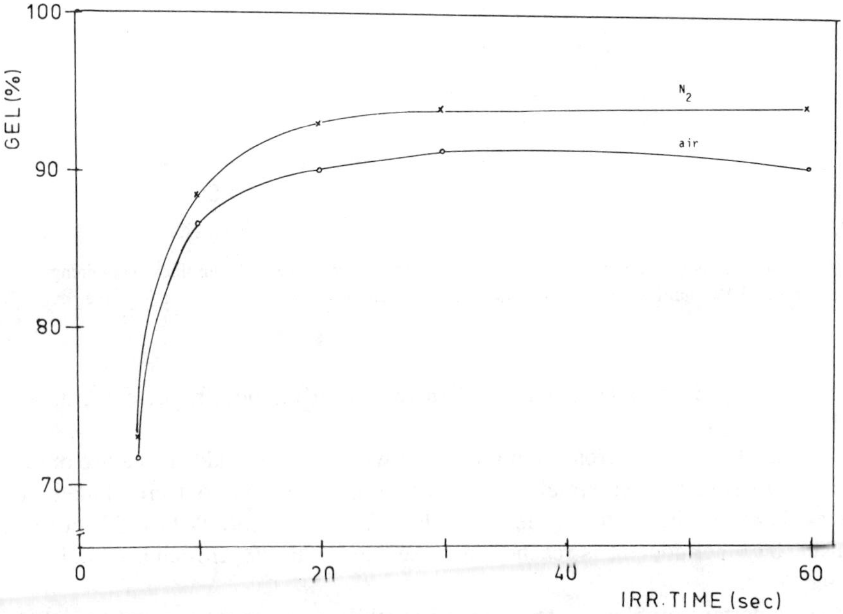

FIG. 12. Gel content of UV-crosslinked linear polyethylene sheets in air and nitrogen atmosphere, respectively, as a function of irradiation time.

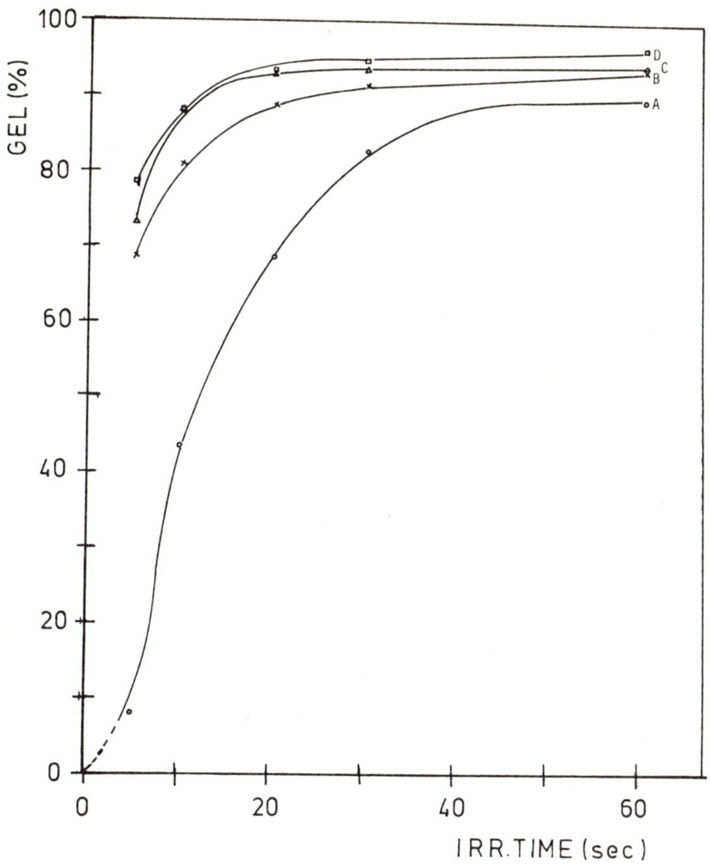

FIG. 13. Gel content of UV-crosslinked linear polyethylene sheets containing 1% sensitizer and increasing amounts of a multifunctional crosslinker (0, 0.5, 1.0, and 2.0% in A, B, C, and D, respectively).

are free radical scavengers which retard or prevent the crosslinking reaction. In addition, the crosslinking reaction *per se* introduces branch points of the chains which means a decrease in thermal stability and photostability. These problems should be studied in more detail.

PERSPECTIVES FOR FUTURE DEVELOPMENT OF POLYMER PHOTOCHEMISTRY

The photochemistry of polymers and polymeric materials is in an early stage of development. The basic mechanisms of the reactions are incompletely understood, and the large-scale industrial applications are rather few and limited to UV curing of some surface coatings and the production of photoresists.

Extensive studies of photochemical processes have been made only in the field of photoresist materials. Most photoresists are still produced using classical positive and negative "image" systems known for several decades. Newer systems, which give resolutions higher than 1 μm in the chips and better defined "image" formation, require the use of far UV irradiation and quartz optic for the lithography.

New basic systems are required for reversible photocrosslinking/photodegradation of resist polymers and for dry-processed resist polymers giving volatile degradation products. Such systems are actively studied in various research programs.

Photocuring of solvent-free organic coatings and printing inks are commercial. They are processed at room temperature without evaporation of volatile compounds which requires less energy and shorter process times. Surface-active initiators and cationic initiators have opened new possibilities for fast and selective polymerization and crosslinking of thin surface layers of polymers in commercial applications. The surface-active initiators increase the reaction rate of curing and may also decrease the effect of oxygen. Cationic photoinitiators are active for rapid polymerization of special monomers and are ready for commercial development.

Rapid crosslinking of polyethylene with UV irradiation to high gel content is possible by adding combinations of sensitizer and crosslinker in small amounts. Similar reactions can be initiated for unsaturated polyesters and for certain rubbers using UV irradiation at low temperatures. UV irradiation applied for this purpose is expected to compete successfully with high-energy radiation (electron beam and gamma rays) and with thermally initiated crosslinking. These developments of photochemical polymer processes show great promise for industrial applications; the work has just started.

The prospects for future development of polymer photochemistry — basic and applied — are bright in many areas. Polymer photochemistry is a promising field of research for the 1990s.

REFERENCES

[1] G. Oster, *J. Polym. Sci., 22,* 185 (1956). G. Oster, G. K. Oster, and H. Moroson, *J. Polym. Sci., 34,* 671 (1959).

[2] C. G. Willson, *"Organic Resist Materials — Theory and Chemistry,"* in *Introduction to Microlithography,* L. F. Thompson, C. G. Willson, and M. J. Bowden, Eds., ACS Symp. 1983, Ser. 219, chap. 3.

[3] M. J. Bowden, "A Perspective on Resist Materials for Fine-Line Lithography," in *Materials for Microlithography — Radiation-Sensitive Polymers,"* L. F. Thompson, C. G. Willson, and M. J. Frechet, Eds., ACS Symp. 1984, Ser. 266, chap. 3.

[4] Cf. C. G. Willson et al., Proc. SPE Regional Tech. Conf., Ellenville NY, November 1982, p. 111.

[5] K. Jain, C. G. Willson, and B. J. Liu, *IBM J. Res. Dev., 26*(2), 151 (1982).

[6] G. G. Willson, "Recent Developments in Organic Resist Technology," paper presented at New Trends in the Photochemistry of Polymers, Stockholm, Sweden, August 26-29, 1985.

[7] S. P. Pappas, (Ed.), *UV Curing: Science and Technology,* Technology Marketing Corp., Norwalk, CN, 1978.

[8] S. G. Wentink and S. D. Koch, (Eds.), *UV Curing in Screen Printing for Printed Circuits and the Graphic Arts,* Technology Marketing Corp., Norwalk, CN, 1981.

[9] C. G. Roffey, *Photopolymerization of Surface Coatings,* Wiley, Chichester, England, 1982.

[10] B. Rånby and S. Göthe, *Free Radical Initiation of Photopolymerization and Photocrosslinking,* IUPAC MACRO 83, Bucharest, Romania, September 1983, pp. 180.

[11] B. Rånby and A. Hult, "Photodegradation and Photostabilization of Organic Surface Coatings," *Organic Coatings, 7,* 137-156 (1984).
[12] A. Hult and B. Rånby, "Photocuring in Air Using a Surface Active Photoinitiator," *Polymer Preprints, IACS Division of Polymer Chemistry, 25*(1), 329 (1984).
[13] A. Hult and B. Rånby, "Primary and Secondary Reactions in Photoinitiated Free-Radical Polymerization of Organic Coatings," in *Materials for Microlithography—Radiation-Sensitive Polymers,* L. F. Thompson, C. G. Willson, and M. J. Frechet, Eds., ACS Symp., 1984, chap. 23, pp. 457-472.
[14] H. F. Mark, N. G. Gaylord and N. M. Bikales, Eds., *Encyclopedia of Polymer Science and Technol.,* Wiley-Interscience, NY, 1966, vol. 4, pp. 340, 364, 390.
[15] U.S. Patent 3,219,566 and British Patent 1,087,408.
[16] B. Rånby and Yong-Lie Cheng, unpublished results.

Bengt Rånby was born on April 45, 1920.

PRESENT POSITION

Professor of Polymer Technology and Chairman, Department of Polymer Technology, The Royal Institue of Technology, S-100 44 Stockholm, Sweden

EDUCATION

Uppsala University, Sweden: B Sc 1940, M Sc 1945, "Fil. lic." 1950, Ph.D., 1952. du Pont Research Fellow, Polytechnic Institute of Brooklyn, NY., USA, 1952/1953.

POSITIONS

Research Assistant 1943–52 and "Docent" of Physical Chemistry 1952–55, Uppsala University, Sweden.
Research Specialist 1956–1958, American Viscose Corp., Marcus Hook, Pa, USA.
Research Professor and Director, Empire State Paper Research Institute 1959–1961, State University of New York, Syracuse, NY USA.
Visiting Professor: North Carolina State University, Raleigh, USA, 1968.
Kyoto University, Japan 1973 Hokkaido University, Japan, 1973.
University of Massachusetts, Amherst, USA, 1976.
Midland Macromolecular Institute, Michigan, USA, 1978.

ACADEMIES

Royal Academy of Arts and Sciences, Uppsala, Sweden, 1960.
Royal Academy of Engineering Sciences, Stockholm, 1967.
Finnish Society of Sciences, Helsinki, 1979.
Royal Swedish Academy of Sciences, Stockholm, 1985.

HONORS

Doctor Honoris Causa, Technical University, Wroclaw, Poland, 1978 and University of Helsinki, Finland, 1982.

PROFESSIONAL ACTIVITIES

Swedish Polymer Society, Chairman, 1984.
Scandinavian Paint Research Institute,
Copenhagen, Chairman of Board 1976–
Science Advisory Council of INCENTIVE
AB, Stockholm
Member 1971–, Chairman 1984–

STEREOREGULATION IN POLYMER SYNTHESIS

PIERO PINO
Swiss Federal Institute of Technology, Institut fuer Polymere,
Universitätsstrasse 6, CH-8092 Zuerich, Switzerland

INTRODUCTION

The crucial role of stereochemistry in determining the chemical and physical properties of polymers was already recognized in the 1920s when Mark and Meyer [1] correctly interpreted the differences between natural rubber and guttapercha by suggesting that the former correspond to the *cis* and the latter to the *trans* configuration of 1,4-polyisoprene. Since then stereoregulation has become a very ambitious goal of polymer synthesis but the monopoly of Nature in carrying out stereospecific synthesis of macromolecules was actually broken only in the 1950s.

In the last decades a very large amount of work has been done in this field all over the world, thousands of papers and patents have been published and therefore the topic can certainly not be reviewed in a short survey. What I shall try to do is to mention very briefly the main tools used for obtaining stereoregulation in polymer synthesis and some of the main results achieved up to now. I shall then discuss some possible further developments in this field which seems still very promising from the point of view of both the development of polymer science and of its practical applications. Of course the choice of the examples is largely influenced by personal experience and preferences and no attempt has been made to carry out a full literature search of the field.

DEFINITIONS

Stereoregulation deals in principle with the preferential or exclusive synthesis of a single type of diastereomeric macromolecules starting with monomers which do not contain sites of stereoisomerism or which contain sites of stereoisomerism either not involved or disappearing during the synthesis. As sites of stereoisomerism we shall arbitrarily consider stereogenic [3] atoms or groups of atoms which can generate enantiomers or diastereoisomers having a sufficient configurational or, in the case of atropoisomerism, conformational stability to allow the observation of stereoisomerism phenomena. Some sites giving rise to stereoisomerism phenomena at room temperature are indicated in Figure 1.

Stereoregulation enables us to synthesize stereoregular polymers which consist of macromolecules containing in the main chain regular sequences of

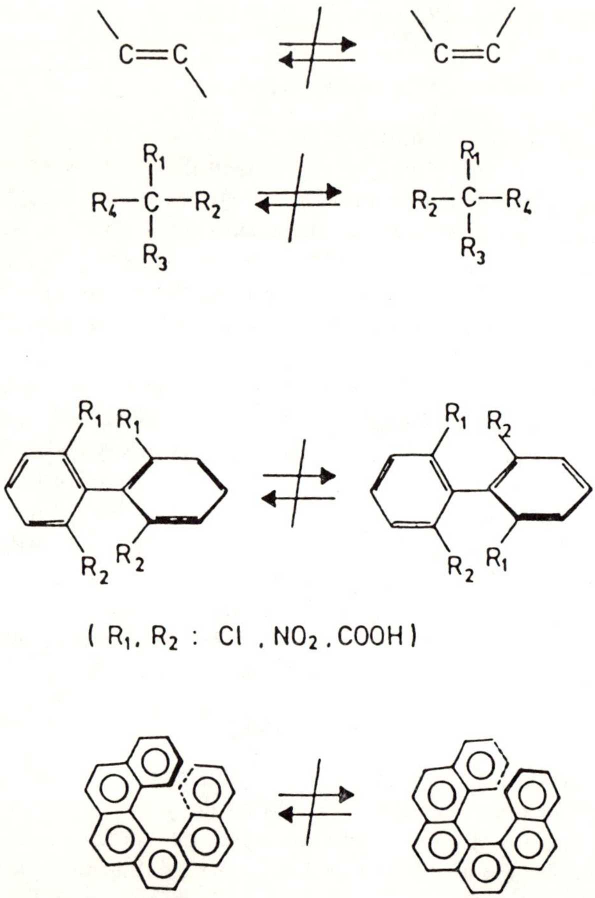

FIG. 1. Some stereoisomerism sites having a substantial configurational or conformational stability at room temperature.

stereoisomerism sites. Some examples of stereoregulation in polymer synthesis starting with monomers not containing stereoisomerism sites, containing stereoisomerism sites disappearing during the polymerization, or stereoisomerism sites not involved in the polymerization are indicated in Figure 2.

By polymerization processes, only macromolecules with relatively simple types of stereoregularity have been synthesized; the most popular examples include sequences of asymmetric carbon atoms with the same absolute configurations (. . . aaaa . .) or with an alternation of the two possible absolute configurations (. . aa′aa′ . . . .) and regular sequences of asymmetric carbon atoms and double bonds (Fig. 3). More recently, examples of sequences of asymmetric sites consisting of asymmetric carbon atoms or double bonds and groups with hindered rotation around single bonds have been mentioned in the literature (Fig. 4).

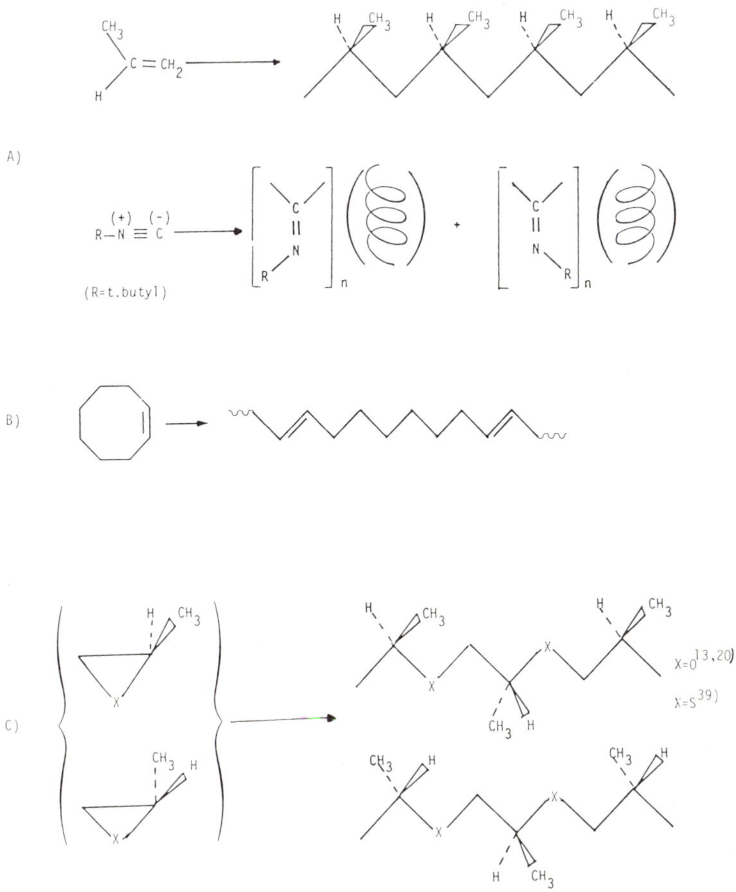

FIG. 2. Some examples of stereoregulation in polymer synthesis: (a) from monomers not containing isomerism sites, (b) from monomers with a stereoisomerism site which disappears during the polymerization; (c) from monomers with a stereoisomerism site not involved in the polymerization.

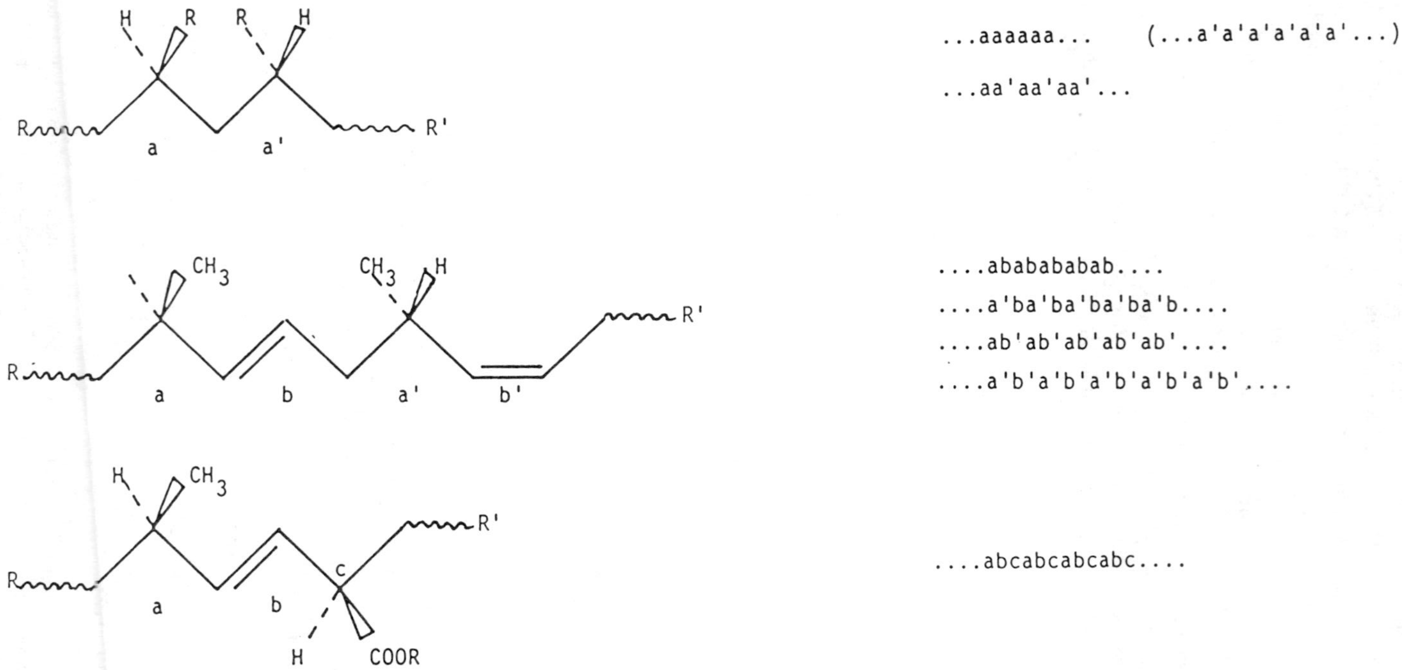

FIG. 3. Some types of ordered sequences of stereoisomerism sites consisting of asymmetric carbon atoms and disubstituted double bonds.

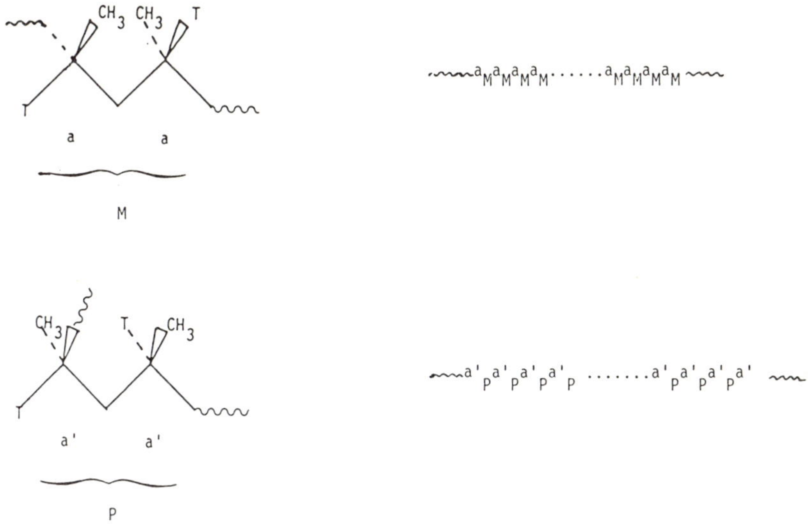

$T = -COOC(C_6H_5)_3$

a_M and a'_P = monomeric units inserted in a M and respectively P helix

FIG. 4. Some types of ordered sequences of stereoisomerism sites including groups with hindered rotation around single bonds.

Polycondensation methods are in principle more suitable for synthesizing more sophisticated types of stereoregularities using stepwise synthesis. However, up to now, sequences mostly of the type (aa′aa′ . . .) have been synthesized starting with *d* and *l* amino acids. In fact, the main aim in the synthesis of polyamino acids has been to obtain ordered sequences of different amino acids with the same absolute configuration to reproduce natural polypeptides.

STEREOREGULATING FACTORS IN POLYMER SYNTHESIS

The main stereoregulating factors in polymer synthesis are (i) the structure of the growing chain end, (ii) the structure of the polymerization catalytic sites, and (iii) the environment of the monomer during the polymerization. In principle, stereoregular polymers can also be synthesized by asymmetric reactions at prochiral centers of already synthesized polymer chains; however this method has not been very successful up to now and will not be discussed further in this section.

STEREOREGULATION BY GROWING CHAIN ENDS

This type of stereoregulation occurs in free radical or free ion polymerization and in some insertion polymerizations with achiral complexes (Table I). In general, the degree of stereoregulation is very poor; in vinyl polymers syndiotactic

TABLE I
Stereoregulation by Growing Chain End

Monomer	Initiator	T(°C)	Solvent	Polymer stereoregularity (mm)	(mr)	(rr)	Ref.
Triphenylmethyl-methacrylate	AIBN	60	Toluene	64	22	14	73
Ethyl-α-chloroacrylate	BPO	70	—	5	29	66	74
Methyl-methacrylate	BPO	50	—	85	315	60	75
Methyl-methacrylate	$LiAlH_4$	−70	THF	7	16	77	76
(Z)-α-ethyl-β-d_1-methyl-acrylate	Fluorenyl Li	−78	THF	5	21	74	77
Diphenylmethyl methacrylate	*n*-Butyl Li	−30	THF	4	12	84	73
2-Chlorobutadiene	Redox system	−40	[a]	[b]			78
Propylene	$V(acac_3/Al(C_2H_5)_2Cl$	−78	Toluene			[c]	79
Propylene	$Cp_2Ti(Ph)_2/$ Methylalumoxane	−75	Toluene	72	25	2	35

[a]Emulsion polymerization.
[b]94–97% 1,4-*trans*.
[c]r diads [81].

diads prevail over the isotactic ones. As expected in the polymerization of unsaturated monomers, stereoregulating capacity increases by decreasing temperatures and by increasing the bulkiness of the substituents as shown in Table I. Only in the polymerization of tritylmethacrylate, in which a very bulky substituent in the lateral chains is present, more than 70% of isotactic diads have been found in the polymer [4]. To our knowledge it has not been investigated if in this case the chain end group is the CH_2- or the $C(CH_3)COOR$ group, as expected, and therefore no reliable hypothesis can be made on the nature of the transition state regulating the stereochemistry of the chain growth.

STEREOREGULATION BY THE POLYMERIZATION CATALYST

In the polymerization of unsaturated and cyclic monomers, chain growing can occur via insertion of the new monomer into a substantially covalent metal to growing-chain bond or into a tight ion couple formed by a growing chain bearing a positive or a negative charge at its end and a counter ion. The significance of the insertion polymerization for the synthesis of structurally regular and stereoregular polymers was first recognized by Ziegler [5] working on the insertion reactions on cumyl potassium; by Morton [6] working on the polymerization of 1,3-butadiene with the "Alfin" catalysts, and by Natta [7] working on the polymerization of α-olefins.

The couple formed by the growing chain end and the counter ion or the metal complex covalently bound to the growing chain is indicated as the "catalytic center." Both moieties of the catalytic centers can contribute to stereoregulation. Depending on the type of monomer and on the structural features of the two moieties, the contribution to stereoregulation by the growing chain end and by the counter ion or organometallic complex bound to the growing chain can be largely different and at times difficult to estimate separately.

Concerning stereoregulation in the polymerization of nonchiral monomers, only unsaturated monomers will be considered. In fact, in the ring-opening polymerization of cyclic monomers only the stereospecific polymerization of *cis*-2,3-epoxyethane [8] having a meso structure and yielding a stereoregular polymer has been investigated, other interesting cases as 3-substituted oxacylobutanes [9] having been largely ignored up to now.

Mono- and disubstituted ethylenes undergo insertion polymerization with organometallic complexes and with ion couples serving as catalysts. The polymers obtained are in general isotactic and the stereoregulating capacity of these systems is generally very high (Table II). In these cases the catalytic center must be able to distinguish among the two enantiofaces of the monomers and therefore must be chiral. Using nonracemic organometallic complexes, optically active polymers have been obtained from nonchiral monomers (Table III).

Stereoregulation by the catalytic sites is also assumed in the insertion polymerization of aldehydes [10] yielding isotactic polymers and of isonitriles [11] yielding stereoregular polyiminomethylenes. The chirality of the catalytic centers produces an interesting type of asymmetric transformation starting with racemic monomers. In α-olefins when the chirality center of the monomer is in α or β position with respect to the reactive group of the monomers instead of random copolymers of the two antipodes, macromolecules prevailingly deriving from one or from the other antipode are formed (Scheme 1). This type of polymerization is indicated as stereoselective polymerization [12] and yields optically inactive polymers. The same phenomenon was observed in the polymerization of racemic epoxides [13].

TABLE II
Some Examples of Stereoregulation by Catalytic Sites

Monomer	Catalytic site	T(°C)	Solvent	Polymer stereoregularity (mm) %	(mr) %	(rr) %	Ref.
Propylene	Ti—C—	60	*n*-Heptane	94.5[a]	3.0[a]	2.5[a]	80
Styrene	Ti—C—	25	Toluene	99.2[b]	0.5[b]	0.3[b]	81
Cyclopentene	W=C<	−30[a]	—		[c]		
Methylvinylether	$[BF_4(C_2H_5)_2O]^-[C—]^+$	−78	Toluene/hexene	100	0	0	83
Methylvinyl ketone	Zn—C—	0	Toluene	85			84

[a]For the *n*-heptane-insoluble polymer fraction.
[b]For the acetone insoluble polymer fraction.
[c]Polypentenamer; 96-97% *trans* monomeric units.

TABLE III
Some Examples of Stereoregulation by Chiral, Nonracemic Catalytic Centers

Monomer	Catalyst	T(°C)	Solvent	Type of polymerstereo-regularity	Ref.
trans 1,3-pentadiene	$MgCl_2/TiCl_4/(-)PMA/Al(iBu)_3$[a]	25	*n*-heptane	1,4-*trans*; isotactic[b]	85
Methyl sorbate	LiBu/(−)menthyl ethyl ether	−40	Toluene	erythro,diisotactic, *trans*[c]	86
Tritylmethacrylate	(−) Sparteine/BuLi	−78	Toluene	Isotactic (M or P)[e]	87
Benzofuran	(−)β-phenylalanine/$AlCl_3$	−75	Toluene	threo di-isotactic[d]	88

[a](−)PMA is (−)menthylanisate.
[b]$[\alpha]_D^{25}$ = +15,1 (*o*-dichlorobenzene).
[c]$[\alpha]_D$ = +6.4 ($CHCl_3$).
[d]$[\alpha]_D$ = −30 to −75° (benzene).
[e]$[\alpha]_D^{20}$ = +363 (toluene); optical activity due to conformationally stable M or P helix.

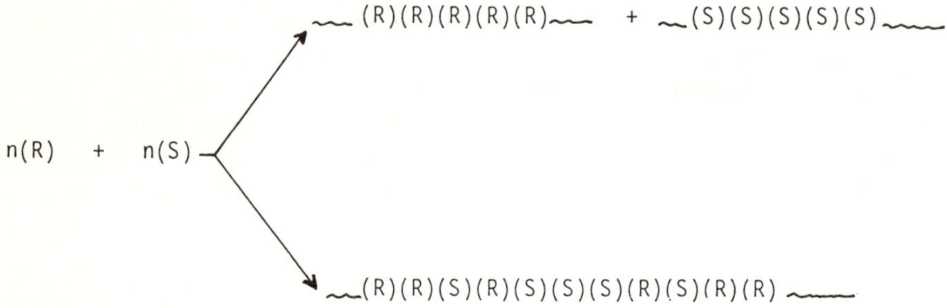

SCHEME 1. Ideal stereoselective polymerization of a racemic monomer.

The origin of stereoselectivity in the polymerization of racemic vinyl monomers is connected with a different reactivity of the two diastereofaces of each antipode [14].

Also with these catalysts, asymmetric polymerizations yielding optically active polymers from nonchiral or from racemic monomers has been achieved (stereoelective polymerization) [15] or asymmetric selective polymerization [13] (Table IV). In the case of the polymerization of cycloolefins via methatesis, the stereoregulating mechanism is still under investigation [16]. In the case of conjugated diene polymerization, double bonds are present in the polymers. Stereoregulation in the case of 1,4-polymerization with metal complexes is believed to be connected with the *syn* or *anti* configuration of the π-allyl complexes formed by the metal atom of the catalyst, with the last monomeric unit of the growing chain (Table V). Finally, in the case of acetylene polymerization only the *cis*-polyacetylene [17] is expected to form according to the *cis* stereochemistry [18] of the insertion polymerization. However, due to the presence of a system of conjugated double bonds in the polymer chains, a very fast *cis-trans* isomerization can take place.

TABLE IV
Some examples of Stereoelective Polymerization

Racemic monomer	Catalytic system	Conversion %	Relative polymerization rate[a]	Ref.
$\underset{\diagdown\;O\;\diagup}{\overset{CH_3}{\overset{\vert}{C}H}-CH_2}$	$Zn(C_2H_5)_2/(+)$ borneol	11.6	3.43	89
$CH_2{=}C(CH_3)-COOCH(CH_3)C_6H_5$	$c-C_6H_{11}MgCl/(-)$ sparteine		[b]	90
$CH_2{=}CH-\underset{\overset{\vert}{CH_3}}{CH}-C_2H_5$	$TiCl_4/(+)\ Zn(CHCH_3C_2H_5)_2$	7.2	1.7	91
$CH_2{=}CH-CH_2-\underset{\overset{\vert}{CH_3}}{CH}-C_2H_5$	$MgCl_2/TiCl_4/AlR_3/MA$[c]	43	1.07	92

[a] Conversion of faster to slower polymerizing enantiomer.
[b] Optical purity of unreacted monomer nearly 100% at about 65% conversion.
[c] (1)-menthyl-*p*-methoxybenzoate.

TABLE V
Stereoregulation in the 1,4-Polymerization of Butadiene by *bis*-(η^3-Allylnickel Trifluoracetate)

Ligand (L)	[L]/[Ni]	% *cis*	% *trans*
—	—	98	1
$P(OR)_3$	1	—	99
C_2H_5OH	1	—	99
CF_3COOH	1	50	49

Racemic heterocyclic monomers undergo stereoselective and stereoelective polymerization [19]. For example, in the case of monosubstituted epoxides [20] and episulfides [21] stereoregularity in the polymer main chain has been obtained starting with an optically pure antipode of the monomer. To obtain an isotactic polymer stereoregulation is needed only when the monomer is not optically pure or racemic. The exact nature of the catalytic sites is unknown also in this case; it is believed that an insertion occurs between a metal atom of the catalyst and the growing chain. The possible stereoregulating role of the growing chain has not been deeply investigated.

STEREOREGULATION BY THE MONOMER ENVIRONMENT

Stereoregulation by the environment is achieved by fixing the monomer molecules in the desired conformation and at the right intermolecular distance before polymerization. In this way a particular diastereomeric transition state for the reaction between growing chain and monomer molecule is favored for geometrical reasons and a stereoregular polymer can be obtained in the absence of stereospecific catalysts. The methods used up to now are the homogeneous and heterogeneous topotactic polymerization [22], polymerization in thin layers in perwoskite-type structures [23], and polymerization in inclusion complexes [24]. Stereoregulation has also been obtained by polymerization in ordered monolayers [25]. The first two methods require a very accurate investigation of the crystalline structure of both the monomer and polymer, and in some cases, the preparation of suitable single crystals [26, 27]. Presently, it is limited to relatively few classes of monomers such as diacetylenes or dienes. The method using inclusion complexes is more flexible and has been successful in the polymerization of various monomers including vinyl chloride, 1,3-butadiene, 1-3 pentadiene, 2-methylbutadiene (Table VI). Also, in this case, a suitable host compound must be found for each type of monomer to be polymerized.

All the above examples belong to the large group of polymerization reactions indicated as solid-state polymerizations; they are in general initiated by radiation. However, in many cases solid-state polymerization fails to produce stereoregular polymers because of local melting processes in the positions in which new covalent bonds between monomers are formed. The melting process destroys the orientation of the monomers which polymerize as they would in a liquid-phase yielding atactic polymers.

TABLE VI
Examples of Inclusion Polymerization

Monomer	Host[a]	Type of stereoregularity	Ref.
$CH_2{=}CH{-}CH{=}CH_2$	Urea	*trans*-1,4	94
$CH_2{=}C(CH_3){-}C(CH_3){=}CH_2$	Thiourea	*trans*-1,4	95
	PHTP	*trans*-1,4	46
$CH_3{-}CH{=}CH{-}CH{=}CH_2$	*trans*, anti, *trans*, anti, *trans* PHTP[a]	*trans*-1,4-isotactic[b]	46
$CH_2{=}CH{-}Cl$	Urea	syndiotactic	94

[a] PHTP = Perhydrotriphenylene.
[b] $[\alpha]_D^{25}$ + 2.5 ($CHCl_3$).

SOME IMPORTANT ACHIEVEMENTS IN STEREOREGULATION

Probably the main achievement in stereoregulation has been the synthesis of isotactic poly-α-olefins [28], which has a very great theoretical and industrial importance. In this case, the degree of stereoregularity achieved can be so high that only a lower limit can be given for its value, which depends on the accuracy of the method used for its determination. The main stereoregulating factor in this case is the chirality of the catalyst (enantiomeric sites control) which enables the catalytic centers to distinguish between the two enantiofaces of the α-olefins. This discriminating capacity is mainly connected with steric interactions between monomer and chiral catalytic centers which, according to a simple stereochemical model [29], easily accounts for the very high regio- and stereospecificity. It is very likely that stereospecificity is enhanced by the presence of a chiral center in the last monomeric unit of the growing chain which is in the β-position or in α-position with respect to the transition metal atom of the catalytic centers when a 1-2 or a 2-1 insertion mode occurs, respectively [30]. The existence of chiral catalytic centers in the α-olefins polymerization catalysts, postulated by Natta and co-workers as early as 1957 [31, 32], was indirectly proved in a number of ways [33] and was finally directly shown by the experiments of Ewens [34] and Kaminsky et al. [35] using nonchiral or chiral soluble transition metal/aluminoxanes catalytic systems. The concept of enantiomeric site control [31, 36], has been very fruitful for stereoregulation in polymer synthesis; stereoselective polymerizations [12] were obtained not only with α-olefins [37] but also with acrylic esters [38] mono- and disubstituted epoxides [21], and episulfides [39]. The attempts to synthesize chiral nonracemic catalysts have been successful both in the case of Ziegler-Natta catalysts [40, 41] and in the case of some ionic catalysts or initiators [42]. In this way, optically active polymers could be prepared both from nonchiral and from racemic monomers. Some examples are given in Tables III and IV. The most spectacular result is probably the synthesis of optically active isotactic tritylmethacrylate, in which, using for example, a Li butyl/sparteine catalytic system, stereoregulation of both the configuration of the asymmetric carbon atoms and the conformation of the polymer main chain could be achieved [42] (Scheme 2).

Remarkable results have also been obtained in the synthesis of stereoregular polymers from diolefins. According to current views stereoregulation is mainly connected with the interaction between the π-allyl metal complex formed by a metal atom of the catalyst with the last monomeric unit of the growing chain and the incoming monomer (Scheme 3). This is shown for the polymerization of 1-deutro-*cis*-1,3-pentadiene [43] in (Scheme 3). Also, in this case the structure of the ligands bound to the transition metal atom of the catalytic site can play an important role [44].

In the field of stereoregulation obtained using suitable monomer environments, the most impressive results have been achieved by Addadi and his group [45], who succeeded in obtaining chiral crystals from nonchiral monomers suitable for solid-state polymerization. Radiation-initiated polymerization in chiral monocrystals yields substantially optically pure oligomers and polymers. This work may be considered as the first absolute asymmetric synthesis occurring with very high optical yield although the prevailing chirality is decided by the type of chiral crystals obtained in the crystallization of the achiral monomer and cannot be predetermined. Also remarkable are the homogeneous topochemical reactions in which a single crystal of polymer can be obtained from a single crystal of monomer, as in the case of diacetylene photopolymerization [27]. Finally, in the field of inclusion complexes, the use of chiral channels [46, 47] has allowed achievement of a satisfactory control both of double bonds and asymmetric carbon atom configuration in the 1,3-pentadiene and 2-methyl-1,3-pentadiene 1,4-polymerizations, a rare case of "through space" asymmetric induction [48].

SOME POSSIBLE DIRECTIONS OF FUTURE DEVELOPMENTS IN STEREOREGULATIONS

Despite the remarkable results achieved in the last two decades further research is needed to understand the structure of the active centers for the poly-

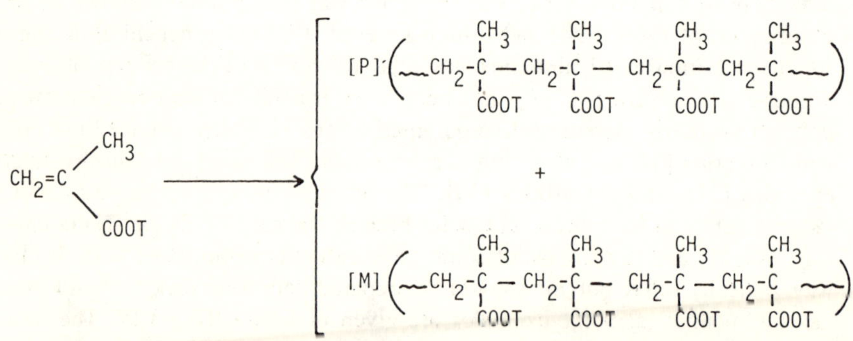

SCHEME 2. Polymerization of tritylmethacrylate to isotactic chains with different helical conformations. T = $-C(C_6H_5)_3$; [P] = right-handed, and [M] = left-handed helical conformation.

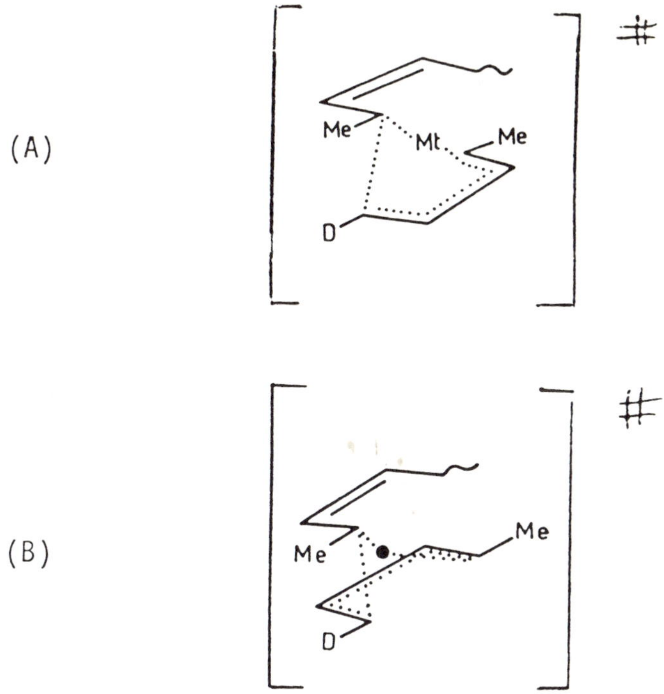

SCHEME 3. Possible schemes for the transition states responsible for the synthesis of a *cis*-1,4-threodisyndiotactic poly-1,3-pentadiene (a) and *cis*-1,4-erythrodiisotactic poly-1,3-pentadiene (b).

merization of unsaturated monomers and for the ring-opening polymerization of heterocyclic compounds. This is particularly true concerning the heterogeneous catalysts which have, in some cases, remarkable industrial importance.

This knowledge is needed to attempt the synthesis of catalysts with centers of the same type with the ultimate aim of increasing the selectivity of the synthesis and to synthesize polymers with predetermined stereoregularity.

Taking the synthesis of the vinyl polymers as an example (Fig. 5), Schildknecht and co-workers [49] first discovered, depending on polymerization conditions, that completely amorphous or partially stereoregular polyvinylethers can be synthesized. A few years later Natta and co-workers [28] discovered that, in the case of α-olefins, polymers having different stereoregularity could be separated and the diastereomeric composition of the polymer mixture obtained in the synthesis could be varied in a large interval depending on the type of catalyst used. More recently Sinn and Kaminsky succeeded in preparing completely atactic [50] or completely isotactic [35] polymers of propylene using soluble catalysts. However, the problem of synthesizing macromolecules, each of which has the same predetermined degree of stereoregularity, is still an important open problem. In fact the obvious solution to controlling stereoregularity using different polymerization temperatures cannot be adopted either with heterogeneous nor with homogeneous catalysts. In the first case the catalytic centers have a different thermal stability and the stereoregulating capacity of the

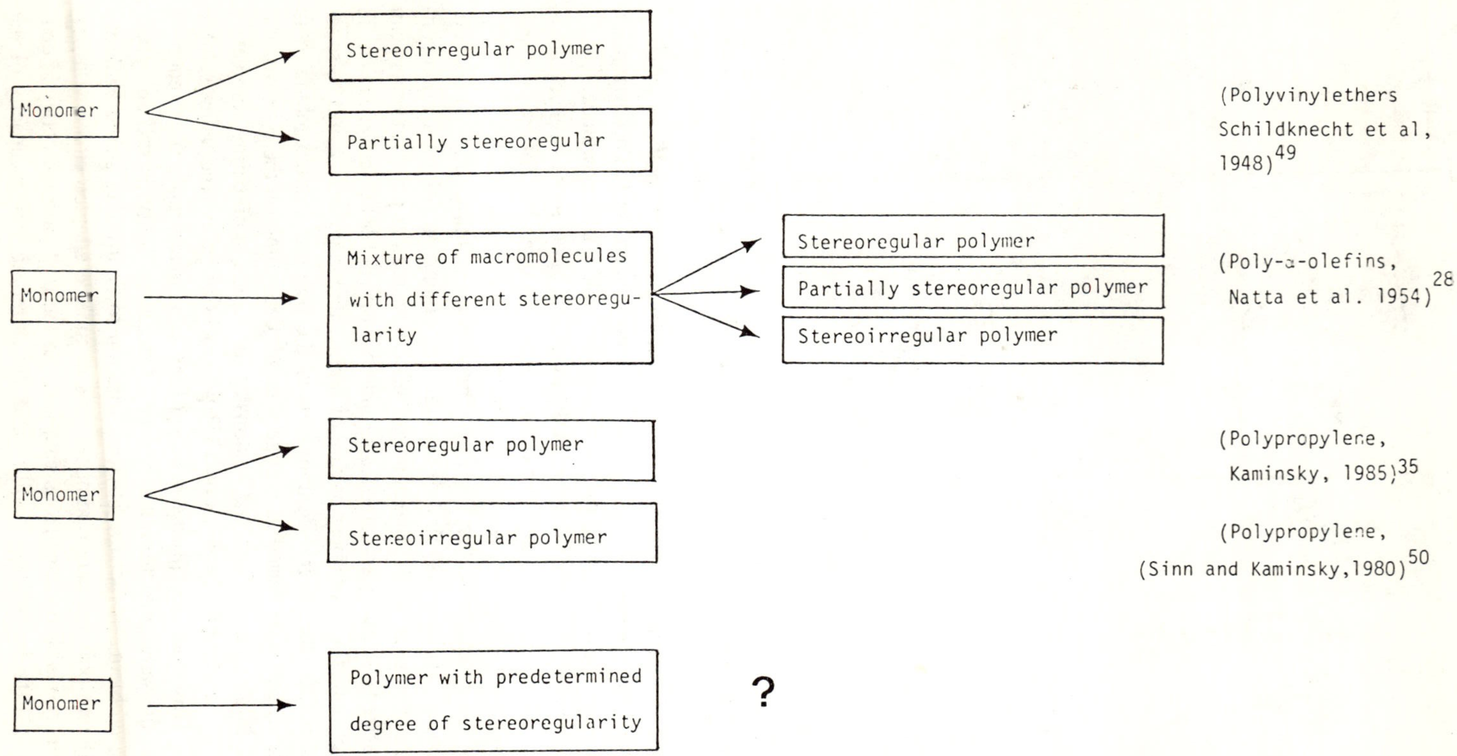

FIG. 5. Synthesis of polymers with predetermined stereoregularity degree.

different centers does not vary in the same way as does the temperature. With the available homogeneous catalysts, stereospecificity decreases by increasing temperature, but regiospecificity may also decrease with temperature, and therefore, producing structurally irregular polymers [51]. The synthesis of vinyl polymers with predetermined stereoregularity not only has a theoretical but may also be of practical interest as the change of the polymer properties with decreasing stereoregularity has not been systematically investigated, and new polymer blends with interesting properties might be prepared using polymers with a predetermined stereoregularity.

A second interesting area is the achievement of a better control of polymer chain conformation. The first attempts to control the conformation of the polymer main chains in solution were made in the 1960s, when in the case of some optically active isotactic poly-α-olefins, the existence of helical conformations in hydrocarbon solution was experimentally proved [52]. The control in this case is connected with thermodynamic factors favoring one of the two screw senses of the low energy helical conformations of the main chain, the percent of helical conformations depending on the structure of the side chains, and on temperature [53].

More recently a similar behavior has been found in isotactic polyvinylketones [54] and in isotactic polyvinylethers [55]. Conformational rigidity connected with kinetic factors, or in other words, atropoisomerism phenomena, in linear polymers was discovered in the 1970s when poly-*t*-butyl-iminomethylene was prepared and separated by elution chromatography in fractions having optical activity with opposite signs [11, 56], a method already used to prove stereoelective polymerization of α-olefins [57]. The optical activity was attributed to the existence of two low energy helical conformations of the polymer main chains separated by an energy barrier sufficiently high to hinder the interconversion of the two conformations at room temperature. A more impressive example is the previously mentioned asymmetric polymerization of tritylmethacrylate by Okamoto and co-workers [42] (Scheme 2). The polymer maintains its optical rotation in solution indefinitely at room temperature. This behavior is interpreted, assuming the existence in solution, of an optically active probably helical conformation of the isotactic main chain which is stable in solution. A transformation of a regular probably helical conformation either in a random coil or in a helical conformation of opposite screw sense can be postulated. However, the existence of a double-stranded helix in methyl methacrylate [58] and the observation of an optical activity in poly(methylmethacrylate) prepared with anionic initiators [59], which disappears at room temperature in few hours, might indicate a more complex "racemization mechanism." If the conformation of methyl methacrylate formed during the synthesis is a double-stranded helix, it is possible that after dissolution in a suitable solvent slow to separation occurs of the rigid double helix into two single-stranded helixes which then rapidly interconvert or give a random coil causing the disappearance of the optical activity.

These two examples of atropoisomerism in macromolecules show that the preparation of block copolymers with rigid and flexible sections is, in principle, conceivable, however, the realization of the synthesis of tailor-made copolymers with planned conformations is still very far in the future.

Concerning control of the conformation of flexible polymer chains during their synthesis, it is in principle possible to stabilize particular conformations (e.g., helical conformation) if crystallization occurs during the polymerization. In this case, optical activity should be lost rapidly by melting or by dissolving the polymer, and interesting information might be obtained on the rotation barriers around single bonds in polymer chains.

Polymers which show optical activity in the solid state have been obtained by Corley and Vogl [60], who polymerized trichloroacetaldehyde with an optically active cationic initiator. However, in this case the polymer decomposes before melting, is insoluble in most of the common solvents, and is not suitable for investigating the racemization rate.

A more interesting case might be the case of isotactic polypropylene and poly-l-butene which have been reported to be optically active in the solid state when prepared with an optically active Ziegler catalyst [61]. Research in this field proceeds slowly mainly due to the lack of suitable methods for measuring the optical activity in the solid state [61].

Stereoregulation in polycondensation has been little investigated up to now. The only important work in this field was carried out by Khorshak and co-workers [63, 64], who succeeded in obtaining a certain degree of conformationally regular polymers starting with sterically hindered bisphenols(*bis*-chloro-2-hydroxyphenylalkanes) with terephthalic acid chloride. Although it has been known for many years that condensation copolymers obtained from a single antipode or from the racemic form of a comonomer have different physical properties [65] no attempt has been published, to our knowledge, to investigate stereoselective polycondensation. In this case starting with a suitable racemic monomer or comonomer it should be possible to obtain macromolecules each of which contains monomeric units prevailingly arising from one of the two antipodes of the racemic monomers. This type of synthesis is, in principle, possible if a chiral racemic catalyst is used.

In the presence of optically active catalysts the synthesis of optically active condensation polymers should be possible starting both from racemic monomers via stereoelective polycondensation or from nonchiral monomers via asymmetric polycondensation. Examples of asymmetric condensations of monofunctional compounds yielding a prevailing one atropoisomer are already known in the literature [66].

Finally, the remarkable progress made in the last decade in the field of asymmetric synthesis and asymmetric catalysis suggest new ways to prepare stereoregular polymers starting from polymers containing "prochiral" sites, that is sites which can be transformed in stereoisomerism sites with suitable reactions. According to the present knowledge of asymmetric catalysis, asymmetric hydrogenations and epoxydation seem particularly promising (Scheme 4). For example, in the case of the hydrogenation of the alternating ethylene carbon monoxide copolymers a strong interaction between the already formed alcohol functions and the catalyst might favor the asymmetric induction and hence the

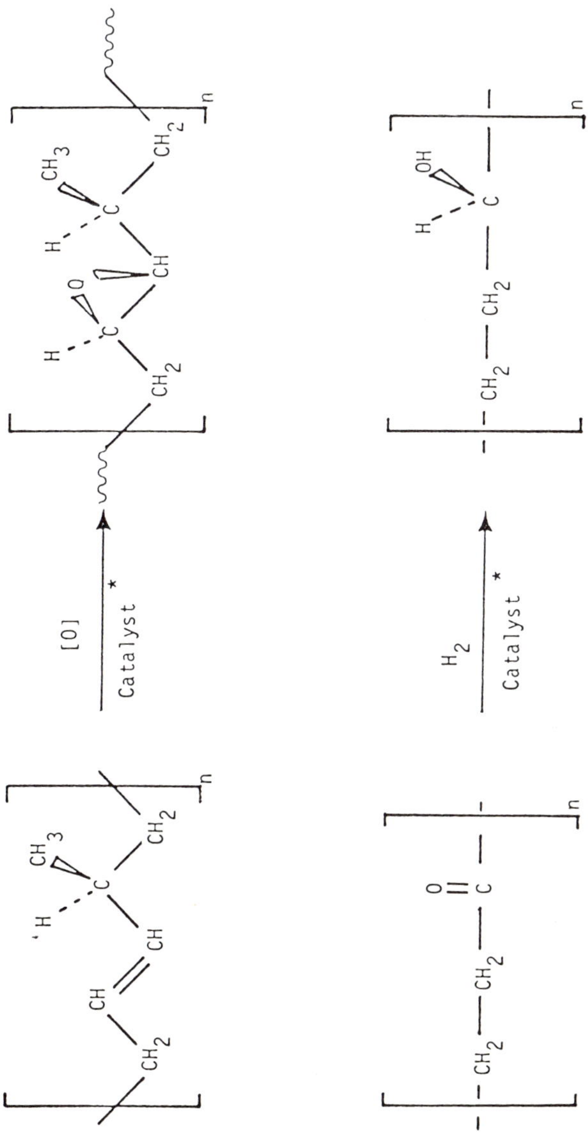

SCHEME 4. Asymmetric modification of polymer chains.

synthesis of a stereoregular polyalcohol. Asymmetric reductions of polyketones are already known in the literature [67].

CONCLUSIONS

After about 30 years of intensive investigations, the field of stereoregulation in polymer synthesis is still far from reaching maturity. The revival of interest in polymer synthesis connected with the rapid development of polymer specialities in the last decade, is further adding increased significance to this topic.

Although spectacular achievements are not expected in the next decade, stereoregulation will continue to offer exciting opportunities in the synthesis of tailor-made macromolecules and materials as well as in the investigation of polymerization mechanisms. Without doubt is will continue to be a uniquely aesthetic and intellectual challenge to synthetic polymer chemists.

I thank Dr. G. P. Lorenzi for carefully reading and revising the manuscript.

REFERENCES

[1] K. H. Meyer and H. Mark, *Der Aufbau der hochpolymeren organischen Naturstoffe,* Akad. Verlagsgesellschaft, Leipzig, 1930, p. 205.
[2] A. Fredga, in *Les Prix Nobel en 1963,* Nobel Found., Ed., Norstedt, Stockholm, 1964, p. 33.
[3] K. Mislow and J. Siegel, *J. Am. Chem. Soc., 106,* 3319 (1984).
[4] H. Yuki, K. Hatada, Y. Kikuchi, and T. Niinomi, *Polym. Lett., 6,* 753 (1968).
[5] K. Ziegler and K. Bähr, *Ber., 61,* 253 (1928).
[6] A. A. Morton, *Ind. Eng. Chem., 42,* 1488 (1950).
[7] G. Natta, *Chim. Ind. (Milan), 46,* 397 (1964).
[8] E. J. Vandenberg, *Polym. Lett., 2,* 1085 (1964).
[9] J. Kops, in *Preparation and Properties of Stereoregular Polymers,* R. W. Lenz and F. Ciardelli, Eds., Reidel, London, 1980, p. 226.
[10] G. Natta, G. Mazzanti, P. Corradini, P. Chini, and I. W. Bassi, *Rend. Accad. Naz. Lincei, 28,* (8) 8 (1960).
[11] R. J. M. Nolte, J. W. Zwikker, J. Reedijk, and W. Drenth, *J. Mol. Catal., 4,* 423 (1978).
[12] P. Pino, F. Ciardelli, and G. Montagnoli, *J. Polym. Sci., Part C, 16,* 3265 (1968).
[13] T. Tsuruta, S. Inoue, and I. Tsukuma, *Makromol. Chem., 84,* 298 (1965).
[14] R. Lazzaroni, S. Bertozzi, C. Bertucci, P. Salvadori, and P. Pino, *Isr. J. Chem., 15,* 63, 1976/77.
[15] P. Pino, F. Ciardelli, and G. P. Lorenzi, *J. Am. Chem. Soc., 85,* 3888 (1963).
[16] H. Höcker, in *Preparation and Properties of Stereoregular Polymers,* R. W. Lenz and F. Ciardelli, Eds., D. Reidel, London, 1980, p. 151.
[17] J. C. W. Chien, *Polyacetylene,* Academic, New York, 1984, p. 46.
[18] J. J. Eisch, A. M. Piotroski, S. K. Brownstein, E. J. Gabe, and F. L. Lee, *J. Am. Chem. Soc., 107,* 7219 (1985).
[19] For a review of this subject see P. Pino and G. P. Lorenzi, in *Preparation and Properties of Stereoregular Polymers,* R. W. Lenz and F. Ciardelli, Eds., D. Reidel, London, 1980, p. 1.
[20] C. C. Price and M. Osgan, *J. Am. Chem. Soc., 78,* 4787 (1956).
[21] N. Spassky, P. Dumas, M. Moreau, and J. P. Vairon, *Macromolecules, 8,* 956 (1975).
[22] For a review on Solid State Polymerization, see G. Wegner, *Pure Appl. Chem., 49,* 443 (1977).
[23] B. Tieke and G. Wegner, *Makromol. Chem. Rapid Commun., 2,* 543 (1981).
[24] M. Farina, *Makromol. Chem. Suppl., 4,* 21 (1981).
[25] H. Bader, K. Dorn, B. Hupfer, and H. Ringsdorf, *Adv. Polym. Sci., 64,* 1 (1985).

[26] M. Hasegawa, *Chem. Rev., 83,* 507 (1983).
[27] G. Wegner, *Chimia., 29,* 475 (1974).
[28] G. Natta, P. Pino, P. Corradini, E. Mantica, G. Mazzanti, and G. Moraglio, *J. Am. Chem. Soc., 77,* 1708 (1955).
[29] P. Pino, B. Rotzinger, and E. von Achenbach, in *Catalytic Polymerization of α-Olefins,* T. Keii and K. Soga, Eds., Kodansha, Tokyo, 1986, p. 461.
[30] P. Corradini, U. Barone, R. Fusco, and G. Guerra, *Eur. Polym. J., 15,* 1133 (1979).
[31] G. Natta, P. Pino, and G. Mazzanti, *Gazz. Chim. Ital., 87,* 528 (1957).
[32] G. Natta, *Inorg. Nucl. Chem., 8,* 589 (1958).
[33] For a review on this topic, see P. Pino and R. Mülhaupt, *Angew. Chem. Int. Ed. Engl., 19,* 857 (1980).
[34] J. A. Ewens, *J. Am. Chem. Soc., 106,* 6355 (1984).
[35] W. Kaminsky, K. Külper, H. H. Brintzinger, and F. W. Wild, *Angew. Chemie, Int. Ed. Engl., 24,* 507 (1985).
[36] R. A. Shelden, T. Fueno, T. Tsunatsugu, and J. Furukawa, *Polym. Lett., 3,* 23 (1965).
[37] P. Pino, A. Oschwald, F. Ciardelli, C. Carlini, and E. Chiellini, in *Coordination Polymerization,* J. C. W. Chien, Ed., Academic, New York, 1975, p. 25.
[38] Y. Okamoto, K. Urakawa, K. Otha, and H. Yuki, *Macromolecules, 11,* 719 (1978).
[39] H. Sakakibara, Y. Takabashi, H. Tadokoro, P. Sigwalt, and N. Spassky, *Macromolecules, 2,* 515 (1969).
[40] G. Natta, L. Porri, and S. Valenti, *Makromol. Chem., 67,* 225 (1963).
[41] G. Natta, M. Farina, M. Donati, and M. Peraldo, *Chim. Ind. (Milan), 40,* 1363 (1960).
[42] Y. Okamoto, K. Suzuki, K. Ohta, K. Hatada, and H. Yuki, *J. Am. Chem. Soc., 101,* 4763 (1979).
[43] L. Porri, M. C. Gallazzi, S. Destri, and A. Bolognesi, in *Transition Metal Catalyzed Polymerization,* P. Quirk, Ed., Harwood, London, 1983, p. 555.
[44] G. P. Durant, F. Dawans, and P. Teyssie, *J. Polym. Sci., Part B, 6,* 760 (1968).
[45] L. Addadi, J. van Mil, and M. Lahav, *J. Am. Chem. Soc., 104,* 3422 (1982).
[46] M. Farina, G. Audisio, and G. Natta, *J. Am. Chem. Soc., 89,* 5071 (1967).
[47] G. Audisio and A. Silvani, *J.C.S. Chem. Comm.,* 481 (1976).
[48] M. Farina, G. Di Silvestro, and P. Sozzani, *Makromol. Chem., Rapid Commun., 2,* 51 (1981).
[49] C. E. Schildknecht, A. D. Zoss, and C. McKinley *Ind. Eng. Chem., 39,* 180 (1947).
[50] H. Sinn and W. Kaminsky, *Adv. Organomet. Chem., 18,* 99 (1980).
[51] P. Pino, P. Cioni, and J. Wei, unpublished results.
[52] P. Pino, *Adv. Polym. Sci., 4,* 393 (1965).
[53] P. L. Luisi and P. Pino, *J. Phys. Chem., 72,* 2400 (1968).
[54] P. Pino, A. Klaus, and A. Allio, *Gazz. Chim. Ital.,* in press.
[55] P. Pino, P. Salvadori, G. P. Lorenzi, E. Chiellini, L. Lardicci, G. Consigulio, O. Bonsignori, and L. Lepri, *Chim. Ind. (Milan), 55,* 182 (1973).
[56] R. J. M. Nolte, A. J. M. Van Beijnen, and W. Drenth, *J. Am. Chem. Soc., 96,* 5932 (1974).
[57] P. Pino, F. Ciardelli, G. P. Lorenzi, and G. Natta, *J. Am. Chem. Soc., 84,* 1487 (1962).
[58] H. Kusanagi, H. Tadokoro, and Y. Chatani, *Macromolecules, 9,* 531 (1976).
[59] D. J. Cram and Y. Sogah, *J. Am. Chem. Soc., 107,* 8301 (1985).
[60] L. S. Corley and O. Vogl, *Polym. Bull. (Berlin), 3,* 211 (1980).
[61] W. Kaminsky, in *Catalytic Polymerization of Olefins,* T. Keii and K. Soga, Eds., Kodansha, Tokyo 1985, p. 300.
[62] For a review in this field see P. Pino, F. Ciardelli, and M. Zandomeneghi, *Ann. Rev. Phys. Chem., 21,* 561 (1970).
[63] V. V. Khorshak, S. V. Vinogradova, and V. A. Vasnjev, *Faserforsch. Textilchemie, 28,* 491 (1977).
[64] V. V. Khorshak, T. A. Baluskwa, S. V. Vinogradova, A. V. Vasilev, and G. K. Senim, *Isv. Akad. Nauk, S.S.S.R. Ser Khim.,* 1921 (1970).
[65] L. B. Sokolov, *Synthesis of Polymers by Polykondensation,* Israel Program for Scientific Translations, Jerusalem 1968, pp. 237–340.

[66] K. Tamao, H. Yamamoto, H. Matsumoto, N. Miyake, T. Hayashi, and M. Kumada, *Tetrahedron Lett.*, 1389 (1977).

[67] E. Selegny, L. Merl Aubry, in *Optically Active Polymers*, E. Selegny, Ed., D. Reidel, London, 1979, p. 68.

[68] G. Natta, G. Dall'Asta, I. W. Bassi, and G. Carella, *Makromol. Chem.*, *91*, 87 (1966).

[69] G. Natta, L. Porri, P. Corradini, G. Zanini, and F. Ciampelli, *J. Polym. Sci.*, *51*, 463 (1961).

[70] G. Natta, L. Porri, A. Carbonaro, and G. Stoppa, *Makromol. Chem.*, *77*, 114 (1964).

[71] G. Natta, M. Farina, P. Corradini, M. Peraldo, M. Donati, and P. Ganis, *Chim. Ind. (Milan)*, *42*, 1361 (1960).

[72] R. S. Cahn, C. Ingold, and V. Prelog, *Angew. Chem.*, *78*, 413 (1966).

[73] H. Yuki, K. Hatada, T. Niinomi, and Y. Kikuchi, *Polym. J.*, *1*, 36 (1970).

[74] R. W. Lenz, *J. Macromol. Sci. (A)*, *9*, 945 (1975).

[75] F. A. Bovey, *J. Polym. Sci.*, *46*, 59 (1960).

[76] T. Tsuruta, T. Makimoto, and H. Kanai, *J. Macromol. Chem.*, *1*, 31 (1966).

[77] W. Fawells, C. Schuerch, F. A. Bovey, and F. P. Hood, *J. Am. Chem. Soc.*, *89*, 1396 (1967).

[78] J. T. Maynard and W. E. Mochel, *J. Polym. Sci.*, *13*, 235 (1954).

[79] Y. Doi, F. Nozawa, and K. Soga, *Makromol. Chem.*, *186*, 2529 (1985).

[80] B. Rotzinger, Diss. ETH Nr. 7590 (1984).

[81] M. Schmid and P. Pino, unpublished results.

[82] G. Natta, G. Dall'Asta, and G. Mazzanti, *Angew. Chemie*, *76*, 765 (1964).

[83] S. Brownstein and D. M. Wiles, *J. Polym. Sci. (A)*, *2*, 1901 (1964).

[84] K. Matsuzaki and S. Ohshima, *Makromol. Chem.*, *164*, 265 (1973).

[85] E. von Achenbach, Diss. ETH Nr. 7970 (1986).

[86] G. Natta, M. Farina, and M. Donati, *Makromol. Chem.*, *43*, 251 (1961).

[87] Y. Okamoto, K. Suzuki, and H. Yuki, *J. Polym. Sci.*, *Polymer Chem. Ed.*, *18*, 3043 (1980).

[88] G. Di Silvestro, M. Grassi, and M. Farina, IUPAC Macro, Florence, 1980, *Preprints*, *2*, 166 (1980).

[89] T. Tsuruta, "Stereospecific Polymerization of Epoxides," in *Stereochemistry of Macromolecules*, A. D. Ketley, Ed., Dekker, New York, 1967, Vol. 2, p. 177.

[90] Y. Okamoto, K. Ohta, and H. Yuki, *Makromolecules*, *11*, 724 (1978).

[91] P. Pino, F. Ciardelli, and G. P. Lorenzi, *Makromol. Chem.*, *70*, 182 (1964).

[92] P. Pino, G. Fochi, O. Piccolo, and U. Giannini, *J. Am. Chem. Soc.*, *104*, 7381 (1982).

[93] Ph. Teyssié, M. Julémont, J. M. Thomassin, E. Walckiers, and R. Warin, in *Coordination Polymerization*, J. C. W. Chien, Ed., Academic, New York, 1975, p. 327

[94] D. M. White, *J. Am. Chem. Soc.*, *82*, 5678 (1960).

[95] J. F. Brown, Jr and D. M. White, *J. Am. Chem. Soc.*, *82*, 5671 (1960).

Piero Pino was born on April 9, 1921.

PRESENT POSITION

Professor of Macromolecular Chemistry at the Swiss Federal Institute of Technology-Zurich (E.T.H.-Z).

EDUCATION

University of Pavia and Florence, Doctor in Chemistry, 1943.

POSITIONS

Lecturer at Florence University 1943–1946;
Assistant of Prof. G. Natta and Lecturer at Politecnic of Milan 1946–1956.
Professor of Industrial Organic Chemistry at the University of Pisa 1956–1968
Professor of Macromolecular Chemistry at the E.T.H.-Z 1968–.
Visiting Professor:
Weizmann Institute Rehovot 1977; Midland Molecular Institute 1982; University of Massachusetts, Materials Research Laboratory, 1984.

PROFESSIONAL ACTIVITIES

President of the Toscana Section of the Italian Chemical Society 1966–1968.
President of the Chemical Society Zurich 1972–1973.
Chairman of the "Ad hoc committee for Polymer Science" of the European Science Foundation; 1979–.
President of the Swiss Polymer Group 1984–.
Corresponding member of the Accademia dei Lincei (Roma).

LANTHANIDES AS COORDINATION CATALYTIC POLYMERIZATION SYSTEMS

BAOGONG QIAN
Wuhan Branch, Academia Sinica, Wuhan, People's Republic of China

FOSONG WANG, ENLE ZHOU, YANG XU, and FUSHEN YU
Changchun Institute of Applied Chemistry, Academia Sinica, Changchun, People's Republic of China

SYNOPSIS

For the polymerization of conjugated dienes, such as butadiene and isoprene, with 4-*f*-orbital lanthanide catalysts, the proposed structure of the catalytic active center was confirmed by the isolation and identification of the bimetallic neodymium complex in soluble, polycrystalline form and as a single crystal. The formation of a *cis*-1,4 or *trans*-1,4 unit depends on the anti/syn isomerization of the metal butenyl group when a new monomer, for example, hexadiene and pentadiene, is incorporated. For the "living" character of Ln-polymerization of butadiene with Nd-complexes at $<-30°C$, the molecular weight of Ln-polybutadiene (Ln-PB) increased with conversion while the number of macromolecules per Nd remained unchanged. For the bulk crystallization of Ln-PB, as well as U-PB, with nearly perfect *cis* content, the rate of crystallization is much quicker than that of Ni- and Co- with lower *cis* content and even with branching. The crystallization rate of Ln-PB fractions depends not only on stereoregularity but also on molecular weight (MW) (e.g., M_η) and molecular weight distribution (MWD); and even on entanglemental MW, $M_e = M_\eta/m$, where m represents entanglemental density. By morphological evidence, the crystallization rate of high MW fractions is much controlled by entanglement. Then the nucleation theory of the overall rate equation for 10% crystallization can be modified by introducing the entanglemental density into the equation. As a structural property, the yield process of Ln-polyisoprene raw rubber under stress is kinetic in character. By comparing the values of activation energy for yielding ($\Delta E_y = 8.0$ kJ/mol) and for disentanglement ($\Delta E_{deN} = 8.4$ kJ/mol), it can be recognized that the occurrence of the yielding process is a loosening process of physical entanglements. There exists a linear relationship between actual yield strength and MW, for example $[\eta]$ or log M_e, and then the value of critical MW for the onset of yielding can be estimated by extrapolation as $M_c(m = 1) = (5.4 \pm 0.2) \times 10^4$ per yield segment. The prospects of lanthanide as a coordination polymerization system are looking up.

INTRODUCTION

The use of catalysts based on rare earth elements or lanthanides (Ln: from La to Lu), which are not rare in China, in stereospecific polymerization of butadiene with high 1,4-*cis* content was reported by our lab for the first time two decades ago, and the monograph on the characteristics of Ln polymerization systems in Chinese [1] and its abstract in English [2] were published in 1980–81. The National Prize for Natural Sciences was awarded to the research scientists in 1982 for their evaluation of the possibility of using the 4-*f*-orbital electrons of rare earth elements to coordinate with ligands in Ziegler-Natta catalyst systems instead of the third orbitals of transition metals generally used, as well as for their elucidation of three "high" structural characteristics of the Ln-polymerized polybutadiene, namely, high stereoregularity, high molecular weight, and high linearity of its macromolecules.

In recent years more systematic investigations were carried on applying the Ln-polymerization to more dienes in order to give stereoregular polydienes with a *cis*-1,4 or *trans*-1,4 structure, clarifying the mechanism of Ln polymerization, and realizing the relationship between structure and properties of the highly stereoregular and linear polydienes with rather high molecular weight and then entangled, especially the influence of entanglement on the bulk crystallization of polybutadiene (Ln-PB) and on the yield behavior of polyisoprene (Ln-PIR).

SOME MECHANISTIC ASPECTS OF 1,4-POLYMERIZATION OF DIENES WITH RARE EARTH-BASED CATALYST SYSTEMS

As reported 1,3-dienes can be polymerized, with *f*-orbital transition metal catalysts, to give stereoregular polymers with a *cis*-1,4 or *trans*-1,4 structure, [1, 4, 5] (Table I). This challenged our interest on clarifying the polymerization mechanism. In recent years some research work has been done in our lab on the structure of active species, the nature of the growing chain, the characteristics of the polymerization process, and on the mode of the monomer insertion.

STRUCTURE OF ACTIVE SPECIES

In our lab's first paper with respect to the polymerization of isoprene, a bimetallic complex containing rare earth and aluminium atoms, which are connected to each other by bridge bonds, has been proposed as a catalytic active species [1] (Fig. 1). This hypothesis was supported further by the kinetic results obtained in both our lab and Prof. Rafikov's [4, 7]. The isolation of a rare earth and aluminium-containing crystalline bimetallic complex active for stereospecific polymerization of dienes from the soluble system

TABLE I
Homopolymers and Copolymers of Dienes Obtained with Rare Earth-Based Catalyst Systems

Monomer	Structure of polymer obtained
butadiene	*cis*-1, 4-polymer
1, 3-pentadiene	*cis*-1, 4-isotactic polymer
ld-1, 3-pentadiene	*cis*-1, 4-erythro diisotactic polymer
isoprene	*cis*-1, 4-polymer
E, E-2, 4-hexadiene	*trans*-2, 5-threo diisotactic polymer
2, 3-dimethyl butadiene	*cis*-1, 4 high crystalline polymer
copolymer of butadiene and isoprene	both butadiene and isoprene units in high *cis*-1, 4 configuration
copolymer of butadiene and 1, 3-pentadiene	both butadiene and pentadiene units in *cis*-1, 4 configuration

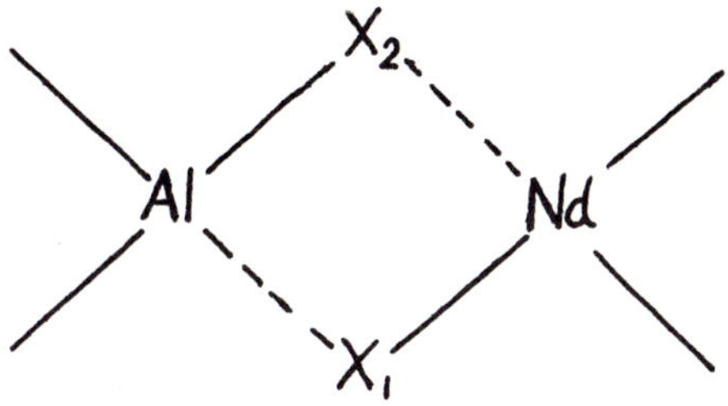

FIG. 1. Bimetallic complex composed of Al and Nd atoms.

$(CF_3COO)_2NdCl \cdot EtOH\text{-}Al(Et)_3$ further confirmed the possible structure of active species proposed previously. For this bimetallic complex we have succeeded in observing either Debye rings, characteristic of polycrystalline samples (Fig. 2), or a clear spot diagram typical of single crystals by electron diffraction technique (Figs. 3 and 4). For the complex isolated from $Nd(i\text{-}C_4H_9O)_3\text{-}Al(C_2H_5)_2Cl\text{-}Al(C_2H_5)_3$ the x-ray diffraction diagram of its single crystal recorded by four-circle diffractometer is shown in Figure 5.

1,4-POLYMERIZATION MECHANISM

It was generally recognized that in the polymerization of conjugated dienes with catalysts based on *d*-orbital transition metals, the bond between the growing chain end and the transition metal is of allylic type, the formation of *cis*-1,4 units depends on the structure of the allyl complex (anti/syn isomerism of the metal butenyl group), and the anti/syn isomerization takes place through the formation of η'-allyl complex as an intermediate compound. Evidently, if the polymerization of dienes with rare earth-based catalysts is similar to that with *d*-orbital transition metals, the polymerization mechanism may be schematized as shown in Figure 6, where K_p and K_{iso} are the propagation rate constant and isomerization rate constant, respectively, mon is monomer. According to the above diagram the percentage of *cis*-1,4 and *trans*-1,4 units in the polymer depends on the rate ratio of reaction “a” to reaction “b.” It may, therefore, be predicted that with the decrease of the monomer concentration, the percentage of synallylic form in the growing chain increases. The results concerning the effect of monomer concentration on the *cis*-1,4 content of polydienes [including polybutadiene (Fig. 7), poly-1,3-pentadiene (Fig. 8) and polyisoprene (Fig. 9)] confirmed the above-mentioned polymerization mechanism.

According to the above scheme if a monomer (E,E)—CHX=CH—CH=CHY yields *trans* units A and *cis* units B, the stereoregular 1,4-polymer that can be obtained from it are those as shown in Figure 10. The results obtained by Wang and Porri [5,8] on the polymerization of (E,E)-2,4-hexadiene(I) and (E,E)-1d-1,3-pentadiene(II) with rare earth-based catalyst are in agreement with the above scheme. For monomer I *trans*-threo-diisotactic polymer of I was obtained, while for monomer II *cis*-erythro-diisotactic polymer of II was isolated. The stereochemistry interpretation of the formation of *cis*-erythro-

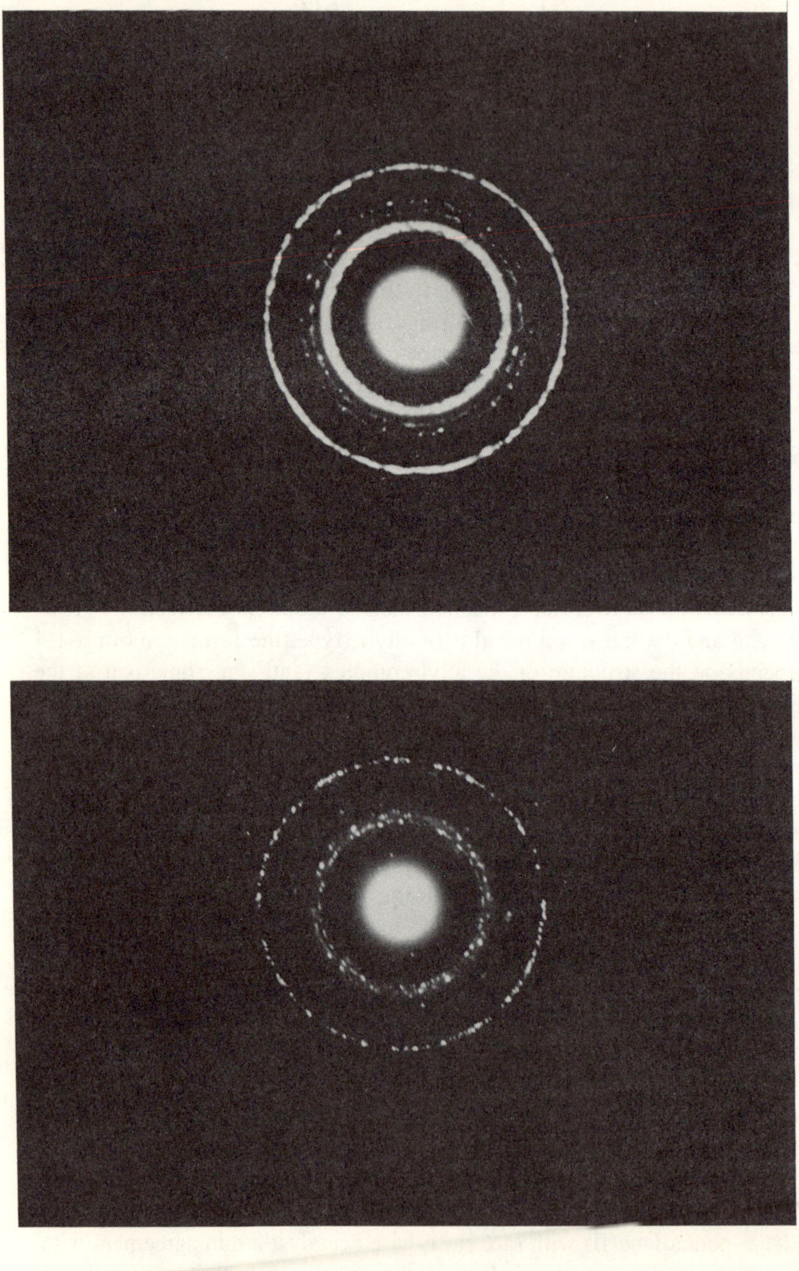

FIG. 2. Debye pattern from electron diffraction of the complex obtained from $(CF_3COO)_2 \cdot NdCL + Al(C_2H_5)_3$

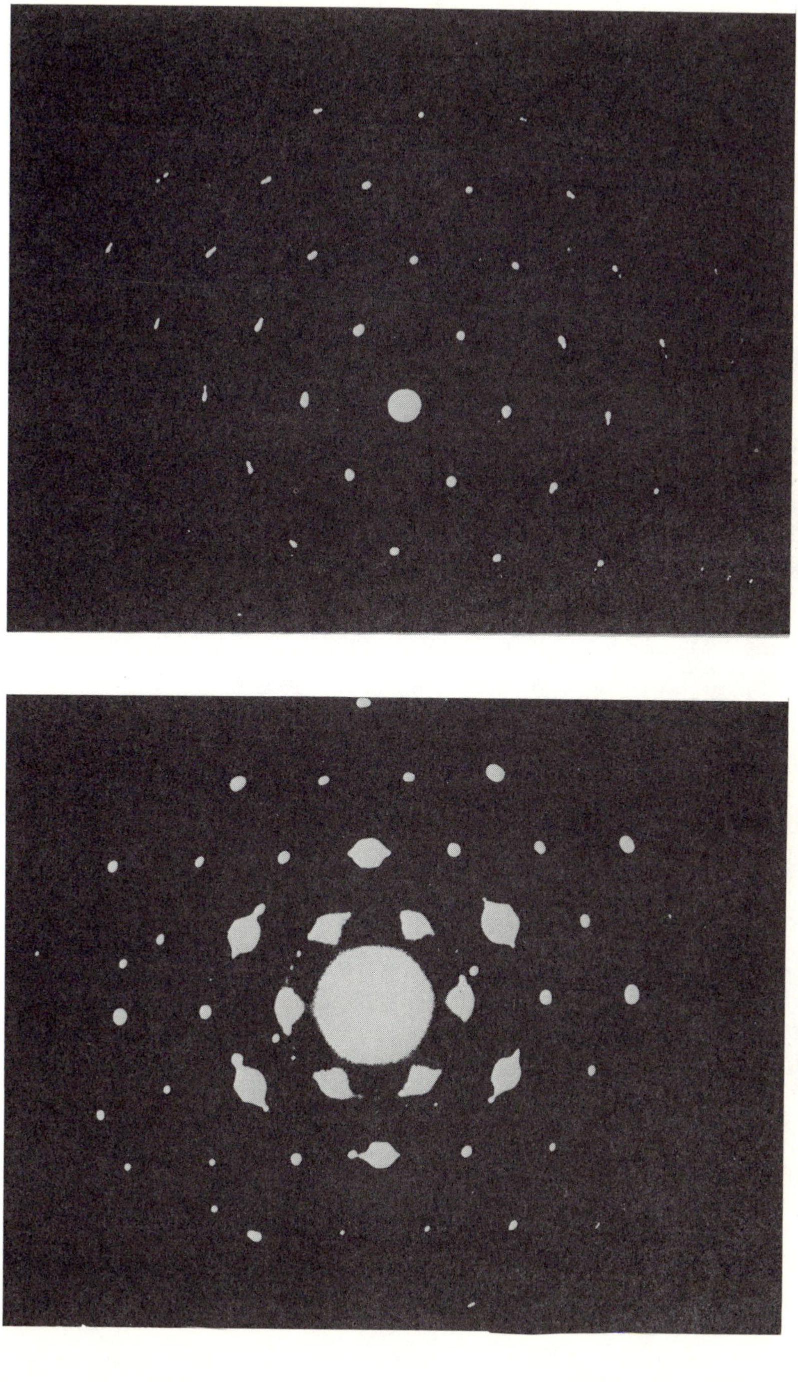

FIG. 3. Spot diagram from electron diffraction of the complex obtained from $(CF_3COO_2)NdCl$ + $Al(C_2H_5)_3$.

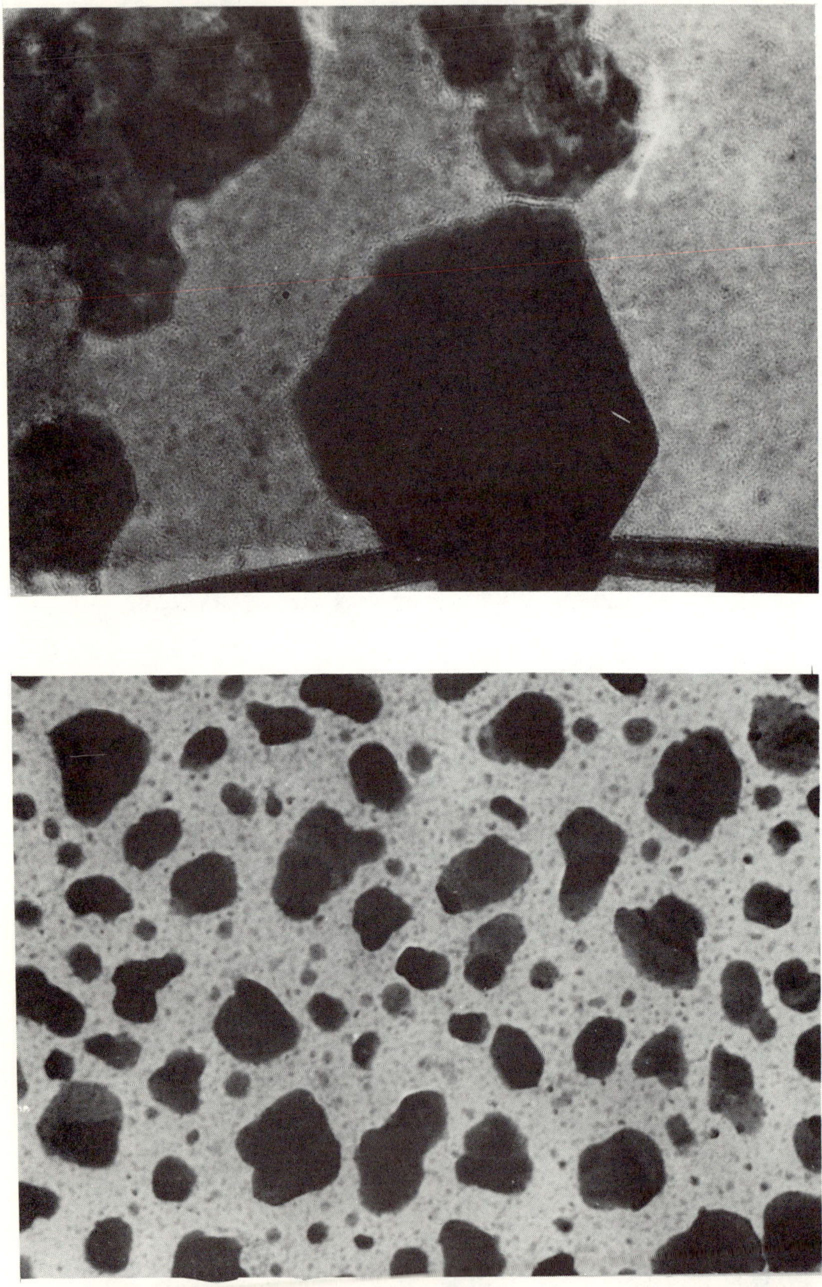

FIG. 4. Single crystal observed by electron microscopy for the complex indicated in Figure 3.

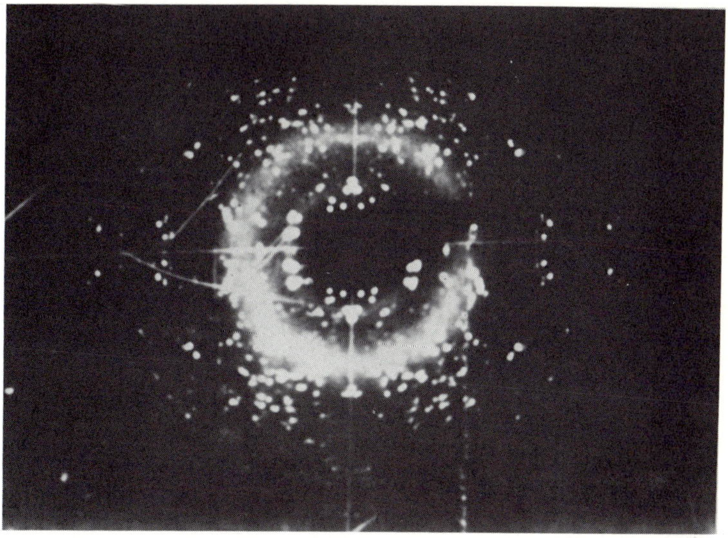

FIG. 5. Spot diagram from four-circle diffractometer of the complex isolated from $Nd(i\text{-}C_4H_9O_3)$-$Al(C_2H_5)_2Cl$-$Al(C_2H_5)_2$. Parameters of a triclinic unit cell: a = 13.92, b = 15.38, c = 15.37 Å, α = 82.95°, β = 89.60°, γ = 85.14°, V = 3255.38 Å^3.

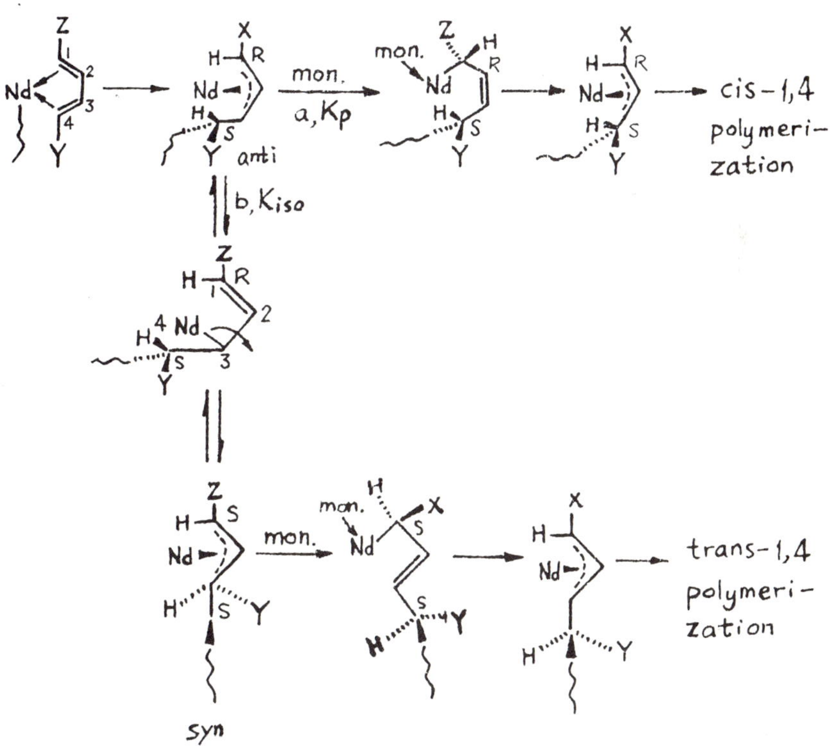

FIG. 6. Possible stereochemistry of polymerization mechanism for 1,4-disubstituted butadiene.

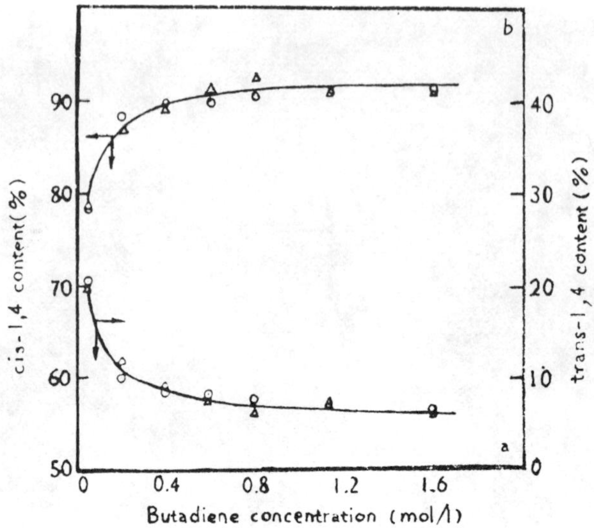

FIG. 7. The influence of butadiene concentration on the microstructure of polymer: (a) $Nd(naph)_3$-$Al_2(C_2H_5)_3Cl_3$-$Al(i\text{-}Bu)_3$ system: Nd = 2.5×10^{-3} mol/L; Al/Nd(molar ratio) = 30; Cl/Nd = 3.0; polymerization in heptane at 50°C for 5 h. (b) $NdCl_3$-C_2H_5OH-$Al(C_2H_5)_3$ system: Nd = 2.5 × 10 mol/L; Al/Nd (molar ratio) = 20. Other conditions are the same as in a.

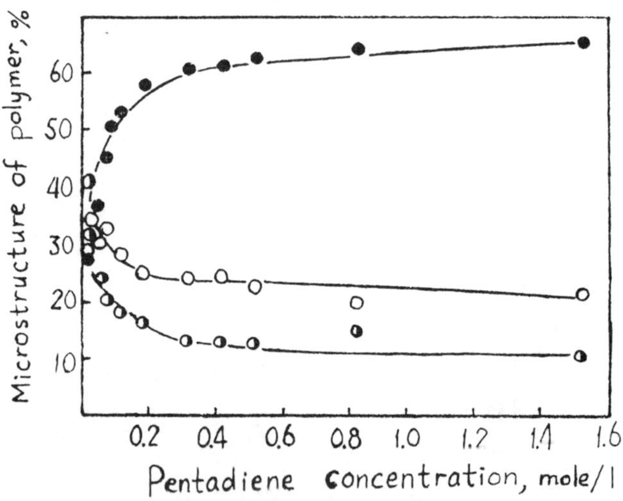

FIG. 8. Effect of the 1,3-pentadiene concentration on the microstructure of polymer: Catalyst system: $Nd(naph)_3$-$Al_2(C_2H_5)_3Cl_3$-$Al(i\text{-}Bu)_3$; Al/Nd = 40, Cl/Nd = 3.0, Nd = 1×10^{-3} M, polymerization at 50°C for 2 h. ●——*cis*-1,4 content, ◑——*trans*-1,4 content, ○——*trans*-1,2 content.

diisotactic polymer with Nd-based catalyst can be illustrated by the following scheme (Fig. 11).

The last unit in the polymer is in the anticonfiguration since a *cis*-1,4 monomeric unit is formed and the new monomer is coordinated with two

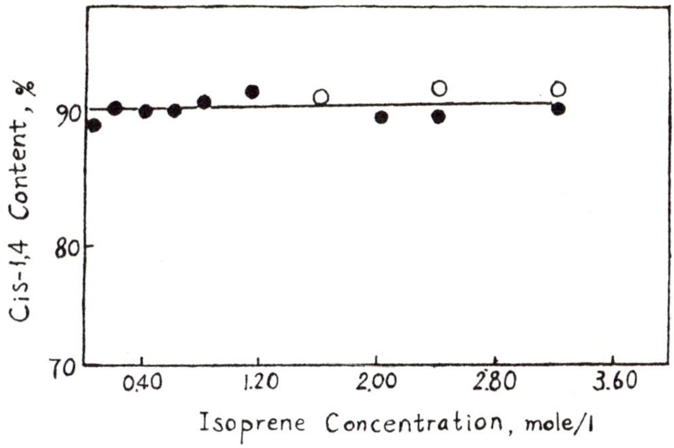

FIG. 9. Effect of the isoprene concentration on the microstructure of polymer. Catalyst system: $NdCl_3$-C_2H_5OH-$Al(C_2H_5)_3$; Al/Nd = 20, Nd = 2.5×10^{-2} M; polymerization at 50°C for 5 h.

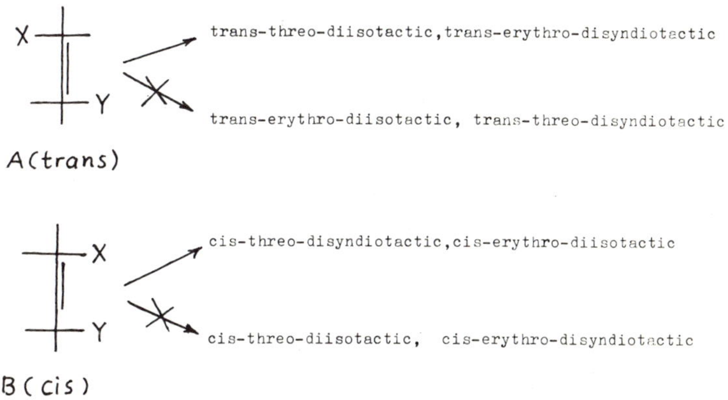

FIG. 10. Types of stereoregular 1,4-polymers that can be obtained from (E,E)-1,4-disubstituted butadiene.

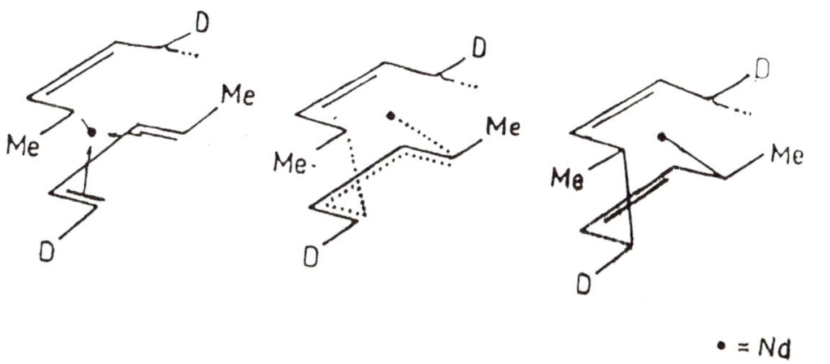

FIG. 11. Possible scheme for the formation of an erythrodiisotactic poly-ld-1,3-pentadiene.

double bonds to give, after insertion, an anti-butenyl group. With mode presentation shown an erythro situation is formed. The new monomer gives, after insertion, a butenyl group of the same chirality as the preceding one. Hence, a *cis*-1,4 diisotactic polymer will result.

"LIVING" POLYMERIZATION CHARACTER

We have already mentioned that under given conditions polymerization of dienes with catalyst system $Nd(naph)_3 \cdot (C_2H_5)_3Al_2Cl_3 \cdot Al\ (i\text{-}C_4H_9)_3$ proceeds by a "living" mechanism. It has been found recently that similar results were obtained for the polymerization of butadiene with catalyst system $NdCl_3$-i-C_3H_7OH-$Al(i\text{-}C_4H_9)_3$. The molecular weight of the polybutadiene increased with polymerization time or conversion, while the number of macromolecules per Nd remained unchanged at polymerization temperatures lower than −30°C (Fig. 12). The gel permeation chromatography (GPC) curve of the polymers obtained shifted to high MW side with increasing conversion (Fig. 13). It reveals that chain transfer reactions and termination of the active centers were not observed under polymerization temperature lower than −30°C and polymerization could proceed with a "living" mechanism.

BULK CRYSTALLIZATION OF Ln-PB

Cis-1,4-polybutadiene prepared with lanthanide catalyst (Ln-PB) contains a very high *cis* content. It can reach nearly 100%, similar to that of U-PB, but higher than Co- and Ni-PB (Table II). Such high stereoregularity provides the linearity of macromolecular chain, as confirmed by the straight linear plot of intrinsic viscosity against weight average molecular weight M_w, while the values

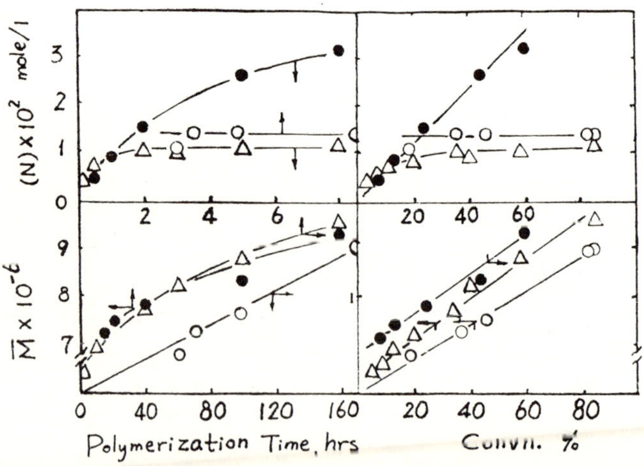

FIG. 12. Molecular weight and number of macromolecules vs. polymerization time and conversion in butadiene polymerization at various temperatures with $NdCl_3$-3i-PrOH-$Al(C_2H_5)_3$ system. ○— −70°C, △—— −30°C, ●——0°C.

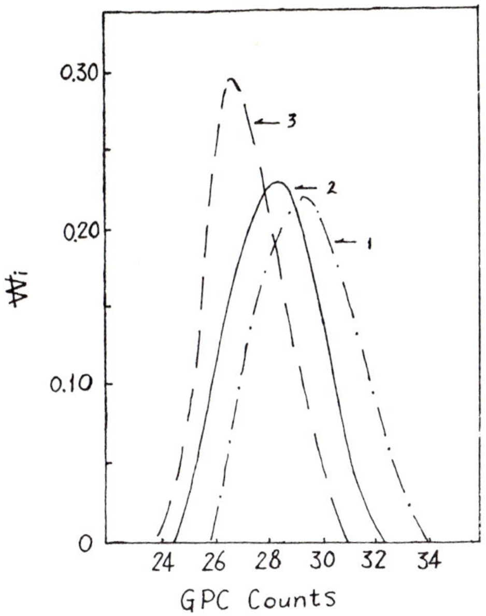

FIG. 13. MWD of polybutadiene with conversion at −70°C; conversion (%): 1-17.9, 2-46.4, 3-85.0.

deviated from the line for the branched Ni-PB (Fig. 14) [9]. In consequence it causes the overall rate of crystallization to proceed rapidly (Fig. 15) [9]. Furthermore, the molecular weight is so high that entanglement should occur.

MORPHOLOGICAL EVIDENCE OF ENTANGLEMENT

It has been found [10] that the low molecular weight fraction crystallized primarily with heterogeneous nuclei [Fig. 16(a)], and its lamellae are not

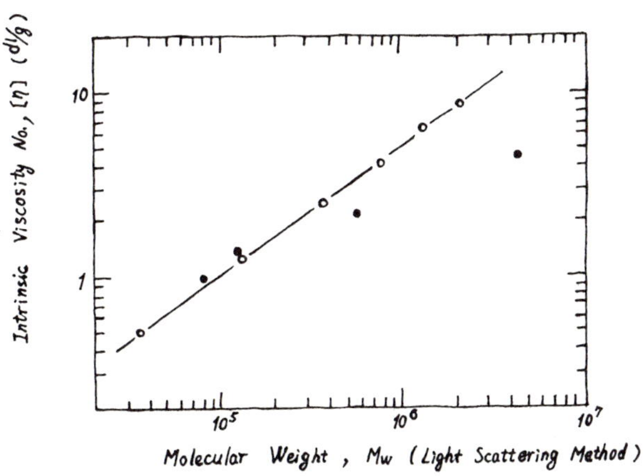

FIG. 14. $[\eta] = KM_w^{\alpha}$, double logarithm plots of $[\eta]$ vs. M_w (○) Ln-Pb, $[\eta] = 3.24 \times 10^{-4}$ $M_w^{0.70}$, dL/g (toluene, 25, 30°C); (●) Ni-PB.

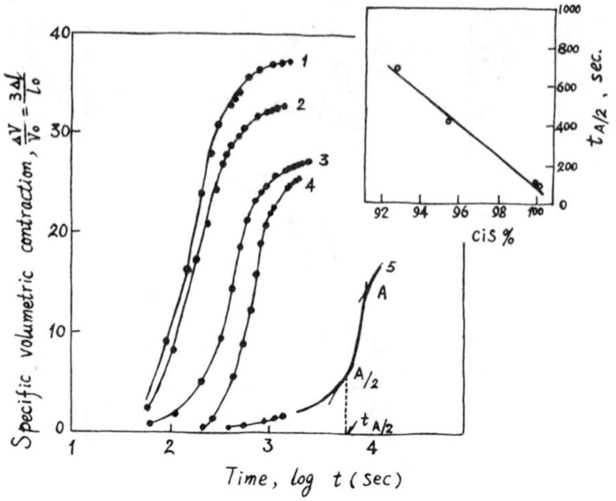

FIG. 15. Crystallization isotherms of *cis*-1, 4-polybutadienes at −26°C.

Sample no.	Degree of crystallinity	Crystallization half-time, $t_{A/2}$
1. Ln-PB	40.5%	170″
2. U-PB	36.0	158
3. Co-PB	29.2	427
4. Ni-PB	27.6	692
5. NR	17.5	6761

entangled [Fig. 16(b)], while the high molecular weight fraction did primarily form homogeneous nuclei [Fig. 17(a)], and lamellar entanglement developed during crystallization [Fig. 17(b)]. Hence a schematic model for the lamellar entanglement around the point of chain entanglement (Fig. 18) can be drawn.

Furthermore, it has been found that there are two types of spherulite in the mixtures of different fractions of Ln-PB (Fig. 19). The growth rates of spherulite of concentric ring and radiation type are 0.12 and 0.04 μm/min, respec-

TABLE II
Some Parameters for *cis*-1, 4-Polybutadienes

Sample	*cis*-1, 4 (%)	[η] (dL/g)	M_w/M_n
1. Ln-PB (ML = 33, our lab)	99.8	3.14	6.0
2. U-PB (Italy)	100 (99)	3.24	5.6
3. Co-PB (Japan, 1220)	95.2		
4. Ni-PB (China, 76.6.26)	92.8	2.69	2.7

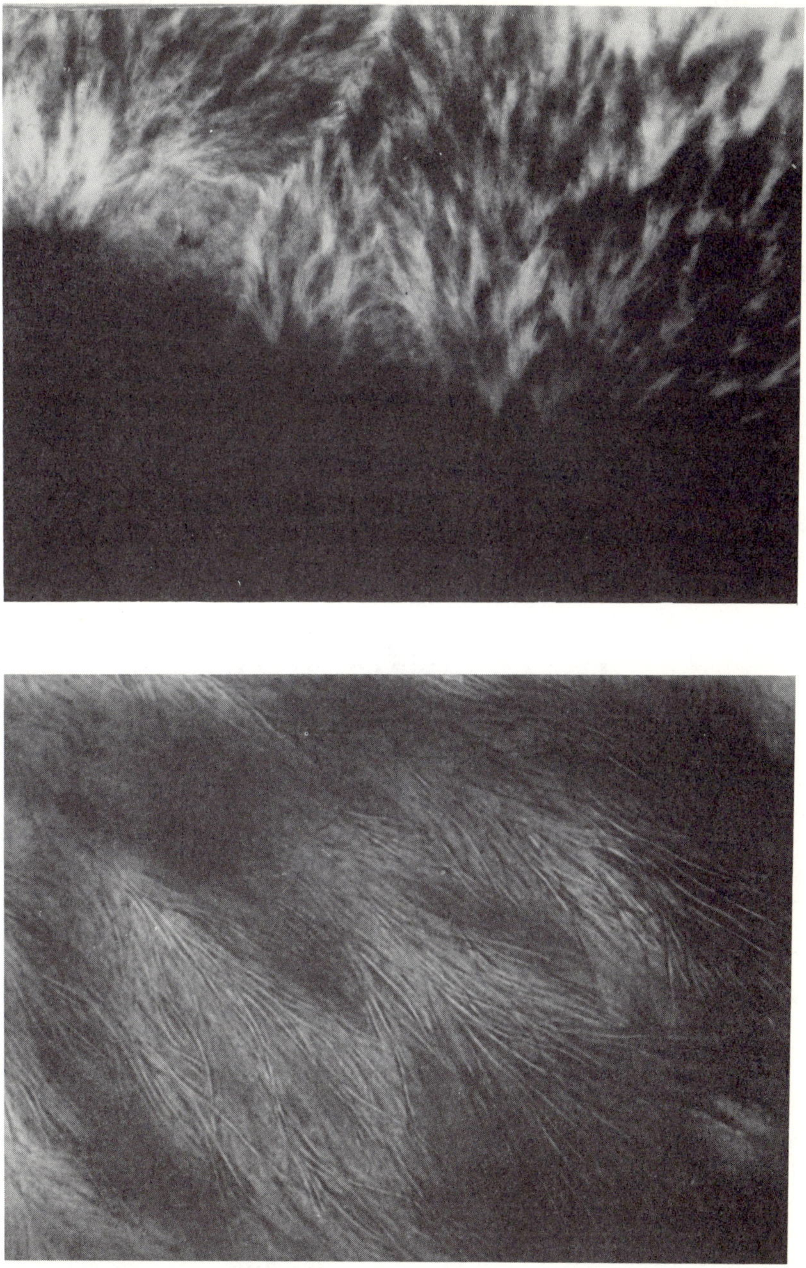

FIG. 16. Low MW fraction of Ln-PB, crystallized at −30°C; $M_\eta < M_{c(m=1)} = 1.27 \times 10^5$. (a) $M_\eta = 0.58 \times 10^5$, nuclei—heterogeneous (×5,000). (b) $M_\eta = 0.58 \times 10^5$, lamellar growth—not entangled (×20,000).

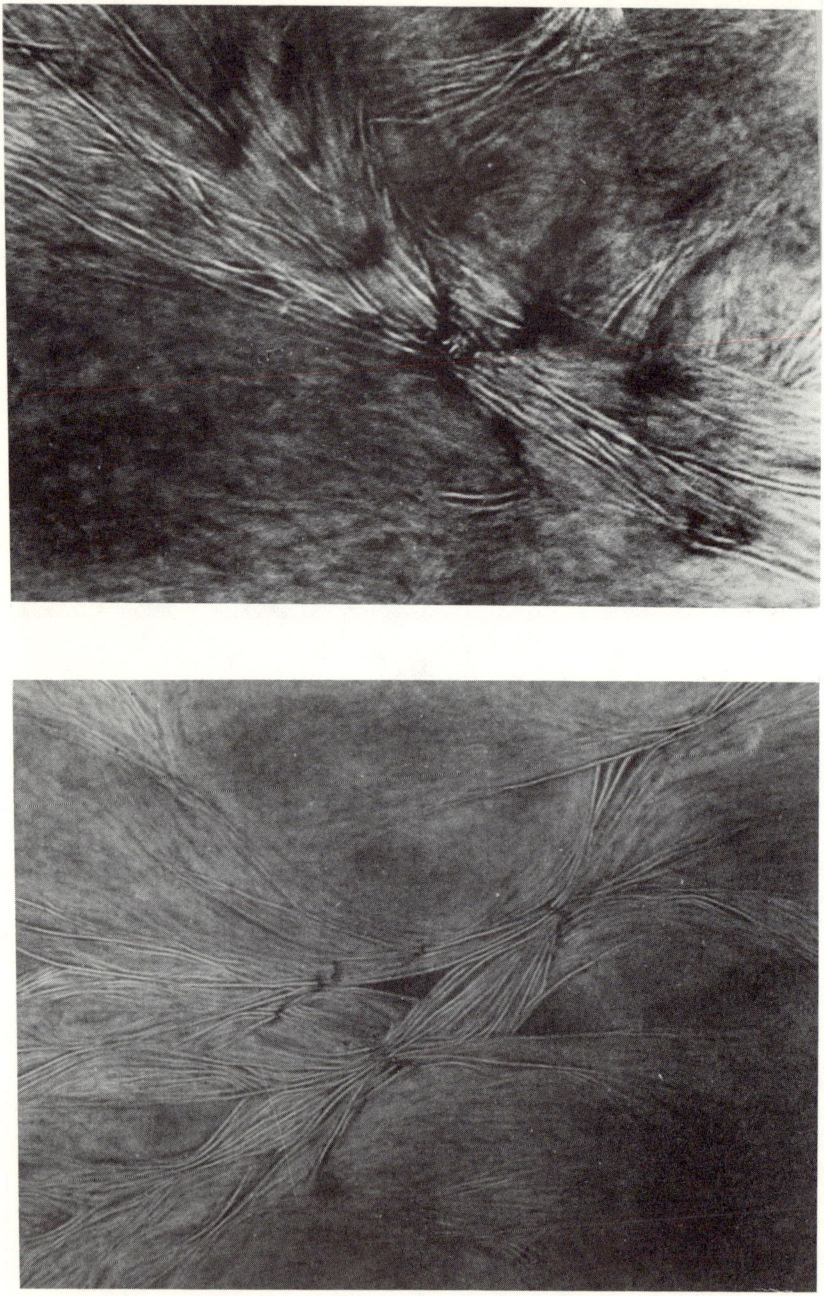

(b)

FIG. 17. High MW fraction of Ln-PB crystallized at $-30°C$; $M_\eta > M_{c(m=1)} = 1.27 \times 10^5$. (a) $M_\eta = 8.44 \times 10^3$, nuclei—heterogeneous (×50,000). (b) $M_\eta = 6.55 \times 10^5$, lamellar—entangled (×30,000).

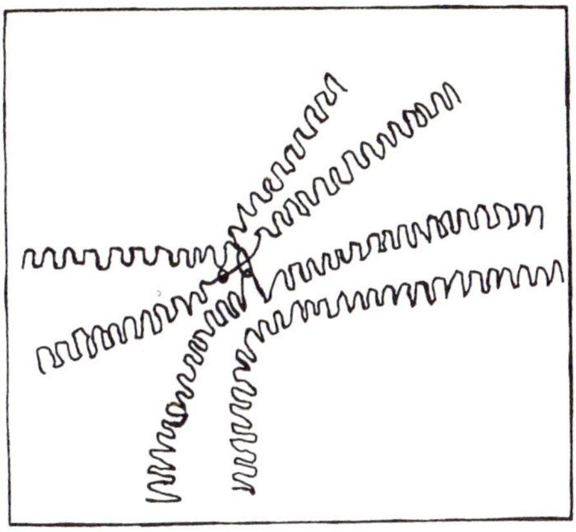

FIG. 18. The schematic model of lamellar entanglement around the point of chain entanglement.

tively. There is no entanglement of lamellae in spherulite of concentric ring type. In the spherulites of radiation fiber type there are many points of entanglement of lamellae with helix growth. The higher molecular weight fractions are found to be of significance in the crystallization of the polymer. Then we can define the average molecular weight of entangled segment as $M_e = M_\eta/m$, where m represents the average number of entangled segments per mole, which can be estimated from relaxation spectrum by Chompff's approximation. From the double logarithmic plots of m vs. M_η (Fig. 20), we obtain

$$m = 1.90 \times 10^{-12} \cdot M_\eta^{2.29}$$

and

$$M_e = 5.1 \times 10^{11}/M_\eta^{1.29}$$

It means that the higher the value of M_η, the larger the m or the lower the M_e. By extrapolating log m to 0 or m to 1, the critical value of molecular weight for entanglement can be estimated as $M_{c(m=1)} = 1.27 \times 10^5$ [11].

DEPENDENCE OF CRYSTALLINITY AND CRYSTALLIZATION RATE ON MOLECULAR WEIGHT, M_η or M_e

The samples of ~99% *cis*-1,4-LnPB were fractionated into several fractions of different M_η, covering a rather wide range from 7.8×10^4 to 5.54×10^6, about 70 times (Table III). Experiments of isothermal bulk crystallization were carried out with linear or volumetric dilatometer at a given temperature T_c.

The typical isothermal crystallization curves are shown in Figure 21. Together with the spherulitic morphologies of different molecular weight fractions of Ln-PB, these results show that the high molecular weight fractions nucleate readily, but with the lowest crystallinity; with the intermediate molecular weight fractions the rates are very slow both in nucleation and in growth of

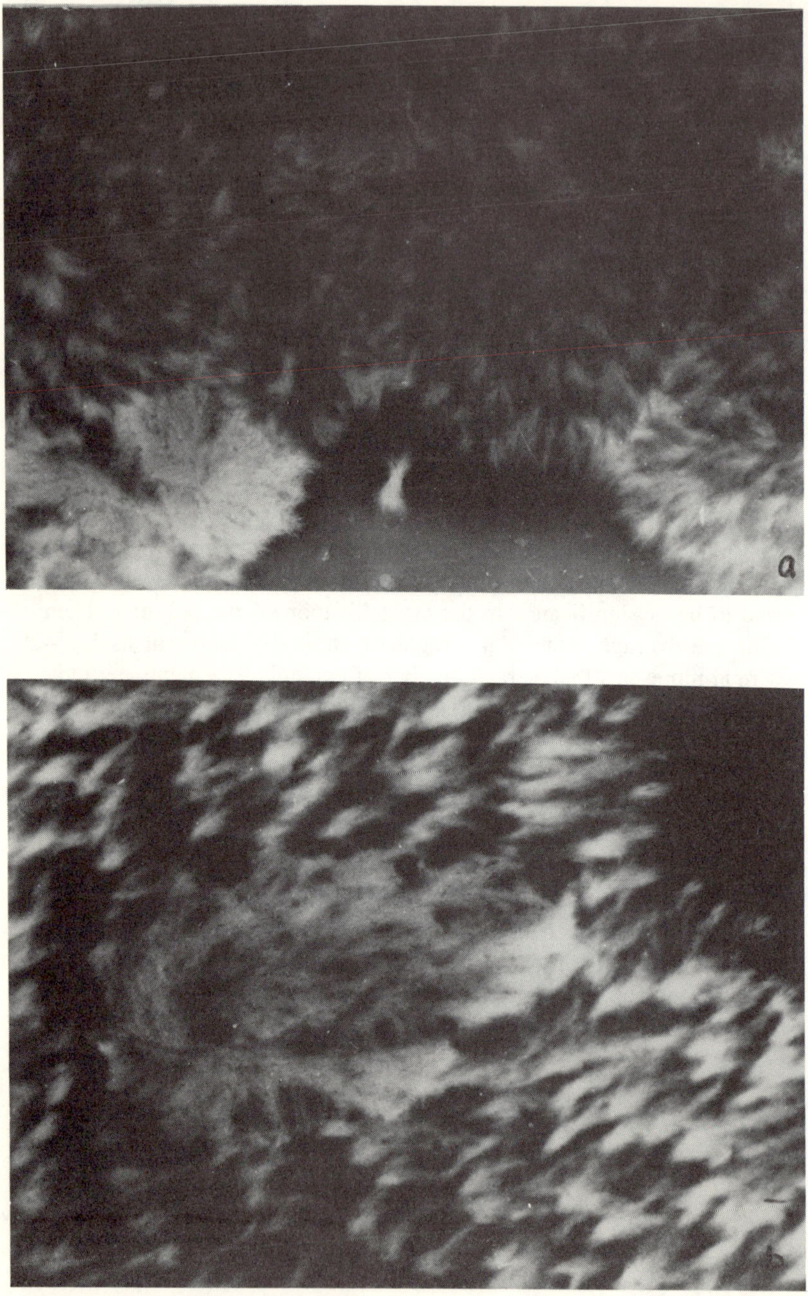

FIG. 19. Two types of spherulites, crystallized at −30°C for mixtures of different MW fractions of Ln-PB: (a) M_η: 0.58 × 10^5, 8.44 × 10^5, 20.70 × 10^5, (×3000). (b) M_η: 0.58 × 10^5, 8.44 × 10^5 (×4000).

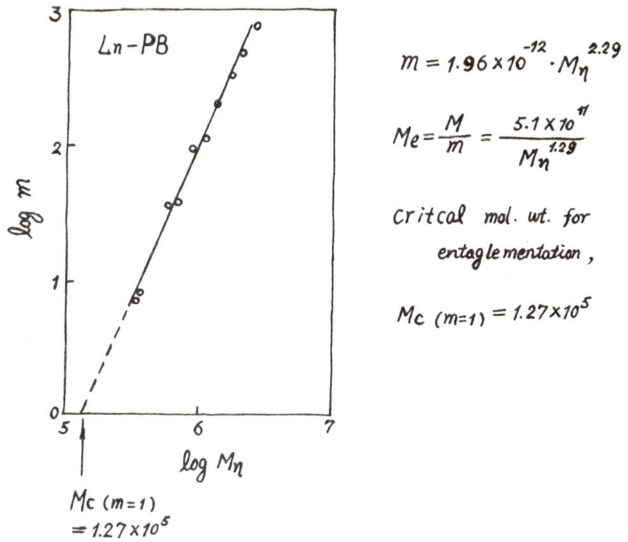

FIG. 20. Estimation of MW of entangled segment of Ln-PB. Average molecular weight of entangled segment, $M_l = M_\eta/m$; M_η: Vis. average molecular weight; m: Average no. of entangled segments per mole, estimated from relaxation spectrum by Chompff's approx.

TABLE III
Characterization Data for Ln-PB Fractions

Sample	*cis-* %	1,2- %	*trans-* %	M_w $\times 10^{-5}$	M_n $\times 10^{-5}$	$\frac{M_w}{M_n}$	$[\eta]$, dL/g	M_η $\times 10^{-5}$
C-2	98.9	0.2	0.9				17.02	55.4
C-3	98.7	0.3	1.0				15.13	46.8
C-4	98.9	0.3	0.8				12.71	36.5
A-2	98.7	0.5	0.8				11.05	29.9
A-3	99.0	0.3	0.7	18.83	10.50	1.79	8.31	19.9
A-5	99.1	0.3	0.6	11.82	9.32	1.27	6.37	13.6
A-7	98.8	0.4	0.8	6.24	4.85	1.29	3.94	6.85
A-8	98.5	0.5	1.0	3.43	2.79	1.23	2.49	3.56
A-9	98.8	0.4	0.8	2.08	1.26	1.65	1.74	2.13
A-10	98.7	0.4	0.9	1.25	0.83	1.62	1.24	1.31
A-11	98.9	0.5	0.6	0.84	0.45	1.89	0.86	0.78

spherulites, but the crystallinity is higher; as for the lower molecular weight fractions, the rate of growth of spherulites is fast, the crystallinity is the highest, but the nucleation is slower. If both M_η and T_c are increased, the crystallinity of Ln-PB is decreased (Fig. 22).

The crystallization rates, $1/t_{0.1}$ of some samples of Ln-PB measured at different temperatures are shown in Figure 23, in which $t_{0.1}$ is the time necessary for 10% conversion of the primary crystallinity. It can be seen that as

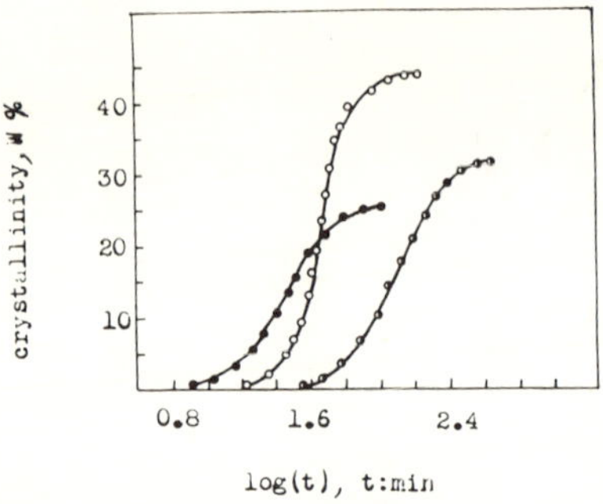

FIG. 21. Plots of crystallinity W against log (t) for crystallization at −13°C. Molecular weight: ● $M_\eta = 3.0 \times 10^6$; ◑ $M_\eta = 6.9 \times 10^5$; ○ $M_\eta = 4.9 \times 10^4$.

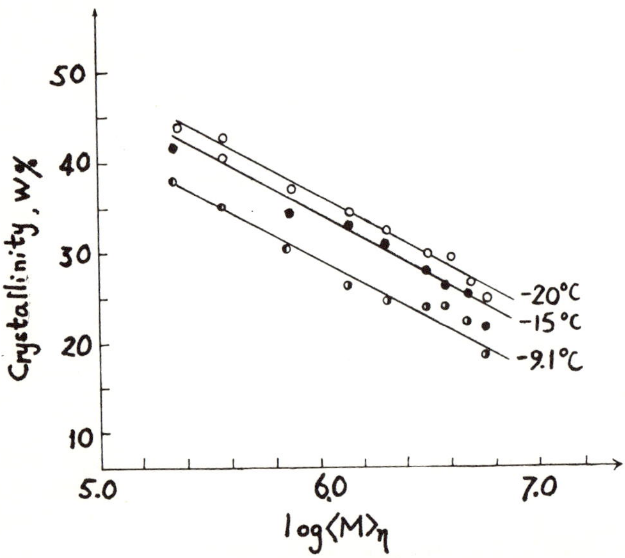

FIG. 22. Plots of crystallinity against log M_η.

molecular weight increases, the crystallization rate decreases at first, reaching a minimum value in the range of $M_\eta = 4.6 \sim 6.7 \times 10^5$; it then increases with the increase of molecular weight until a maximum value occurs in the range of $M_\eta = 3.0 \sim 3.8 \times 10^6$; finally it decreases with the increase of molecular weight again. It seems that this curve has produced the controversial results reported in the literature [12] of the changes of crystallization rate of a polymer with its molecular weight.

The molecular weight of Ln-PB corresponding to minimum crystallization rate (about 5.6×10^5) and the critical molecular weight of entanglement segment ($M_{c(m-1)} = 1.27 \times 10^5$) evaluated from relaxation spectrum are comparable, we can then take entanglement [13] or M_e [14] into consideration. If log ($1/t_{0.1}$) is plotted against log (M_e), a three-section zigzag line in Z type

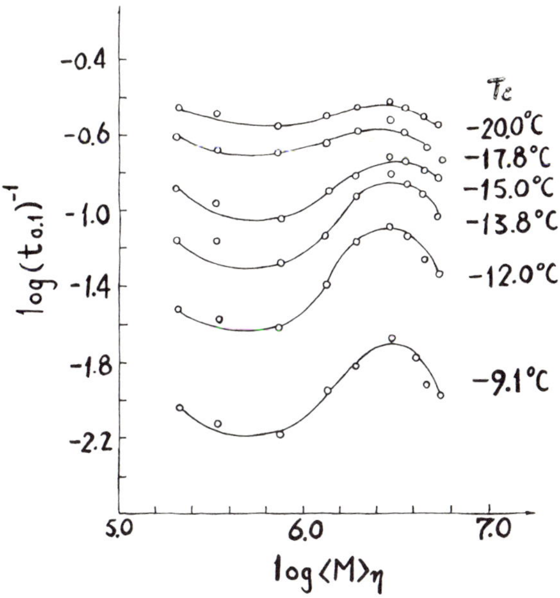

FIG. 23. Double logarithmic plots of crystallization rate $1/t_{0.1}$ vs. molecular weight M_η.

is obtained (Fig. 24), with a maximum value occurring at $M_e = 2.26 \times 10^3$ and a minimum value occurring at $M_e = 1.51 \times 10^4$. For section A and section B, $1/t_{0.1} = K \cdot M_e^\alpha$, the crystallization rate increases with the increase in M_e, but $\alpha_A > \alpha_B$ and $\alpha_A/\alpha_B = 2.5 \sim 4.2$. The difference between α_A and α_B means that the crystallization mechanisms of A and B are somewhat different. Section C is the additive transition section.

In conclusion, it can be said that the bulk crystallization of high molecular weight fraction of Ln-PB is rather controlled by entanglement, so we can modify the nucleation theory of crystallization by introducing the entanglement density m into the overall rate equation for 10% crystallization (Fig. 23/24) [14].

YIELD BEHAVIOR OF Ln-PIR RAW RUBBER

For the whole sample of Ln-PIR raw rubber, yield behavior is characterized by nearly perfect linearity of macromolecular chain in comparing the values of $\langle g \rangle_\eta$ with Ti-PIR samples (Table IV), which may be an indication of branching if $\langle g \rangle_\eta < 1$.

YIELD BEHAVIOR [14]

Since the yield process of a bulk rubber is kinetic in character, our experiments were carried out at constant rate of stretching and various temperatures (T = 80 to −100°C). The activation energy of yield behavior ΔE_y may be ob-

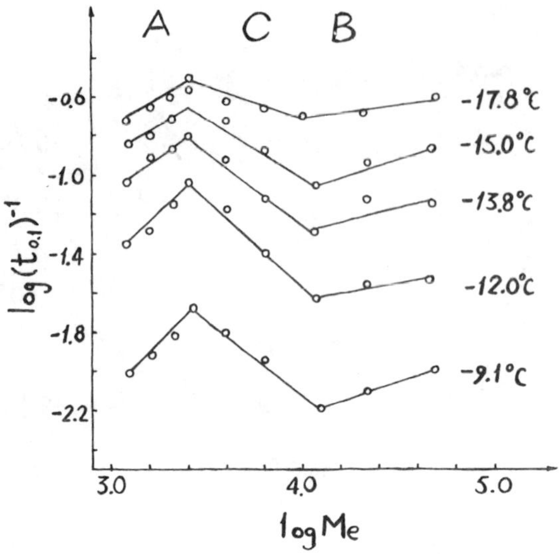

FIG 24. Plots of $\log(t_{0.1})^{-1}$ vs. $\log M_e$.

$$M_e = M_\eta/m = 5.1 \times 10^{11}/M^{1.29}$$

$$1/t_{0.1} = KM_e^\alpha$$

$T_{c,°C}$ \ α	α_A	α_B	α_A/α_B
−17.8	0.62	0.15	4.13
−15.0	0.74	0.30	2.47
−13.2	0.58	0.18	3.22
−12.0	0.84	0.20	4.20
− 9.1	0.73	0.21	3.48

TABLE IV
Molecular Characteristics of PIR Whole Samples

PIR Sample	Gel[a] %	GPC[b] $M_w \cdot 10^{-6}$	$/M_n \cdot 10^{-6}$	$= M_w/M_n$	Intrinsic Viscosity[c] $[\eta]_{exp}$	$/[\eta]_{GPC}$	$= \langle g \rangle_\eta$
Ln-(J)	0.9	2.44	0.81	3.02	466	465	1
Ti-(A)	4.8	3.46	1.23	2.81	326	585	0.56
Ti-(B)	4.2	2.94	1.66	1.77	416	554	0.75
Ti-(C)	4.5	2.35	1.27	1.85	374	469	0.80
Ti-(D)	4.8	1.99	0.81	2.74	326	403	0.81

[a] Toluene-insoluble portion of the sample.
[b] Calculated from the chromatogram as linear polymer and corrected for instrumental spreading.
[c] Toluene solution, mL/g.

tained (Table V) from an Arrhenius plot of true yield strength log $(\lambda_y \sigma_y)$, where λ_y represents yield elongation ratio and σ_y nominal yield strength, against 1/T (Fig. 25). For Ln-PIR in rubbery state, $\Delta E_y = 8.4$ kJ/mol, in

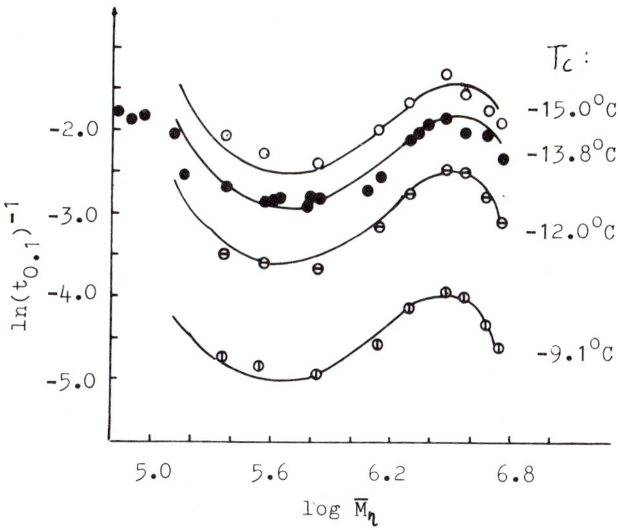

FIG. 23/24. Plots of $\mathrm{Ln}(t_{0.1})^{-1}$ vs. $\log \overline{M}_\eta$. Entanglement modification for the 10% crystallization overall rate $(t_{0.1})^{-1}$ equation based on nucleation theory (14A),

$$\mathrm{Ln}(t_{0.1})^{-1} = -27.09 + 3214\,\overline{M}_\eta^{-0.5} + 2.29\,\mathrm{Ln}\,\overline{M}_\eta - \frac{750}{130 + T_c - T_g} - \frac{5.972\sigma_e T_m^\circ}{T_c(T_m^\circ - T_c)};$$

where: $T_m^\circ = 278.4 - 1.4 \times 10^5/\overline{M}_\eta$

$T_g = 173.0 - 8.700 \times 10^4/\overline{M}_\eta$

$\sigma_e = 20.37 - 3.122 \times 10^{-7}\,\overline{M}_\eta + 2.172 \times 10^{-13}\,\overline{M}_\eta^2$

while $m = 1.96 \times 10^{-12}\,\overline{M}_\eta^{2.29}$ (Fig. 20). ——Calculated curve; ○●⊖⦶ Experimental values.

TABLE V
Activation Energy ΔE_y and Activation Volume V_y for Yielding Process:
Arrhenius, $\lambda_y \sigma_y = K_y \cdot \exp(\Delta E_y/RT)$
Eyring, $d\varepsilon_y/dt = K_y' \cdot \exp(V_y \lambda_y \sigma_y/RT)$

ΔE_y or V_y	Sample	Rubbery (20°C)	Glassy (−80°C)
Activation energy, ΔE_y (KJ/mol)	Ln-PIR	8.4	6.8
	Ti-PIR	9.6	
	NR	11.7	
Activation volume, V_y nm³	Ln-PIR	15.4 (X_y = 136)	0.17 (X_y = 1.5)

comparison with the value of activation energy of disentanglement $\Delta E_{\mathrm{deN}} =$ 8.0 kJ/mol, as obtained independently with stress relaxation method. Then it can be considered that the yield of raw rubber under stress is actually a process of loosening of physical entanglements among macromolecular chains.

On further quenching to the glassy state, the yield strength increased abruptly as expected, while ΔE_y decreased approximately by 20%. Here the potential

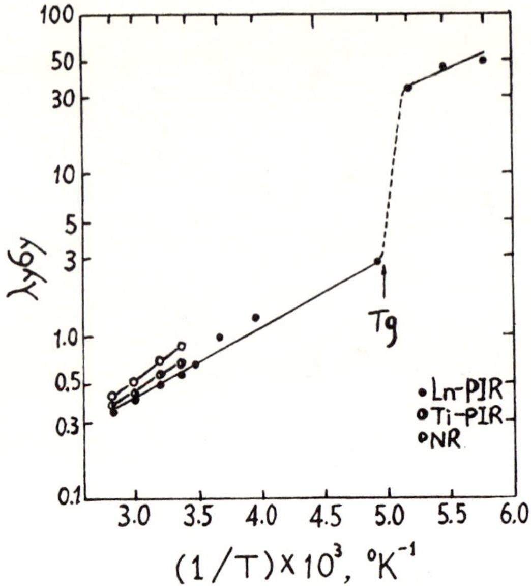

FIG. 25. Arrhenius plots of true yield strength $\lambda_y\sigma_y$ vs. temperature $(1/T)$ for PIR.

barrier to be overcome may be a very small domain instead of the original one in the rubbery state.

With experiments at constant temperature and various rates $\dot{\varepsilon} = 0.27 \sim 0.0027\ \text{sec}^{-1}$) of extension, according to Eyring's viscosity theory for yield process, the activation volume V_y can be obtained by plotting strain rate log $\dot{\varepsilon}$ against strength $\lambda_y\sigma_y$ (Fig. 26). Assuming this to be the volume of moving segment during the occurrence of yield deformation, the corresponding number of yielding segments can be estimated by this equation $X_y = V_y\sigma N/M_o$; where σ is the density (0.91 g/mL), N is the Avogadro's number and M_o is the molecular weight of monomeric unit. Activation volume of Ln-PIR rubber is found to be 15 nm^3, corresponding to 136 monomeric units (Table V) being disturbed on yielding at room temperature. When the specimen was quenched from rubbery state to glassy state, it affected only a very small domain, corresponding to one and a half units.

DEPENDENCE OF YIELD STRENGTH ON MOLECULAR WEIGHT OR $[\eta]$

It has been found that, in the case of PB raw rubber, there exists a linear relationship between yield strength and molecular weight [11,15]. It is further confirmed in the case of PIR rubbers (Fig. 27, Table VI),

$$\lambda_y\sigma_y = A[\eta] + B \qquad (\sigma_y \text{ in MP}_a,\ [\eta] \text{ in dL/g})$$

for Ln-PIR, A = 0.14 and B = −0.05. By extrapolation to σ_y approaching to zero, the critical value of molecular weight for the occurrence of yield process,

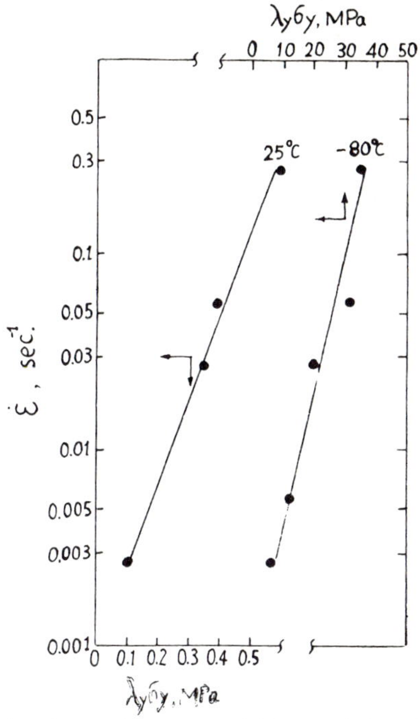

FIG. 26. Eyring plots of strain rate ($\dot{\varepsilon}$) vs. true yield strength $\lambda_y\sigma_y$ for Ln-PIR below and above T_g.

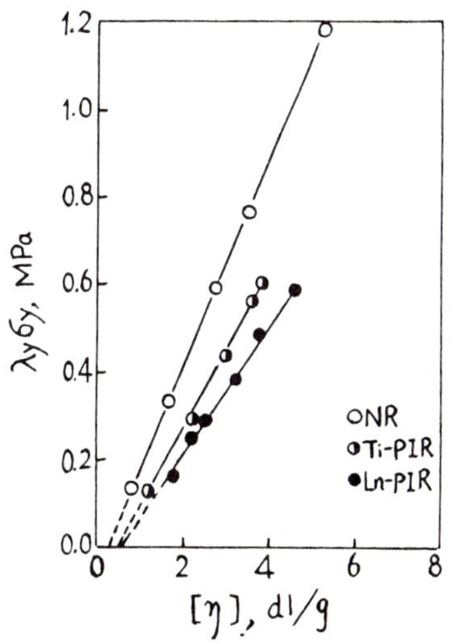

FIG. 27. Dependence of true yield strength $\lambda_y\sigma_y$ (25°C, 500 mm/min) on intrinsic viscosity [η] (30°C, toluene).

$M_{c(\sigma_y=0)} = 5.7 \times 10^4$ may be obtained, equivalent to 840 monomeric units per yielding segment. This value agrees well with the critical value of molecular weight of entangled segment $M_{c(m=1)} = 5.28 \times 10^4$, which was estimated with stress relaxation method.

TABLE VI
Dependence of Yield Strength σ_y on Molecular Weight M_η or $[\eta]$ for PIR's
$\lambda_y \sigma_y = A[\eta] + B$

PIR	A	$-B$	$[\eta]_c$, dl/g	Critical Value as $\sigma_y = 0$: $M_c \times 10^{-4}$	$X_c = M_c/M_o$
Ln-PIR	0.14	0.05	0.37	5.7[a]	840
Ti-PIR	0.18	0.08	0.45	7.6[a]	1100
				4.0[b]	590
NR	0.24	0.06	0.27	1.4[c]	210

[a] $[\eta] = 1.73 \times 10^{-4} \cdot M_w^{0.70}$. (This paper, 1981).
[b] $[\eta] = 2.0 \times 10 \cdot M_w^{0.728}$. (Beattie, W. H., and Booth, C. J., *Appl. Polym. Sci.*, *7*, 507, 1963).
[c] $[\eta] = 5.02 \times 10 \cdot M_w^{0.67}$. (See, e.g., Fraga, D. W., *J. Polym. Sci.*, *41*, 522, 1959).
[d] M_o of $C_5H_8 = 68$.

However, such a relationship exists only in rubbery state, but not in glassy state, as shown in Table VII, the nominal yield strengths at −80°C for all three Ln-PIR samples with different molecular weights are all equal to 20 kg/cm^2.

Since the yield strength is dependent on $[\eta]$, and $[\eta]$ is additive in nature, σ_y should also be additive [15], that is, $\sigma_y = \Sigma W_i \sigma_{yi}$. When two samples of PIR rubber with different molecular weights and yield strengths are blended in different weight proportions (W_i), as expected, the experimental results agree very well with the calculated line as shown in Figure 28.

ENTANGLEMENT AND YIELD BEHAVIOR

The molecular weight of Ln-PIR is so high that entanglement between macromolecules should be taken into consideration. The number of entanglement segments m per mole in bulk polymer can be estimated from the relaxation spectrum according to Chompff's approximation [11]. The calculated results show that the value of m increases exponentially with the increasing of M_η. By double logarithmic plotting m vs. M_η (Fig. 29), one may obtain the value of e for the following equation

$$m = K_e M_\eta^e$$

for Ln-PIR e equals 1.48 as listed in Table VIII. The critical molecular weight M_c for the onset of entanglement coupling can be obtained by extrapolating to

TABLE VII
Yield Strength σ_y for PIR in Rubbery and Glassy State

$[\eta]$, dL/g	Nominal yield strength, kg/cm^2: Rubbery (25°C)	Glassy (−80°C)
4.6	1.7	20
6.8	2.8	20
9.5	4.4	21

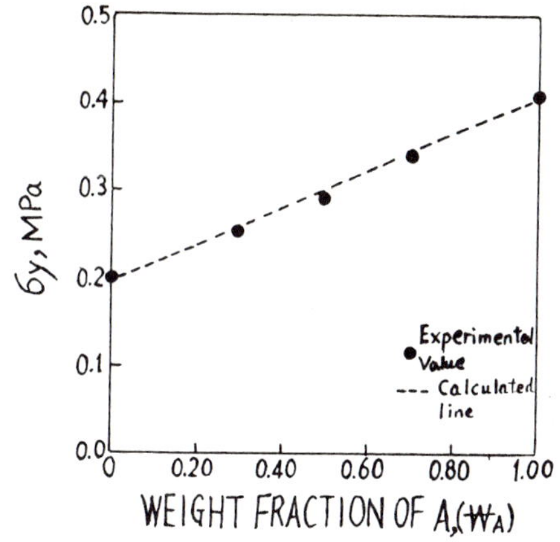

FIG. 28. Additivity of yield strength σ_y for Ln-PIR.

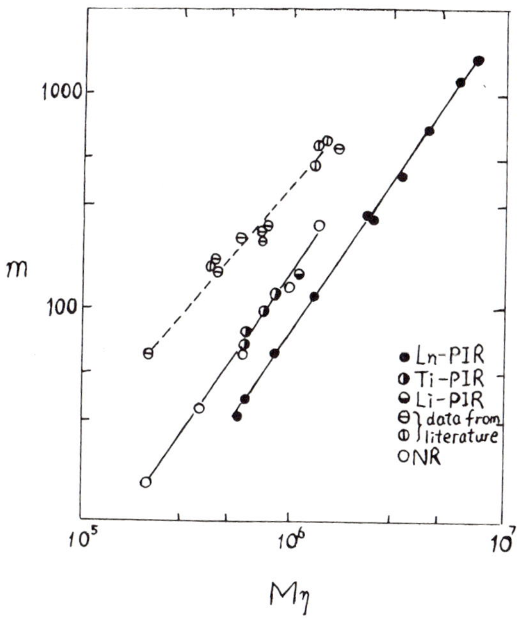

FIG. 29. Double logarithmic plots of number of entangled segment m vs. viscosity average molecular weight M_η.

$m = 1$, $M_{c(m=1)} = 5.28 \times 10^4$ for Ln-PIR, equivalent to $X_c = M_c/M_o = 778$ monomeric units per entanglement segment, nearly 1.7 times that for Ti-PIR and NR.

Such entanglement should be physical in nature and therefore dependent on temperature, it can then be disentangled by increasing temperature. By Arrhenius plotting of log m vs. $1/T$ (Fig. 30), one can obtain the activation energy of disentanglement ΔE_{deN}, which equals 8.0 kJ/mol.

TABLE VIII
Dependence of Number of Entangled Segments m on Viscosity Average Molecular Weight M_η
$m = KM_\eta^e$

Sample	e	K_e	$M_{c(m=1)}$	$X_c = M_c/M_o$
Ln-PIR	1.48	1.05×10^{-7}	5.28×10^4	778
Ti-,NR	1.43	3.70×10^{-7}	3.14×10^4	462

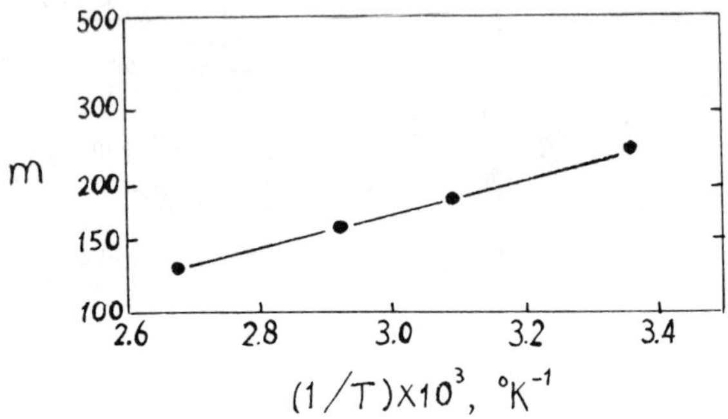

FIG. 30. Arrhenius plots of number of entangled segments m vs. temperature (1/T) for Ln-PIR.

CRITICAL MOLECULAR WEIGHT M_c FOR YIELDING

According to the theory of rubber elasticity, the equilibrium modulus due to entanglement E_{eN} of the rubbery plateau can be represented by

$$E_{eN} = 3g_n\sigma RT$$

The molecular weight of entangled segment M_e can be estimated by proper choice of the value of g_N, which usually ranges from 0.5 to 1.0, but Pearson (1974) [16] obtained a value of 1.22 for PB. By plotting log $M_e(g_N = 1)$ against $(\lambda_y\sigma_y)$, the straight lines in Figure 31 may be expressed as

$$\text{Log } M_e = C(\lambda_y\sigma_y) + D$$

The results are listed in Table IX, for the Ln-PIR (J-O ~ 5) series, C = −0.99 and D = 4.64. By taking $g_N = 1.22$ and extrapolating as $\sigma_y = 0$, we obtain the critical value $M_{c(\lambda_y\sigma_y=0,\, g_N=1.22)} = 5.4 \times 10^4$, just lying between the values of $M_{C(m=1)} = 5.28 \times 10^4$ by relaxation method and $M_{c(\sigma_y=0)} = 5.7 \times 10^4$ by yield strength method, so the average value of M_c may be taken as $(5.4 \pm 0.2) \times 10^4$ for Ln-PIR raw rubber.

PROSPECTS

Based on the characteristics of Ln polymerization systems and some preliminary experimental results, we would like to develop these systems into a new stage during the next decade:

1. Beside the solution Ln polymerization method, bulk polymerization was carried out on butadiene and isoprene, where no solvent was required. The pro-

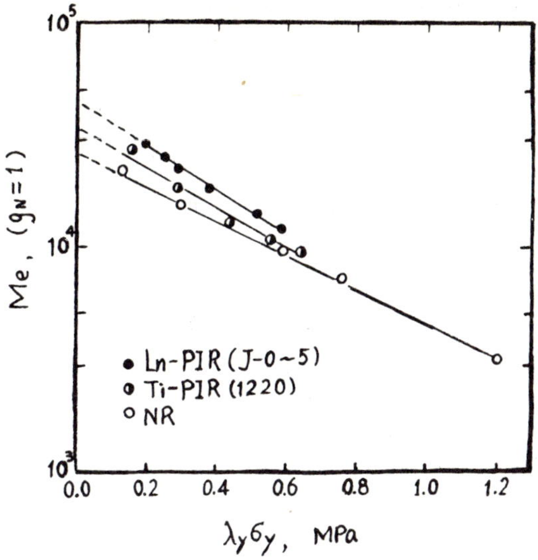

FIG. 31. Plots of molecular weight of entangled segments $M_{e(gN=1)}$ vs. true yield strength $\lambda_y \sigma_y$ for PIR.

TABLE IX
Critical Molecular Weight M_c for Yielding Estimated from Log M_e vs. $\lambda_y \sigma_y$ Plot for PIRs
$M_e = 3g_N \rho RT/E_{eN}$, $\log M_e = C(\lambda_y \sigma_y) + D$

Sample	$-C$	D	$M_c(\sigma_y = 0, g_N) \times 10^{-4}$ $g_N = 1.0$	$g_N = 1.22$
Ln-PIR	0.99	4.64	4.4[a]	5.4($X_c = M_c/M_o$ = 800)
Ti-PIR	0.97	4.49	3.1[b]	3.8 (560)
NR	0.79	4.40	2.5[c]	3.1 (450)

[a,b,c] Same as Table VI.

cessability of their raw rubbers was superior to that of solution polymerized, so that the mechanism of the bulk Ln polymerization and the relationship between structure and property of such polymerizates would be reexamined.

2. For ethylene and mono olefins, as well as styrene and vinyls, the Ln polymerization catalysts are also effective. The specificity and efficiency of Ln catalysts for a given monomer would be recognized, especially, the role of monomers in Ln-copolymerization would be investigated in more detail.

For example, the copolymerization of ethylene and 1,3-butadiene with different Ln catalysts was investigated. A copolymer containing 8% *cis*-1,4-butadiene and long ethylene-ethylene sequences is obtained [17]. The activity of individual Ln catalyst is in the order Pr > Nd > Gd > Dy > La > Er. This order is the same as that of Ln polymerization of ethylene homopolymer, but different from that of *cis*-1,4-polybutadiene for which Nd > Ce > Pr > Gd [3].

3. Acetylene also can be polymerized with lanthanide catalysts [18]. The *cis*/*trans* isomerization rate of polyacetylene (Ln-PA) is slower than that of Ti-

PA, and the air stability of Ln-PA is far better than that of Ti-PA in ambient conditions [20]. This provided a better catalyst for the polymerization of acetylenes, including substituted acetylenes.

Furthermore, PA can be obtained with scandium (Sc) catalyst [21], similar to Ln-PA, Sc-PA possesses a high *cis*-content (90–100%) with metallic sheen of film, and the electrical conductivity of iodine-doped Sc-PA is 11–12 orders of magnitude higher. Thus it can be said that scandium played a role as an active member of the rare earth elements in coordination of catalytic polymerization.

REFERENCES

[1] T. Sheng, et al., *Sci. Sin., 13,* 1339 (1964).

[2] F. Wang, Q. Sheng, et al., *Collected Papers on Synthetic Rubbers by Rare-Earth Catalysts,* Science Press, Beijing, 1981.

[3] Shen Zhiquan, Qian Baogong, et al., "The characteristics of lanthanide coordination catalysts and the cis-polydienes prepared therewith," presented at the ACS/CSJ Chemistry Conference, Honolulu, Apr. 5, 1979, *J. Polym. Sci.,* (Polym. Chem. Ed.), *18,* 3345–3357 (1980).

[4] The Fourth Laboratory, Changchun Institute of Applied Chemistry, *Sci. Sin., 17,* 656 (1974).

[5] F. Wang, A. Bolognesi, A. Immizi, and L. Porri, *Makromol. Chem., 182,* 3617 (1981).

[6] Z. Gong, B. Li, G. Yang, and F. Wang, *Gaofenzi TongXun* (Polymer Bulletin), 116 (1983).

[7] Yu. B. Monokov, S. R. Rafikov et al., *Doklady Akad. Nauk USSR, 234,* 1125 (1977).

[8] A. Bolognesi, S. Destri, L. Porri, and F. Wang, *Makromol. Chem., Rapid Commun., 3,* 187 (1982).

[9] Baogong Qian, et al., *Proceedings of China-US Bilateral Symposium on Polymer Chemistry and Physics,* Oct. 5–10, 1979, pp. 155–170; Beijing, Science Press, Beijing; Reinhold, New York, 1981; *Kexue Tongbao* (Monthly Journal of Science, English ed.), *27(3),* 278–284 (1982).

[10] Enle Zhou, et al., *China-Japan Bilateral Symposium on the Synthesis and Materials Science of Polymers* (Oct. 21–24, 1984, Beijing), Preprints p. 159, Academia Sinica, China, The Society of Polymer Science, Japan (1984).

[11] Yanshou Zhang and Baogong Qian, Collected Papers of the Changchun Institute of Applied Chemistry, Academia Sinica, *17,* 36 (1981) presented at the Japan-China Bilateral Symposium on Polymer Science and Technology, Tokyo, October 26–29, 1981.

[12] (a) A. D. Chirico, *Die Makromol. Chim., 175,* 2029 (1974); (b) B. Wunderlich, *Macromolecular Physics,* Vol. 2, Sec.6.1.5 Academic, New York, 1976; (c) E. Turska, *Polymer (Lond.), 12,* 629 (1971); (d) R. Alamu, *Polymer, 23,* 374 (1982); (e) E. Ergos, *Macromolecules, 5,* 147 (1972); (f) E. G. Lovering, *J. Polym. Sci., A-2, 8,* 747 (1970).

[13] B. Wunderlich, *Macromolecular Physics,* Vol. 2, Sec.6.1.7, Academic, New York, 1976.

[14] Xu Yang, Yu Fushen, Zhou Enle, and Qian Baogong, *Chinese J. Appl. Chem., 1* (4), 51–56 (1984) presented at the IUPAC Polymer 85, Melbourne, February 11–14, 1985. (a) Xu Yang, Doctoral dissertation, Changchun Institute of Applied Chemistry, Academia Sinica, 1985. (b) Qian Baogong (Chien, Pao-kung), Yu Fushen, Chen Rongshi, Ruan Meina, Zhang Wanjun, Li Shentian, Zhan Yanshou, and Yang Yuhua, "Characterization of *cis*-1,4-polyisoprene polymerized with Lanthanide System (Ln-PIR)," presented at the Japan-China Bilateral Symposium on Polymer Science and Technology (Tokyo, Oct. 26–29, 1981); *J. Appl. Polym. Sci., 30,* 375–390 (1985).

[15] Fushen Yu, Yuhua Yan, Shentian Li, and Baogong Qian, Collected Papers of the Changchun Institute of Applied Chemistry, *18,* 1 (1982).

[16] Pearson, D. S., Shutnik, B. J., and Bohm, G. G. A., *J. Polym. Sci., Polym. Phys. Ed., 12,* 925 (1974).

[17] S. Wang, Z. Li and F. S. Wang, *Polym. Bull., 6,* 420 (1984).

[18] Z. Shen, et al., *Sci. Sin. (B), 26* (8), 785–798 (1983).
[19] F. S. Wang and X. J. Zhao, China-Japan Bilateral Symposium on the Synthesis and Materials Science of Polymers, October 21–24, 1984, Beijing, preprints, p. 109.
[20] J. Yang and F. S. Wang, China-Japan Bilateral Symposium on the Synthesis and Materials Science of Polymers, October 21–24, 1984, Beijing, preprints, p. 113.
[21] Z. Shen, et al., China-Japan Bilateral Symposium on the Synthesis and Materials Science of Polymers, October 21–24, 1984, Beijing, preprints, p. 107.

Baogong Qian was born on March 18, 1916.

PRESENT POSITION

Professor of Polymer Science,
President of the Wuhan Branch
Academia Sinica
Wuchang, Wuhan City
P. R. China.

EDUCATION

Polytechnic Institute of Brooklyn, M.S., 1949.

POSITION

Research Associate, Professor,
Deputy Director of Changchun Institute of Applied Chemistry, 1950–1980.
Visiting Professor:
Jilin University, 1950–1980.
Wuhan University, 1981.

PROFESSIONAL ACTIVITIES

Member of the Division of Chemistry, Academia Sinica, 1981–
Second National Prize on Natural Sciences, 1982.

NEW DESIGNS FOR CONDUCTIVE POLYMERS

NAOYA OGATA
Department of Chemistry, Sophia University, 7-1 Kioi-Cho, Chiyoda-Ku, Tokyo, Japan 102

SYNOPSIS

This paper deals with new synthetic designs for both electron and ion conductive polymers in terms of graphite polymers and amorphous solid electrolytes.

Much interest has been focused on electrically conductive polymers and many papers on the topic have been published during the past ten years. Electrical conduction is carried out by either electrons or ions through materials. Highly electron-conductive polymers are formed by the addition of either electron donors or electron acceptors to conjugated polymers such as poly(acetylene) or poly(*p*-phenylene). However, electroconductivities of these doped polymers are not stable enough owing to the release of dopants or other side reactions such as oxidation. Since a long life of conductive polymers is desirable, there is a need for relative stability of electroconductivities.

Graphite has a high conductivity by itself without benefit of a doping process and polymers having graphite structure represent a potential approach to obtaining stable conductive polymers. Polymer pyrolysis or a high temperature polycondensation is a simple procedure which results in a graphite structure of relatively high electrical conductivity.

On the other hand, ionic conduction is carried out by the movement of ions in electrolyte solutions in which solvents are limited to low molecular weight compounds. However, certain kinds of polymers such as poly(ethylene oxide) or poly(propylene oxide) dissolve inorganic salt to form homogeneous ion-containing polymers which show an ionic conduction. These solid ion conductors or polymeric electrolytes have high potential for electric application such as solid-state batteries.

This paper deals with new synthetic designs for both electron- and ion-conductive polymers in terms of graphite polymers and amorphous solid electrolytes.

GRAPHITE POLYMERS

Pyrolytic polymers have not been well appreciated as a synthetic metal, partly because their chemical and electronic structures cannot be assessed very easily. A basic unsolved question may be whether the structure and properties of the pyrolytic polymers bear any relationship to the molecular structure of the starting polymers.

Based on this premise, various types of condensation polymers were subjected to pyrolysis reaction at elevated temperatures and the electronic properties of these pyrolyzed polymers were investigated in terms of electric conductivities and graphite structure. Among these polymers, poly(*p*-phenylene-1,3-oxadiazole) (POD), which was obtained in the form of a thin film with high crystallinity resulted in a highly graphite-like structure with a relatively high electrical conductivity by means of the pyrolysis reaction at elevated temperatures.

Figure 1 shows the dependence of the room-temperature conductivity, σ_{RT}, on T_p, where each POD film was kept at T_p for 60 min after being heated at a rate of 10°C $\min^{-1}$. The figure also includes data for polyimide (Kapton), pyrolyzed under the same conditions. The pyrolyzed POD exhibits σ_{RT} of 68, 350, 410, and 510 S cm^{-1} for T_p = 800, 1,000, 1,200, and 1,400°C, respectively. The pyrolyzed POD had σ_{RT} more than twice as high as that of pyrolyzed polyimide.

Figure 2 indicates the temperature dependence of the electrical conductivity of the POD pyrolyzed at T_p = 800, 1,000, and 1,200°C. Another characteristic feature of POD is a temperature-independent conductivity at low temperatures, and the σ vs. T curves as a whole are expressed by the equation, $\sigma(T) = \sigma(0) + \sigma_1(T)$, where $\sigma(0)$ is the zero-temperature limit conductivity.

Powder X-ray diffraction patterns (with Cu-K radiation) of the pyrolyzed POD films are shown in Figure 3, which indicates that the pyrolyzed POD has

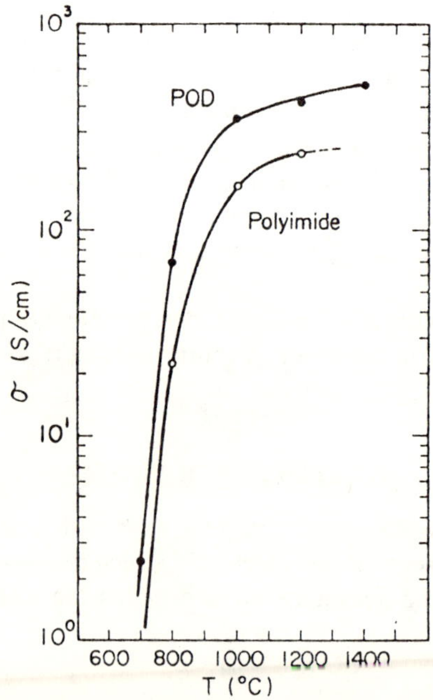

FIG. 1. Electrical conductivity at room temperature vs pyrolysis temperature for POD (●) and polyimide (○).

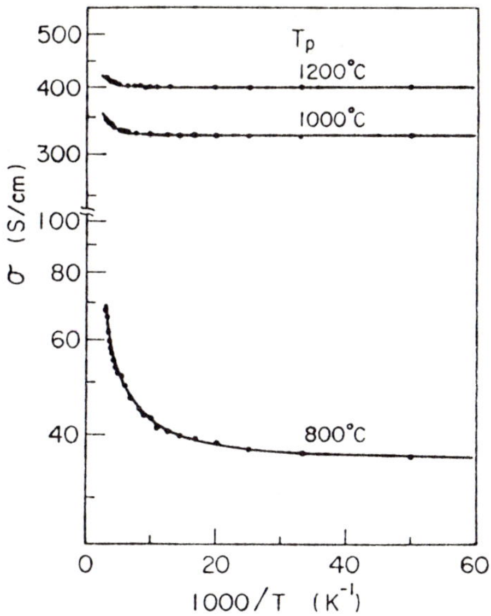

FIG. 2. Temperature dependence of conductivity for POD pyrolyzed at 800, 1,000, and 1,200°C.

an increased ordering in the (002) direction of a graphite structure at higher T_p. The growth of the graphite structure is also suggested by X-ray photoelectron spectroscopy (XPS) shown in Figure 4.

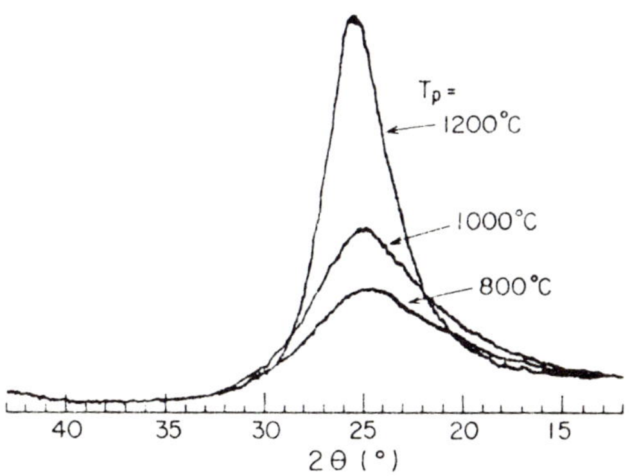

FIG. 3. Powder X-ray diffraction patterns for POD pyrolyzed at 800, 1,000, and 1,200°C.

The pyrolyzed POD has two conducting channels: one is a metallic channel with extended electronic states for which the conductivity has no temperature dependence, and the other is of a disordered system where the electron motion is described by the three-dimensional variable range-hopping model. The X-ray diffraction and XPS analyses suggested that the pyrolyzed POD has a graphite structure in which nitrogen-containing heterocyclic rings are included.

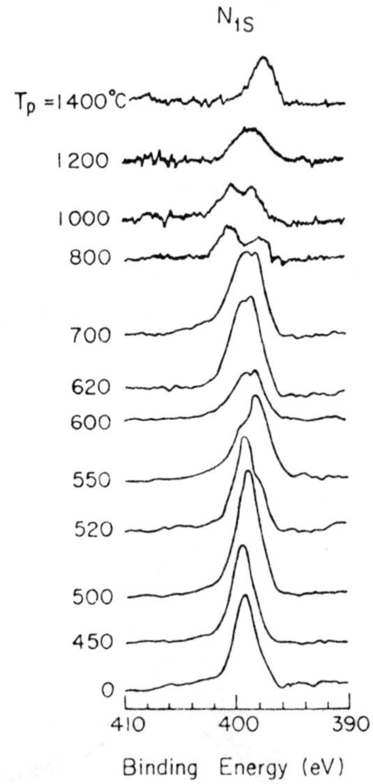

FIG. 4. X-ray photoelectron spectroscopy peaks in the N_{1s} region for POD pyrolyzed at various temperatures.

It had been expected that "one-dimensional graphite" of fused aromatic rings might have distinctive physical properties such as metallic conduction or even high-temperature superconductivity. It was found that very fine whiskers were grown on a pellet of 3,4,5,10-perylene tetracarboxylic dianhydride (PTCDA) when the pellet was heated above 520°C in an inert atmosphere.

The whisker had a rectangular cross-section of about 0.2 to 0.4 μm on a side and a length of 10–15 mm, as shown in Figure 5.

Chemical analyses of the whiskers confirmed that the structure of the whiskers consisted of poly(naphthalene) as shown below:

Poly(naphthalene) (PPN) was formed through a pyrolytic polymerization of PTCDA in a vapor phase, which might proceed through the formation of perylene tetraradicals. The electrical conductivities of the whiskers are summarized in Figure 6, which indicates that the electrical conductivity of the PPN whisker grown at 530°C was 0.2 S cm^{-1} at room temperature with metallic temperature dependence at lower temperature ranges, while the whiskers grown at 1,200°C had a high conductivity of 1,100 S cm^{-1} with almost no temperature dependence.

It is expected that the PPN whiskers which were one-dimensional graphite will lead to the manufacture of composite materials with high electrical conductivities and various applications to electronic devices will be realized.

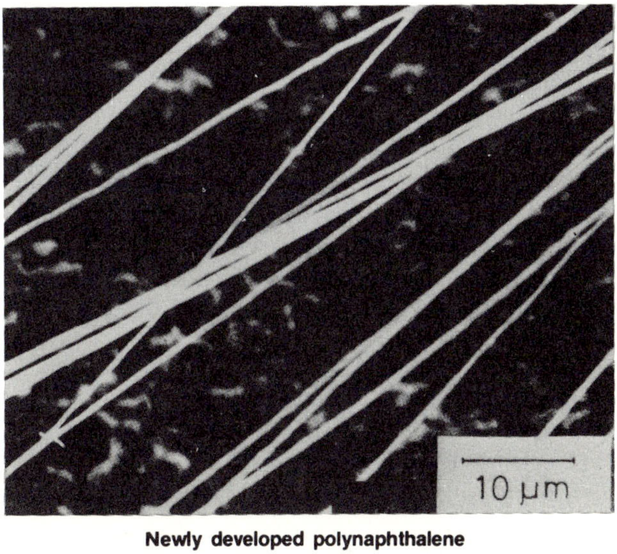

Newly developed polynaphthalene

FIG. 5. Electron microscopic picture of whiskers.

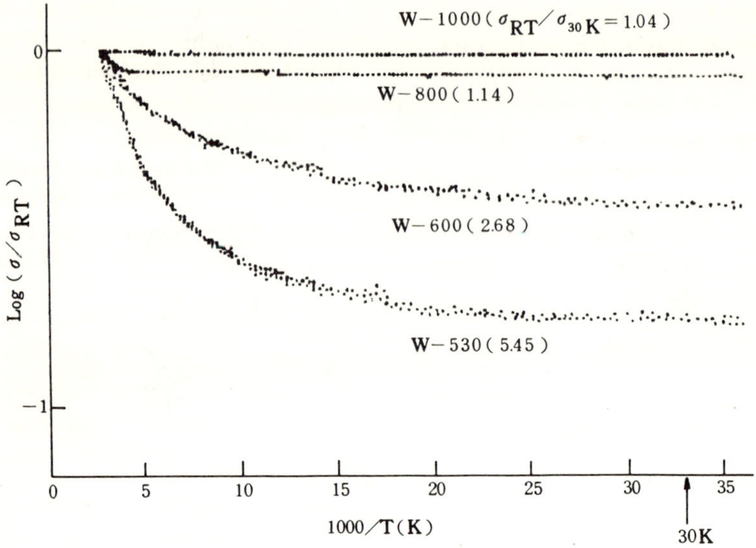

FIG. 6. Temperature dependence of conductivities of the PPN whiskers.

ION CONDUCTORS

A relatively high ionic conductivity has been attained recently in polymer complexes formed by alkali metal salts and polyethers such as poly(ethylene oxide) or poly(propylene oxide).

Triol-type poly(propylene oxide) (PPO) and tolylene-2,4-diisocyanate (TDI) were reacted to form a network of polyurethane-containing PPO segments. The network films were obtained by solvent casting and various alkali metal salts were migrated into the films.

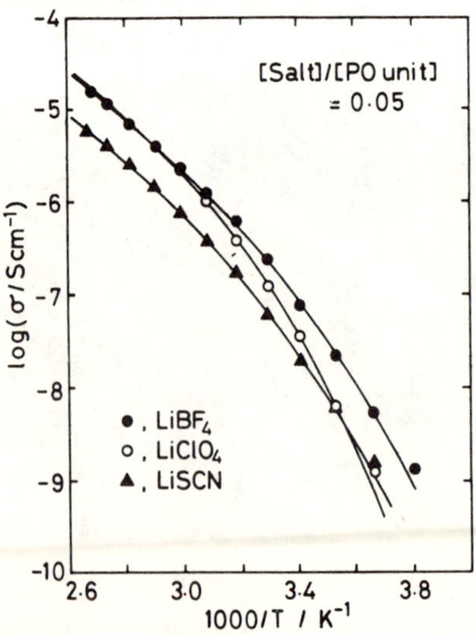

FIG. 7. Relation between logarithm of ionic conductivity and reciprocal of absolute temperature.

Figure 7 indicates the temperature dependence of ionic conductivity of the films, which shows that the temperature dependence did not obey the Arrhenius plots with a temperature-independent activation energy. The PPO complexes were transparent, amorphous, and elastic materials, and their glass transition temperatures were lower than the measuring temperature range.

It was confirmed that alkali metal salts were dissolved in amorphous domains of PPO segments and the temperature dependence of σ could be expressed by the WLF-type equation as shown below:

$$\log \frac{\sigma(T)}{\sigma(T_g)} = \frac{C_1(T - T_g)}{C_2 + (T - T_g)}$$

Various types of amorphous polymers containing polyether segments and also aliphatic polyesters were synthesized and their ionic conductivities were measured. Among these polymers polyurethane-containing PPO segments and poly(β-propiolactone) had a relatively high ionic conductivity of 10^{-4} S cm^{-1} at room temperature.

Naoya Ogata was born on March 30, 1932.

PRESENT POSITION

Professor of Polymer Science
Department of Chemistry
Sophia University
Chiyoda-Ku, Tokyo
Japan 102.

EDUCATION

University of Tokyo, Ph.D., 1961.
Research Associate, 1961–1962
Polytech. Inst. Brooklyn
Research Associate, 1962–1963.
State Univ. New York at Syracuse.

POSITIONS

Toyo Rayon Company (present Toray Ind.), 1954–1966.
Associate Professor, Sophia Univ. 1966–1968.
Professor, Sophia University, 1968–.
Visiting Professor:
University of Massachusetts, 1979–1980.

PROFESSIONAL ACTIVITIES

SPSJ (Society of Polymer Science, Japan): Chief Editor of "KOBUNSHI," 1975–1977.
SPSJ: Committee of Society Activities Chairman, 1980–1982.

ADVANCED POLYMERIC MATERIALS FOR THE FUTURE

NORBERT M. BIKALES
National Science Foundation, Washington, DC 20550

SYNOPSIS

It has been said that "while materials comprise only a small part of the matter in the universe, they are the stuff from which man makes things, so that they become the working substance of civilizations" (M. Cohen of MIT). Increasingly, polymers have become the basic working substance of late twentieth-century civilization! This is due, in no small measure, to the enormous influence and contributions of Professor Herman Mark. My assignment in the symposium honoring this great man is to present a somewhat more materials-oriented talk, to look at the future of advanced polymeric materials.

INTRODUCTION

Although polymers of various types had been in extensive use in industry for many years, the petroleum supply crises of the 1970s provided the impetus for an expanded scope in their usage (Fig. 1). In particular, advanced composites received major attention because of their attractive mechanical properties, often coupled with energy savings in manufacture or in use, as compared to metals. This paper will describe the current status of composites utilization in the air-

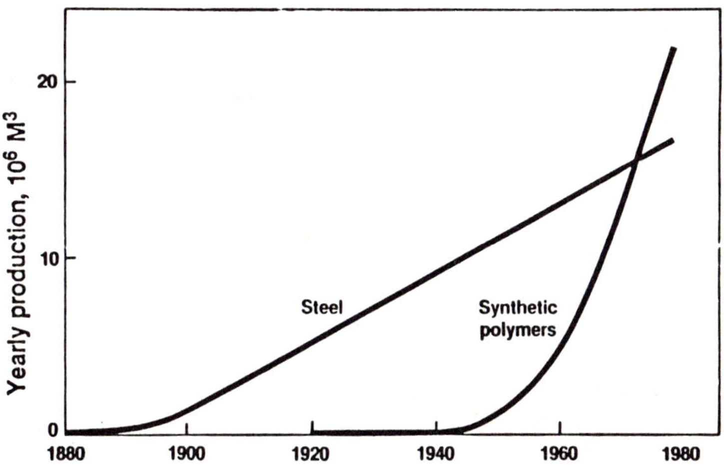

FIG. 1. Annual production of steel and synthetic polymers, in m^3, over a period approximating that of Professor Mark's lifetime. Courtesy National Research Council.

Published by John Wiley & Sons, Inc.

craft and automobile construction industries [1–4]. It will enumerate the research areas that need further advances, and some of the organizational structures that have been devised in the United States to help effect solutions to these scientific and technological challenges.

SOME USES OF ADVANCED COMPOSITES

In Aircraft

The U.S. aircraft industry began making significant use of advanced polymeric composites in the 1970s, with construction of a series of military aircraft. The airplanes known by the designations F-14, F-15, F-16, F/A-18, and AV-8B have 0.8%, 1.6%, 2.5%, 10%, and 26%, respectively, of their structural weight in composite materials (Fig. 2). The AV-8B, in particular, makes extensive use of graphite-epoxy and other composites for the wing-box skins and substructure, the forward fuselage, the horizontal stabilizers, the elevators, the rudder, the overwing fairing, and the ailerons and flaps. Altogether about 600 kg of composites are utilized in this design, at a considerable saving in weight and in the number of parts, as compared to a similar aircraft built from conventional materials.

The U.S. civilian aircraft industry is now also making increasing utilization of such composite materials, in an effort to improve performance, lower the weight, reduce the fuel consumption, and provide greater flexibility in design. The Lockheed L-1011, the McDonnell-Douglas DC-10, and the Boeing 727, 737, and 747 all make use of some composites. However, the first major uti-

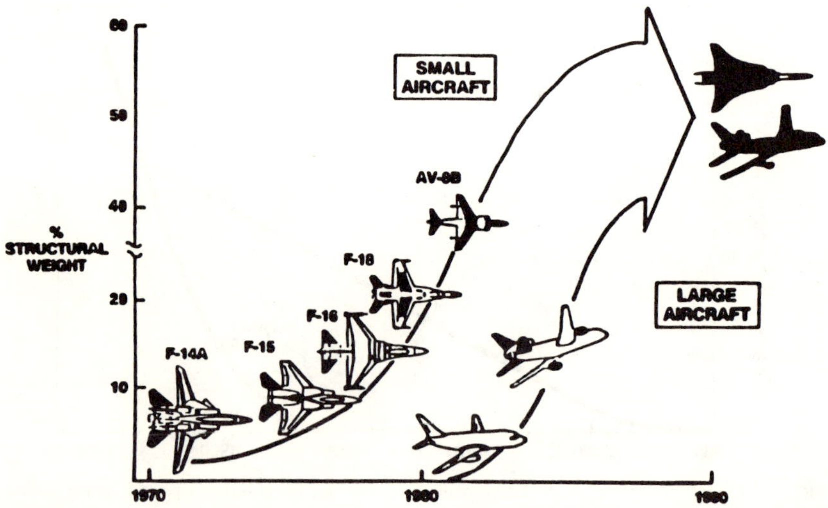

FIG. 2. Usage of advanced polymeric composites in military (small) and civilian transport (large) aircraft over a span of two decades. The dark areas represent the composites in the structures. In future aircraft, composites may be the principal structural material. Courtesy National Aeronautics and Space Administration.

lization came in the new generation of large transport planes known as the Boeing 757 and 767; still more widespread use in airline construction is expected in the late 1980s and early 1990s. Thus, whereas the older 747 has only some 1% of its structural material as composite, the new 767 has 3%, of which 2.5% represents advanced composites reinforced either with graphite or aramid fibers [3]. Virtually the entire structure of the small Lear Fan 2100 is built of composites: some 600 kg of graphite-epoxy in the aircraft's primary structure (fuselage, wings, empennage) and some 135 kg of aramid-reinforced epoxy in the secondary structure. Another aircraft making extensive use of polymeric composites is the new Beech Starship 1. It has been estimated that the large subsonic airplanes to be built in the next ten years can potentially have as much as 65% of their weight made up of polymeric composites.

Before composites can make this major impact, some very significant obstacles will have to be overcome. Better (tougher) resins must be provided; manufacturing costs must be reduced, design flexibility improved, properties must be made more consistently reproducible, and there must be improvements in impact, environmental, and several other properties. Defect detection and repairability are also major concerns.

As will be discussed below, these goals cannot be met solely by applied research and development, but much basic research will be needed in all aspects of polymer chemistry, polymer morphology, surface science, rheology, mechanics, and other disciplines.

In Automobiles

Automotive applications of polymers represent a sizable market for many kinds of polymers [1, 4]. Because of the large number of vehicles produced annually, this market utilizes far greater quantities of polymers than the aircraft industry. In the 1980s, the U.S. consumption of polymers in automobiles has been on the order of 680,000 metric tons annually! About 50% of the total rubber output in the United States is used in automobile tires.

In rough terms, the average U.S.-built automobile now utilizes about 100 kg of polymers, which represents an increase of about 87 kg over twenty years ago (i.e., before the great rise in energy prices). In the next ten years, that weight is expected to double, as major components currently made of metals will be constructed of fiber-reinforced composites in a continuing effort to conserve fuel (Table I).

Among the current applications of polymers are instrument panels, inner door panels, trim, seats; fender extensions, grilles, lenses, filler panels; fuel tanks, vapor emission canisters; fans, pulleys, gears, air-cleaner parts, and reservoirs for brake fluid. In one mass-produced station wagon, the reinforced-polymer tailgate weighs one third less than a metal component it replaces, and consists of 2 pieces rather than 17. A dramatic demonstration of the utility of advanced polymeric materials has been the experimental automobile built by the Ford Motor Company in the late 1970s with the body, chassis, and even powertrain components made of graphite-fiber-reinforced composite.

TABLE I
Typical Weight Savings (lbs.) by Materials Substitution in Automotive Components [1]

Metal part	Substitution material	Potential weight saved (lb)
Wheel	Fiberglass composite	6
Light truck single leaf spring	Fiberglass composite	20–30
Hood	Fiberglass composite	6
Transmission support crossmember	Fiberglass composite	3
Deck lid	Fiberglass composite	8
Bumper	Fiberglass composite	12
Tailgate	Fiberglass composite	14
Seat	Thermoplastic	5
Fuel tank	Thermoplastic	9
Connecting rod (engine)	Graphite-fiber composite	1

Source: Courtesy National Research Council.

A further major advance, introduced by the General Motors Corporation beginning in 1984, has been the mass production of a two-seat car with a complete plastic exterior. The innovation in the Fiero is not only the greater utilization of polymeric materials, but also in the remarkable manufacturing technology. The exterior plastic panels are mounted ("hung") on a metallic frame that provides the structural integrity. The panels are fabricated from sheet-molding compound, reaction-injection-molded (RIM) polyurethane with or without reinforcement, and polyolefins. Adhesive bonding is widely used, and special coatings and molding techniques produce an enhanced esthetic appearance.

Some other uses of polymeric composites have been in springs for sports cars and trucks and in truck driveshafts, to name only a few instances. Even some high-performance racing car engine components (pushrods, parts of pistons, valves) are made of composites. Experimentally, flywheels have been constructed that could potentially improve fuel economy in automobiles, buses, and trains, by more efficient engine operation and regenerative braking.

Major advances are needed in the automotive field to improve appearance and durability, to decrease costs of production, and to reduce the cycle time in processing by the use of better resins and new methods of composite manufacture.

RESEARCH NEEDS

A number of studies have been conducted under the auspices of various U.S. governmental agencies to establish a list of future needs in the materials area. Recurring themes include a call for strengthening of basic research in polymer science and engineering and a greater effort to improve the basic research–

applications linkage [1, 2]. In practice, this will require greater involvement of U.S. university laboratories in research of relevance to industry as well as greater cooperation among academic and industrial researchers. The following compilation of research needs is derived from a study of the National Academy of Sciences [1] and other sources.

Materials. Synthesis of new polymeric systems, improved thermal stability, more rapid cure (especially at low temperatures), tougher matrix materials, thermoplastic matrix materials, lower-cost high-modulus fibers, self-reinforcing composites. Chemistry of thermosets, and relation to properties. Fiber-surface treatments and understanding and control of the fiber-resin interface. Further development of reaction-injection moldable polymers. Structure-property-morphology relationships in polymer systems, including alloys. Transparent polymers with high surface hardness. Polymers with reduced flammability and less toxic byproducts on heating.

Manufacturing Technology. Kinetics of cure reactions. Rheology of flow in molds. Reaction process technology for direct conversion of monomers to shaped objects, utilizing a broad range of polymers. Higher speed processing: curved pultrusion, pull-forming, alternative heating methods. Automation technology and hardware. Greater reproducibility and freedom from defects. Improved repair methods. On-line processing controls. Prediction of part performance based on composition and morphology and fiber-orientation distribution. Prediction of properties based on fabrication parameters. Modeling of molding and forming processes. Heat transfer and mixing in molding.

Adhesion and Bonding. Surface science. Adhesive chemistry developments. Better performance under extreme conditions. Bonding of plastics to plastics and metals. Automated high-speed application technology. Nondestructive evaluation of bond strength.

Coatings. Low-temperature curing, nontoxic, low-emission paints. Solventless coatings. Stability of colloids. Theory of wetting and adhesion. Greater basic understanding in this largely empirical field.

Impact Resistance. Energy absorption of designed structures for impact strength, damage from flying stones, crash worthiness, passenger protection. Knowledge of failure mechanisms.

Design Data Base. Long-term performance under various service conditions, e.g., moisture, solvents, ultraviolet radiation, aging, particularly in combination with stress. Methodology for rapid calculations for stress analysis and failure prediction. Response to different load conditions. Computer-based modeling for design, engineering, and manufacture.

Destructive and Nondestructive Evaluation. Greater reliability at high rates and low cost. Correlation of defects with end-use performance. Failure prediction under static and cyclic stresses. Failure mode analysis. Reproducible test procedures for interlaboratory comparisons.

Energy Storage. Materials, design, and manufacturing technology for high-efficiency flywheels.

ORGANIZATIONAL IMPROVEMENTS

An important finding of a recent study by a joint committee of the National Academies of Science and Engineering [2] is that there are fewer than 30 U.S. universities where research on composite materials is being conducted and that only some 40 faculty members are involved! It is clear that the latter number will have to be increased sharply in the near future.

The National Science Foundation is taking several steps to encourage greater university participation in research of potential importance to industry. Among recently instituted organizational structures are provisions for basic research projects jointly conducted by individual university and industrial researchers, as well as for research centers that are jointly funded by NSF and industry but located at universities. Some of these centers are concerned with polymers, namely, those at the Massachusetts Institute of Technology, the University of Massachusetts, and Case Western Reserve University.

NSF-funded materials research laboratories and groups are designed to encourage multidisciplinary basic research. One such group, at Professor Mark's own Polytechnic University, deals with the aging processes in polymeric blends. A recent innovation is the establishment of university-based engineering research centers to focus the attention of groups of university engineers on problems of pressing national need. In April 1985, the first of these centers became operational, among which is the Center for Composites Manufacturing Science and Engineering at the University of Delaware. Together with the Center for Adhesion Science, funded by the Office of Naval Research at the Virginia Polytechnic Institute, and several others, these new research centers are expected to make significant contributions to the types of materials problems that have been discussed above.

Finally, let me express my belief that a new field of research is emerging: it is a fusion of polymer science and engineering, metallurgy and ceramics science, with components of chemistry and physics, mechanics, rheology, and other disciplines. That field is the new discipline of *materials science*. It is my conviction that many of the difficult problems in materials research and utilization can be solved only by a joint effort of materials scientists in industrial, government, and university laboratories. As Professor Mark has taught us, this cooperation must even extend across national boundaries!

The views expressed here are solely the author's and do not represent those of any other parties. The author wishes to thank a number of colleagues for contributing valuable background information for this paper, and especially C. F. Bersch, T. J. Mao, S. Newman, and J. C. Seferis. The references cited below were also most useful.

REFERENCES

[1] *Polymer Science and Engineering: Challenges, Needs, and Opportunities,* National Academy Press, Washington, 1981.

[2] Research Briefings 1984; *Report of the Research Briefing Panel on High-Performance Polymer Composites,* National Academy Press, Washington, 1984.

[3] J. C. Seferis and L. Nicolais (Eds.), *The Role of the Polymeric Matrix in the Processing and Structural Properties of Composite Materials*, Plenum, New York, 1982.
[4] S. Newman, "Automotive Applications," in *Encyclopedia of Polymer Science and Engineering*, 2nd. ed., J. Kroschwitz Ed., Wiley-Interscience, New York, Vol. 2, pp. 117–143, 1985.

Norbert M. Bikales is currently the Director of the Polymers Program at the National Science Foundation in Washington. Prior to assuming this position in 1976, he was Professor of Chemistry and Director of Continuing Education in the Sciences at Rutgers University. He obtained his doctorate in organic and polymer chemistry at the Polytechnic Institute of Brooklyn (now Polytechnic University) and has worked in industry (American Cyanamid Company) and as an industrial consultant. With Professor Mark, he edited the *Encyclopedia of Polymer Science and Technology* and serves with him on the editorial board of the *Encyclopedia of Polymer Science and Engineering*. He has a number of other books to his credit, including *Cellulose and Cellulose Derivatives* and *Water-Soluble Polymers*, and he is the author of about seventy papers and some thirty patents.

In 1984, Dr. Bikales was a visiting scientist at the Centre de Recherche sur les Macromolécules in Strasbourg. He was elected a Fellow of the New York Academy of Sciences in 1977 and of the American Association for the Advancement of Science in 1986. Dr. Bikales was the recipient of the Great Medal of the City of Paris in 1985 and of the Distinguished Alumnus Award of Polytechnic University in 1986. In 1983, he served as Chairman of the Division of Polymer Chemistry of the American Chemical Society, and has also been active in the International Union of Pure and Applied Chemistry.

RADIATION EFFECTS ON POLYMERS

JAMES H. O'DONNELL
Polymer and Radiation Group, Department of Chemistry
University of Queensland, Brisbane 4067, Australia

SYNOPSIS

Irradiation can produce large changes in the properties of polymers for the absorption of small amounts of energy, due mainly to the marked dependence of properties on molecular weight. The chemical changes induced by radiation commence with ionization and excitation and proceed by a variety of free radical and ionic reactions. The main features of the radiation degradation of polymers are chain scission and crosslinking, formation of small molecular products and structural changes in the polymers. Fundamental research studies utilize a variety of modern instrumentation including IR, UV, NMR, ESR, MS and pulse radiolysis. Industrial applications of irradiation of polymers have increased rapidly with increasing availability of electron accelerators and isotope sources. Electron beam lithography in the microelectronics industry utilizes the sensitivity to chain scission of polymers with appropriate molecular structures. Exposure to radiation, as in sterilization of medical items, in space environments, or in nuclear facilities, may cause significant deterioration in properties, and there is a need for the development of improved polymer materials for these applications. New processes and new materials based on radiation-induced chemical reactions in monomer/polymer systems offer great potential for future technology based on knowledge of the relevant, fundamental radiation chemistry.

INTRODUCTION

I am privileged to have been invited to lecture at this Symposium on "Polymer Science in the Next Decade" which honors Herman Mark on his 90th birthday. I am also privileged to have been a Post-Doctoral Fellow at the Polytechnic Institute of Brooklyn (now the Polytechnic Institute of New York) with Professor Herbert Morawetz in 1963, when I enjoyed the spirit of "Brooklyn Poly" and Professor Mark's stimulating lectures.

International interaction is vital for the development of polymer science and Herman Mark has been a catalyst for communication between members of the world polymer community. He has always shown a broad, interdisciplinary interest in polymers, which has been an example to us all.

Radiation effects in polymer systems were being studied at Brooklyn in the 1960s as part of the program on solid-state polymerization. This has been just one area of the radiation chemistry of monomers and polymers—a field which has expanded steadily in both fundamental research and technological innovation. In this lecture I have attempted to survey some of the likely developments during the next decade.

HISTORY OF RADIATION CHEMISTRY

Radiation chemistry research commenced with the discovery of X-rays from discharge tubes in 1985 by Roentgen and of natural radioactivity from a uranium salt by Becquerel, who observed the chemical changes produced in a photographic emulsion. It is an interesting coincidence that 1895 was the year that Herman Mark was born, and therefore his life has spanned the entire history of radiation chemistry.

The effects of radiation on chemical compounds was a field of little more than academic curiosity for many years. However, it has grown rapidly into an important area of polymer technology as a result of the increasing availability of radiation sources, and recognition of the significant changes in chemical structure and material properties that radiation produces in polymers.

HIGH ENERGY RADIATION

Radiation is a term which covers electromagnetic and particulate radiation from gamma-rays with wave lengths of picometers to radiowaves with wave lengths of hundreds of meters. High energy radiation is understood to mean radiation which can cause ionization in chemical compounds. This requires a minimum of about 10 eV, corresponding to 1,000 kJ per mole and a wave length of 100 nm. Radiolysis contrasts with photolysis, when the photons have only sufficient energy to produce excitation, but not ionization. Characteristics of the radiation which is mainly used to irradiate polymers are given in Table I.

RADIATION SOURCES

Studies in radiation chemistry were limited while the only sources of radiation were naturally occurring radioactive isotopes. Two developments made high energy radiation available in much larger quantities and enabled radiation chemistry to develop. They were:

1. Nuclear reactors, which have been constructed in increasing numbers since 1943, generate high levels of neutron and gamma radiation. There has been continuing interest in the effects of radiation on the materials of construction of the reactors, including polymer materials, such as electrical insulation, lubricants, adhesives, and elastomers. Large amounts of fission products are produced in the fuel elements, which have been utilized as radiation sources. Specific isotopes, such as ^{137}Cs, can be separated by processing. Radioactive isotopes can also be prepared by irradiating stable isotopes with neutrons in a

TABLE I
Characteristics of Radiation Used in Polymer Science and Technology

Radiation	Source	Energy	Dose rate (Mrad/hr)	Penetration
Electron beams	Accelerators Field emission	20 kV-10 Mv	100-2000	μm-mm
Gamma rays	^{60}Co ^{137}Cs	1 Mv	0.1-10	dm
X-rays	Generator Synchrotron	50-200 kV	0.1	cm

reactor. An example is the conversion of ^{59}Co to ^{60}Co. This isotope is widely used as a radiation source; it produces gamma rays of energy 1.1 and 1.3 MeV with a half life of 5.2 years.

2. X-ray generators were produced for industrial and medical purposes, and electron accelerators were developed for research programs in nuclear physics. These radiation sources were utilized for fundamental research in radiation chemistry and adapted for technological applications. An important use of accelerators has been for the treatment of cancer. Recently, electron accelerators have been used for crosslinking polymer films and polymerization of surface coatings. Electrons have much lower penetration than gamma rays, but the accelerators give much higher dose rates (typically by a factor of 1,000) than ^{60}Co gamma-ray sources, and they are especially suitable for irradiation of film or sheet. A big safety advantage is that they can be switched off.

The main commercial interest in radiation chemistry initially was related to operation of reactors in the nuclear power industry, and to the biological area of cancer therapy. Fundamental research led to the discovery of solvated electrons and the utilization of the radiolysis of water to produce hydroxyl radicals. Radiation can be considered as an especially convenient reagent to produce these chemically reactive intermediates.

RADIATION CHEMISTRY AND POLYMER CHEMISTRY

The effects of high energy radiation on monomers and polymers are much greater than for most chemical compounds. The explanation is the substantial amplification factor. Thus, when a monomer is irradiated, each reactive species produced—free radical, anion, or cation—can initiate a chain reaction in which as many as a million monomer molecules may be converted to polymer. The large effect of radiation on polymers is due to the sensitivity of the material properties to molecular weight. Only a few scissions or crosslinks per polymer molecule are sufficient to change markedly the flexural and tensile strength, elongation to fracture, solubility, melting point, and glass transition temperature, and so on, especially if the molecular weight of the unirradiated polymer is in a critical range.

The chemical use of high energy radiation has increased exponentially over the last 40 years, and a large proportion of the technological applications have been in the area of polymers. The next decade will see the 50th anniversary of modern radiation chemistry and industrial applications will continue to increase.

The future development of radiation chemistry requires an interdisciplinary approach. Polymer chemistry has many special features and future advances are dependent on collaboration between chemists and physicists, material scientists and engineers. However, there is a well developed field of radiation chemistry. This expertise has been neglected over recent years by polymer chemists and much greater interaction is essential.

RADIATION EFFECTS ON POLYMERS

Modification of Properties

When a polymer is exposed to high energy radiation, the material properties of the polymer are affected. An understanding of these changes in properties is important if polymer materials are to be subjected to irradiation either in a deliberate, controlled treatment to modify a particular property, or as a result of their use in a radiation environment, as in nuclear reactors. Some property changes initiated by radiation can be utilized for technological processes. For example, in microlithography the difference in the rate of dissolution of irradiated and unirradiated polymer enables a circuit pattern to be transferred from a computer to a metallic mask or to a silicon chip.

Radiation must be absorbed by the molecules before chemical changes can occur in the polymer. Some of the energy causes a rise in temperature of the polymer and is conducted away to the surroundings. The chemical changes in the polymer molecules produce the changes in material properties. The sequence of events is illustrated in Figure 1.

CHEMICAL CHANGES

The chemical changes which occur in polymers on exposures to high energy radiation can be classified into (1) scission and crosslinking, (2) formation of small molecules, (3) structural changes in the polymer. The properties of polymer materials are highly dependent on molecular weight and so the radiation chemical yields for scission and crosslinking are particularly important.

The yield for any radiation-induced chemical change is expressed as the *G* value. This is defined as the number of events, for example, chain scissions, per 100 eV or 16 aJ of energy absorbed. The radiation dose is measured in gray (Gy) or rad (1 Gy = 100 rad = 1 J/kg).

The evolution of gases and modifications of the molecular structure of the polymer can also markedly change the properties of polymer materials. Morphological changes, particularly loss of crystallinity, are also important al-

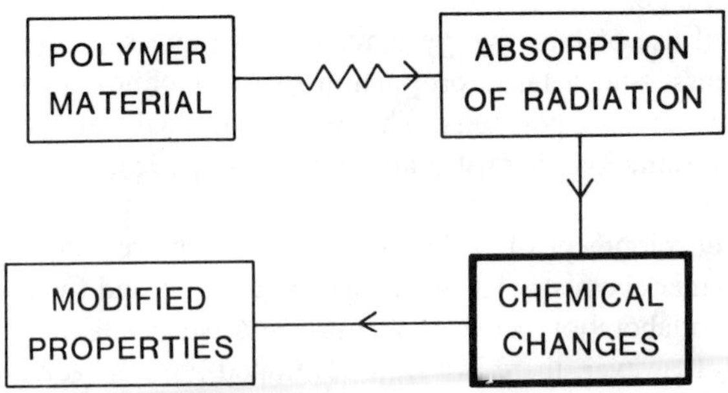

FIG. 1. The sequence of events involved in the modification of the properties of polymer materials by exposure to high energy radiation.

though they usually require large radiation doses. A variety of experimental techniques are utilized to characterize radiation degradation of polymers and some of the most important are listed in Table II.

RADIATION PARAMETERS

The effect of radiation on polymers is determined by a number of parameters, which are listed in Table III. These can be divided into three categories depending on whether they relate to (1) the radiation, (2) the environment, or (3) the polymer. The relationships between each of these parameters and the changes in the polymers need to be determined quantitatively. The parameters are not independent and the total matrix of variables which determines the effect of radiation on polymers is extremely complex.

The most important relationship is between molecular structure and radiation sensitivity. Much has been learned, but a great deal more remains to be discovered and quantified. The objective must be to establish a table of sensitivity factors for chemical groups similar to the chemical shift rules for nuclear magnetic resonance (NMR).

INITIAL CHEMICAL REACTIONS

The characteristic feature of the interaction of high energy radiation with chemical compounds is ionization and this distinguishes it from photolysis. An electron is ejected from the polymer molecule, leaving the parent radical

TABLE II
Chemical Changes Produced in Polymers by High Energy Radiation and Techniques Used for Characterization

1. Chain scission and crosslinking
    - Molecular weight distribution and averages
        - viscometry
        - gel permeation chromatography
        - osmometry
        - light scattering
        - sedimentation velocity
        - sedimentation equilibrium
    - Mechanical properties
        - flexural strength and modulus
        - tensile strength and modulus
    - Soluble fraction
        - extraction and separation
2. Formation of small molecules
    - gas chromatography
    - mass spectrometry
    - high performance liquid chromatography
3. Modification of molecular structure and morphology
    - nuclear magnetic resonance
    - infrared spectroscopy
    - X-ray diffraction

TABLE III
Parameters Which Influence the Nature and Extent of Degradation of Polymers by High Energy Radiation

1. Radiation
    - type
    - energy
    - dose rate
    - dose
2. Environment
    - temperature
    - atmosphere
    - solvent (solutions)
    - additives
    - stress
3. Polymer
    - molecular structure
    - stereochemistry
    - morphology

cation. If the electron is trapped by a chemical group with a high electron affinity, then a radical anion will be formed. These radical ions can undergo unimolecular decompositions and ion-molecule reactions. Neutralization, including recombination of geminate ion pairs ($P^{+\cdot} + e^{-\cdot}$), produces excited molecules, which can decompose by heterolytic bond scission to ions, or homolytically to neutral radicals. Further reactions produce permanent chemical changes in the polymer. Many of the radiolysis reactions in polymers are attributed to the free radical intermediates, which have been identified by ESR spectroscopy. The initial reactions involving the formation of ions and radicals are shown in Figure 2.

RADIATION-SENSITIVE GROUPS

Radiation does not cause random cleavage of chemical bonds in molecules, despite the apparent enormous excess of energy in the radiation compared to the strength of chemical bonds. Thus 1 MeV radiation is equivalent to 10^8 kJ/mol compared with bond energies of 200–600 kJ/mol. However, there is increasing evidence that radiation causes highly specific bond cleavage in molecules. The energy is not initially absorbed preferentially—there are no selective chromophores as in photolysis. The energy must be able to migrate either intramolecularly or intermolecularly. Some chemical groups in the molecule and certain bonds within these groups have a particularly high probability of being broken. Some of these groups are listed in Table IV; various combinations have been used to enhance the radiation sensitivity of polymers.

Permanent scission of the backbone chain in polymer molecules is unfavorable due to the difficulty in separating the fragments formed in the scission reaction. However, some polymers undergo main-chain scission with high G values and the reasons for this are still uncertain.

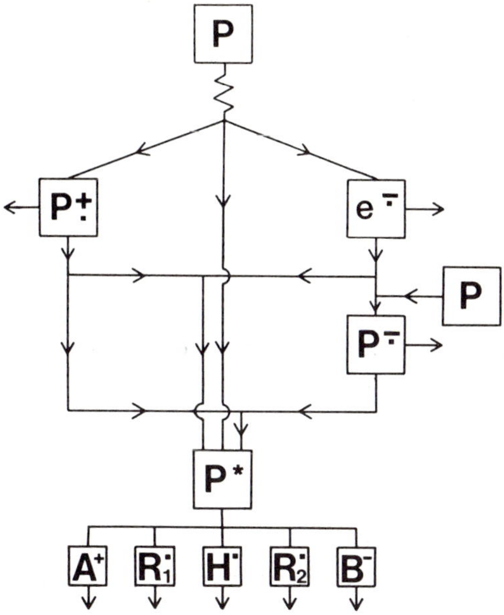

FIG. 2. The initial formation of chemically reactive species in polymer molecules (P) resulting from the absorption of high energy radiation.

TABLE IV
Radiation-Sensitive Groups Which May Influence the Effect of High Energy Radiation on Polymers

—C(R)(R)—	—C—X—C—	
—COOH	$—CH_2—CHR—SO_2—$	
—C—Hal	C=C	$—NH_2$

APPLICATIONS

Some of the technological applications of radiation effects on polymers are listed in Table V. They can be classified into:

1. Treatment of polymers by radiation (e.g., sterilization) in which changes in material properties are incidental, and usually undesirable
2. Treatment of polymers by radiation in order to change their properties (e.g., crosslinking of polymer films)
3. Use of polymers in radiation environments, such as nuclear reactors and space, where deterioration will occur but must be within the constraints of the required service lifetime
4. Polymerization processes, including curing of surface coatings and graft copolymerization

TABLE V
Some of the Technological Applications of Radiation Effects on Polymers

1.	Microelectronics
2.	Crosslinking and branching
3.	Sterilization
4.	Graft copolymerization
5.	Radiation-resistant polymers
6.	Surface coatings
7.	Adhesives

Radiation effects on polymers encompass a wide field of different situations and objectives. In the most general sense, the effect of radiation on polymers is to cause degradation, where degradation is taken to mean any change in the chemical structure of the polymer, and hence eventually in the material properties. Frequently, degradation is used in the sense of deterioration of properties. It may also be applied specifically to a reduction in molecular weight. However, radiation is used to improve some properties (usually at the expense of others). Also, an increase in molecular weight may accompany a decrease in some material properties. Therefore, the more general interpretation of degradation is desirable.

Some of the likely developments during the next decade will be considered in more detail.

MICROLITHOGRAPHY

The utilization of radiation-sensitive polymers in the microlithographic process for the manufacture of silicon chip-integrated circuits is fairly well known. Scission or crosslinking in a polymer film alters the rate of dissolution of the irradiated polymer compared to the unirradiated polymer. The polymer of lower molecular weight may be selectively removed, either in the irradiated areas for positive resists (scission), or in the unirradiated areas for negative resists (crosslinking).

Ultraviolet (UV) visible light in the wavelength range 350–450 nm is used in conventional microlithography and the dimensions of the components are limited by the wavelength of the light, due to diffraction. Electron beams give narrower lines (although scattering now becomes a limiting factor) and circuit patterns with a resolution of less than 1 μm can be written directly from a computer. Electron beam lithography is now in use—especially to make masks and also for direct maskless fabrication of integrated circuits.

The process of electron beam microlithography with a positive polymer resist film is shown in Figure 3. Typical polymers used are poly(1-butene sulfone) and poly(methyl methacrylate). New electron resists with higher radiation sensitivity are being sought and modifications of the chemical structures are being made to enable plasma etching instead of dissolution of the exposed resist. Submicrometer lithography will lead to greater numbers of components on a single chip and the time required to transfer a complete circuit pattern by a traversing

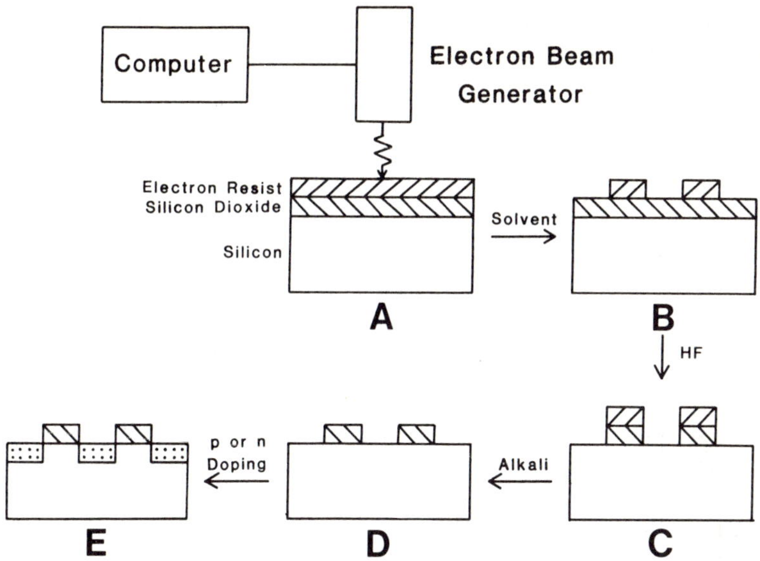

FIG. 3. Schematic diagram of the process of electron beam lithography using a positive resist film.

electron beam then becomes excessive. X-irradiation is being investigated as a possible method combining the short wavelength (and high resolution) of high energy radiation with simultaneous irradiation of entire chips or even of wafers. There are many challenges in this field, including the discovery of sensitive X-ray resists and suitable masks.

Perhaps the greatest developments in integrated circuits will come in the production of sensors on a chip able to measure pressure, temperature, etc., and to analyze chemical composition. Already, simple devices are in mass production for automobiles, and prototypes of more sophisticated sensors are being tested. There are likely to be major advances in the medical field; for example, implanted sensors to control the release of drugs. Collaboration between chemists, pharmacologists, physicists, engineers, and medical scientists can be expected to lead to incorporation of highly sophisticated biological sensors and microprocessors on the same chip.

STERILIZATION

Sterilization is essential in the medical/health field and is an area where the effects of radiation on polymers are of considerable importance. Chemical sterilization is becoming increasingly suspect because of possible toxicity and carcinogenic problems. Heat sterilization has limitations for packaged items and polymer materials. Also, many plastics suffer significant changes in shape and properties during thermal sterilization. Radiation is almost certain to become the main method of sterilization. However, the relatively low radiation dose of 2.5 Mrad required for sterilization can, surprisingly, cause significant changes in the properties of some polymer materials.

There are two approaches to this problem: (1) material selection (requiring knowledge of the effects of radiation on polymers) and (2) control of the properties of the irradiated polymer by additives, thermal treatment, and so on. For example, it is difficult to sterilize unsaturated elastomers by radiation because of the radiation sensitivity of the C=C bond. PTFE would be unsuitable because it is highly radiation sensitive—although extremely resistant to heat and to chemicals. There are also difficulties in the sterilization of polymers in aqueous solution—which may become an important field in the development of targetted drugs.

RADIATION-RESISTANT POLYMERS

Polymer materials are required for use in radiation environments, such as nuclear power plants. Another very interesting radiation environment is space. For example, satellites in geosynchronous orbit at 36,000 km above the earth's surface are situated in the 2nd Van Allen radiation belt and receive large doses of electron and proton irradiation.

It is interesting to consider the molecular features which characterize high radiation resistance. Radiation resistance is provided, primarily, by aromatic structures; for example in polystyrene, polysulfones, and polyimides. The radiation resistance is mainly due to the dissipation of energy in the aromatic rings without chemical change. There may be other molecular or physical structures which will provide radiation resistance, but little has been reported.

There is usually a marked difference between the resistance of polymers to irradiation in vacuum (or an inert atmosphere such as nitrogen) and in air. Radiation-induced oxidative degradation proceeds readily even in aromatic polymers and the material properties deteriorate accordingly.

The objective in selecting radiation-resistant polymers is to achieve adequate service lifetime. Carbon-fiber-reinforced composites are being used for making structures for use in space. Composites are favored because of their high strength to weight ratio and their dimensional stability. Epoxy resins are currently used for the matrix as a natural extension of the composite technology of the aerospace industry. However, a limitation is that a minimum of volatiles must be produced in the high vacuum of space. Also, fabrication of epoxy composites is labor-intensive because they are thermosetting resins. The next generation of carbon-fiber-reinforced plastic composite structures for use in space will undoubtedly be made with thermoplastic resins. However, the thermoplastics must withstand severe temperature cycling. The properties of high temperature engineering plastics, such as poly(arylene sulfone)s, polyetherketones, poly(phenylene sulfide), and polyimides, are currently being investigated as matrices for composites.

SURFACE COATINGS

Surface coatings are traditionally prepared by polymerization of monomer or evaporation of solvent from a polymer solution. Large quantities of solvent are involved in both processes, creating environmental problems—and also limiting the rate of formation of the coating. There has been increasing interest in radiation curing of oligomers with both UV and high energy radiation.

Irradiation of monomers results in polymerization—by free radical, cationic, or anionic chain growth mechanisms, depending on the monomer and the conditions of irradiation. Radiation-initiated polymerization is industrially attractive because of the high amplification factor—one radiation-created radical or ion can result in the polymerization of perhaps 10,000 monomer molecules.

Radiation curing has developed rapidly in recent years, and will continue to be a growth area. High conversions are obtained at high polymerization rates. Solvent-free systems are used to reduce toxicity problems, energy requirements, and the cost of recovery of the solvent. Oligomers, rather than monomers, enable higher initial viscosities and smaller volume contractions. UV initiation is widely used, but is inhibited by oxygen and inert gas environments do not work well in industry. Electron beam irradiation can give very high polymerization rates, even in air. An advantage of radiation over UV is that it works equally well for opaque polymers, such as with carbon black and other pigments.

There is increasing interest in the polymerization of Langmuir-Blodgett layers of oriented polymerizable groups (i.e., double bonds) in solid surface coatings. This is closely related to the research of the 1960s into solid-state polymerization. These layers are particularly amenable to radiation-initiated polymerization.

Perhaps the greatest devolopments will be in the modification of the properties of polymer surfaces, especially of thermoplastics, by irradiation, which can greatly enhance strength and temperature resistance by crosslinking. However, an outstanding problem for commercial application is the need to minimize effects of oxygen.

ADHESIVES

Adhesives can be considered as a special type of surface coating. Normally an adhesive is either a monomer, which can be polymerized by a step-growth or a chain-growth mechanism (examples are epoxies and cyanoacrylates, respectively), or a polymer dissolved in a solvent which must be evaporated.

There is an increasing demand for high performance adhesives with the characteristics shown in Table VI. These requirements can probably all be achieved by utilizing radiation-initiated curing of suitably formulated adhesives. Extremely fast setting of stable adhesives is possible. Radiation-initiated crosslinking of thermoplastic adhesives may also be an important development in the future.

One of the advantages of high energy radiation compared with UV is that it can penetrate an opaque material to initiate a reaction below an overlaying material. Radiation curing also avoids the problem of thermal degradation of the material which can occur during hot curing of adhesives. It is also useful when the substrates are nonabsorbent and solvent cannot escape.

Electron beam energies can be selected which result in enhanced energy absorption at a particular depth in the substrate, as shown in Figure 4. New

TABLE VI
Desirable Characteristics of High-Performance Adhesives

Elimination of:
evolution of volatiles
toxic solvents and monomers
Resistance to:
sustained loads
high temperatures
high humidity

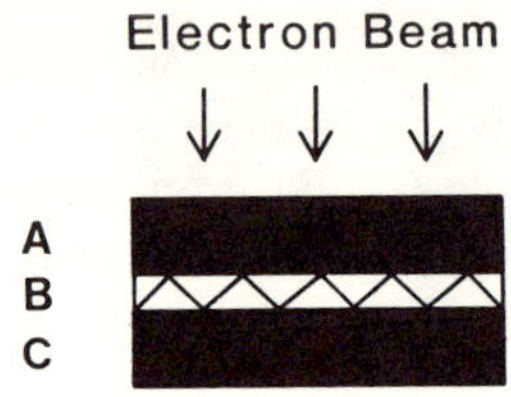

FIG. 4. Utilization of the penetration of high energy radiation to cure an adhesive. A, C = opaque substrates, B = radiation-sensitive adhesive.

monomers and oligomers can be designed and synthesized with high absorption coefficients for electrons.

Another advantage of radiation-cured adhesives is that a stronger "bond" to the substrate is produced. This may involve crosslinking (covalent bonding) between the adhesive and the substrate or "setting" of the adhesive into a mechanical lock with the surface of the substrate.

NEW MATERIALS

Irradiated polymers can exhibit properties that are significantly different from those that are obtained by treatment with chemical reagents. They are in effect new materials. Much research is required to provide an understanding of the unique properties resulting from irradiation. Areas where our basic scientific knowledge is inadequate include:

1. Crosslinked polymers: the crosslinks may be H links or Y links; they may be clustered or randomly distributed: they may occur in amorphous regions, in defects in crystalline regions, or at amorphous/crystalline boundaries.

2. Branched polymers: the frequency and distribution of branches of different lengths have significant effects on the material properties, but characterization of the branches and correlation with properties is needed.

3. Grafted copolymers: graft copolymerization of a monomer onto a base polymer may result in a true (T) graft or chain scission and formation of a block copolymer. The base polymer may be linear or crosslinked (before or after irradiation). Uses for graft copolymers include viscosity modifiers, ion-exchange resins, enhanced oil recovery, and reactive polymers (homogeneous and heterogeneous).

Radiation is not the only way to prepare these polymers, but it frequently produces polymers with new properties.

Plasma polymerization and plasma effects on polymers have great potential for future technology. Plasmas modify the chemical structures and properties of the surfaces of polymers. Crosslinking and dehydrogenation are important processes, but it is difficult to characterize these materials.

INSTRUMENTAL ANALYSIS

Advances in fundamental knowledge and technological utilization of radiation effects on polymers will depend on increasing utilization of highly sophisticated analytical instrumentation. It has been said that the advancement of science is determined by the discovery of new instruments. We can anticipate remarkable advances in the characterization of polymers during the next decade. Advanced instrumentation which is already available includes (1) XPS (ESCA), SIMS, etc., (2) FTIR, (3) NMR, including solid-state and two-dimensional NMR, (4) gas chromatography/mass spectrometry (GC/MS), (5) high-performance liquid chromatography (HPLC), (6) ESR. There will be great advances in these techniques during the next decade. A variety of other techniques are currently crossing the boundary from research specialization to routine analysis.

CONCLUSIONS

The next decade of polymer science will be particularly exciting. On one hand, a wide variety of new polymers and new processes will become commercial; on the other hand, most current commodity polymers will be chemically modified to improve their properties for particular applications. High energy radiation will be used increasingly to achieve both of these objectives.

The author's research into the radiation chemistry of polymers has been supported by the Australian Research Grants Scheme, the Australian Institute of Nuclear Science and Engineering, the Australian Atomic Energy Commission, and NASA Langley, and stimulated by many people, all of whom are gratefully acknowledged.

James M. O'Donnell was born on September 3, 1934.

PRESENT POSITION

Reader in Physical Chemistry
Department of Chemistry
University of Queensland
Brisbane, Queensland 4067
Australia

EDUCATION

University of Western Australia; B.Sc. (Hons.), 1955.
University of Leeds, England; Ph.D., 1963.
Polytechnic Institute of Brooklyn; Postdoctoral Fellow, 1963.

POSITIONS

ICI Australia, Melbourne
Factory Development Chemist 1955–1956,
Research Chemist 1957–1958.
University of Queensland
Lecturer-Reader, 1964–.

Visiting Positions:
University of Manchester, UK, 1971.
University of Leeds, UK, 1978.
University of Lancaster, UK, 1981.
Virginia Polytechnic Institute and State University, USA, 1981.

PROFESSIONAL ACTIVITIES

Royal Australian Chemical Institute Vice-President, 1985.
H. G. Smith Medallist, 1983.

Polymer Division
Chairman, 1972–1973.
Battaerd-Jordan Medallist, 1982.
Royal Society of Chemistry, UK
Australian Lecturer, 1984.

POLYMER SCIENCE AND MEMBRANE BIOLOGY

HELMUT RINGSDORF
University of Mainz, D-6500 Mainz, Federal Republic of Germany

Polymer chemists as poachers in foreign grounds? Why not? It even will be one of the needs for Polymer Science in the next decade. Macromolecular chemistry has become a mature science with all the advantages and handicaps of well-established scientific disciplines: many heights are conquered and the harvest is abundant, but adventures and the future might be elsewhere. Where? Future polymer chemistry cannot be limited to repetitive improvement of already successful mass polymer but rather has to expand into neighboring fields of material science and especially life science.

The architecture and topography of biological membranes is still not matched by any synthetic model. Can polymer chemists really contribute to the understanding or even mimicking of cell membrane functions and cell-cell-interactions? During the last years, polymeric liposomes have shown to combine high stability with the orientation of phospholipids in biomembranes. In contrast to the more labile vesicles from natural lipids they offer potential for studying additional aspects in biomembrane research. Thus, hole formation in partially polymerized and phase separated liposomes or "corked liposomes" was possible. Also the patching and capping reaction occurring in biomembranes e.g. during recognition processes could be mimicked by using vesicles consisting of polymerizable lipids and natural lecithin.

Sugars play an important role as hydrophilic headgroups in cell membranes and are involved in such processes as cell adhesion, cell recognition, immunity and malignant transformation. The simulation of recognition processes with polymeric liposomes carrying sugar headgroups has already been mimicked using concanavalin A. It is important to note that those vesicles are not destroyed during the recognition and precipitation process. Functionalized polymeric liposomes of the next generation thus offer potential e.g. in the field of cell-liposomes adhesion, chemical signaling between cells and especially to study the cellular basis of immunity.

Whereas the lipid bilayer determines the basic functions of biological membranes, proteins are responsible for most of the specific membrane activities, serving as receptors, enzymes or carriers, etc. Can they be inserted into longterm stable polymeric membranes without loss of activity? They can! ATP-synthetase was already used and stayed active. In addition the chromoprotein bacteriorhodopsin has been incorporated into liposomes of fully synthetic, polymerizable lipids. These light-driven polymer embedded proton pumps show the expected long-term stability over months. Jointly inserted little ATP-factories on a cellular level will be available. The field is open: the information on specific membrane proteins and their different functions is abundant. Thus, the incorporation of assemblies exitable by specific chemical or physical stimuli into spherical and planar organized systems may lead to a new generation of polymers useful for energy storage problems, as biosensors and as specific and switchable separation membranes.

First hesitant steps to bridge the gap between polymer science and membrane biology have been made. We are only limited by our imagination and may be even more so by our fear not to be able to verify the new concepts. It is much easier to be sceptical than to be creative. Macromolecular chemistry is still far from being able to provide ultimate solutions for specific problems in membrane biology or life science in general. However in "going for it" membrane biologists and polymer scientists can only benefit from the many chances to learn. The key attitudes to enjoy this adventure are courage to enter neighboring fields and, especially, the willingness to cooperate.

MACROMOLECULES AND EVOLUTION*

FREDERICK R. EIRICH
Polytechnic University, Brooklyn, New York 11201

SYNOPSIS

Macromolecular chemistry, physics, and technology and their major offsprings, molecular biology and bio- and genetic engineering, play ever increasing roles in our daily lives, views, and plans, materially and intellectually. To understand what this means and looking ahead to what might yet come, it might be best to view what happened, and happens, as part of our planet's ongoing evolution which began some 5 billion years ago and has about as many years to continue. It had its slower and faster phases, during all of which macromolecules played a large role.

In the past, macromolecules turned themselves into microbes and evolved into Man. Now Man makes macromolecules faster and more varied than they could have otherwise arisen on this planet. Re-phrasing the question, whether nucleic acids make individuals so that individuals make nucleic acids, or vice versa, one might ask whether Man is just a means for macromolecules to boost their own development.

The first inorganic, macromolecules appeared on this Planet at the time of rock formation. As the temperature dropped and liquid water appeared, water soluble carbon compounds evolved prebiotically, among them the building blocks of life: oligomeric, or low polymeric condensates of carbohydrates, phosphates and sulfur compounds, of amino acids and nucleotides, as well as oligomeric hydrocarbons and their derivatives. Polymers in mixed solution tend to form coacervates, or will aggregate to clusters. From there we theorize that temperature, humidity and drying cycles, plus intense radiation and energy from the heterotrophic degradation of other organics, fostered intensified cyclic, catalytic and auto-catalytic, processes, among them the syntheses of a large variety of peptides and nucleotides which mutually modified their functions and more than replenished what the aggregates lost by degradation. Eventually, one cluster reached steady state and an identity becoming surrounded by a semipermeable mixed peptide and phospholipid membrane, to become the ancestor of life. By further multiplication and diversification, the beginnings of a biosphere were created, a development which deeply affected the evolution of our planet's outer shell, and thereby again further life. Thus, by allowing the blueprint of chemistry inherent in the atomic properties to become realized up to macromolecular dimensions, and then evolve into selfsustaining systems, our Earth entered the life-sustaining phase of planetary and biomacromolecular evolution.

The polymers of the systems that clicked into life were surely much simpler than today's constituents, but it needed macromolecular complexity and versatility to fulfill the many exacting structural and conformational requirements of catalysis and regulation on which the living process is based. Dr. Mark, early on, pointed out the unimaginably large number of different molecules existing in a batch of any polymer, counting those due to MW distribution and isomers, tactic, optical, branched and/or head-tail, even for homo-, and more so for copolymers. In fact, it is unlikely that in all of polymer manufacture ever two exactly alike molecules were produced. Nor had 2 individuals, from the first cells to Man, the same set of macromolecules (except perhaps for cloned individuals which do not contribute the evolutionary changes that characterize life). Biological

*One observer's attempt to see Evolution in terms of his understanding of polymer and colloid science. Apologies are made for the extrapolations into biology, neurology, and information science.

evolution, with its short life spans for individuals that entails generational chemical changes serving the planetary chapter of cosmic evolution, requires the syntheses of huge numbers of slightly different macromolecules of the same general structure and function, a requirement that is readily delivered by macromolecules.

The occurrance of life in all its improbability has prompted mathematicians and physicists to treat it as the result of fluctuations or bifurcations within fluxes from chemical energy to heat, creating sustained dissipative structures and chemical cycles. For the chemist, it seems more appropriate to consider the occurrance of life as akin to the ignition of reactive material, in this case within an in itself improbable, but possible, macromolecular aggregation. This point of change in regime, where the autocatalytic process started to branch, was soon followed by another incident which allowed photosynthesis to produce the compounds whose degradation supplied the energy of the living processes.

Further development of a selfmaintaining clusters must then have changed from chemical evolution, based on innate molecular structures (structural information), to Darwinian selection through adaptation to conditions of need and supply, i.e. changes in nutrients, oxygen, or salinity, and to competition. Those who become responsive to these signals and developed locomotion, completed the acquisition of the basic requirements for living.

The family trees of proteins have been followed for most of the living families by amino acid sequencing of "old" proteins, like cytochromes and ferredoxins. One finds that, for all their individual changes with time (a "molecular clock"), these proteins have conserved their essential composition and structure in all the kingdoms, confirming by their internal relation the all-embracing brotherhood of life. The circulatory systems of transportation and chemical messages, and the nervous systems of receiving and responding to outer and inner stimuli, also remained essentially the same since the formation of multicellular organisms that allowed a division of labor and specialization of functions. In particular, the mode of stimuli transport, that of a tangential soliton wave of membrane charge has remained essentially the same and is a most remarkable cooperation between millions of Na-K-ATP-ase molecules embedded in the phopholipid bilayer of the nerve axons.

What did change in the further course was the emphasis on changes and differentiation of higher functions. We rely here on findings on today's living creatures and we base conclusions on the hypotheses that today's hierarchy reflects the history of evolution. The development, it is believed worked its way from the perception of stimuli and responses, whose record represents what we call memory. Memory was a revolutionary form of information. In contrast to structural information, it is the impression of a non-random array (template) by an external event onto a macromolecular system that, up to then, served only metabolic functions, whereas the new impression serves as a guide for recall. The whole process may be thought to be akin to, e.g., the imprints of patterns of light and dark on the photographic plate, or the newly imprinted charge distributions electrically inscribed on semiconductors. Just as the nearly infinite redundancy of energetically and conformationally interconvertible states, as well the size and multifunctionality of macromolecules are the key to their performances, so are the nearly infinite *p-n* states of semiconductors the source of their performances and memory capability.

The availability of a memory had profound effects. In response to a new stimulus of a given kind, memory permits not only a much faster and more efficient response, but also a welter of combinations with earlier related responses. The acquisition of further memory, and of the memory of past utilization, constitutes the process of learning and makes choices available, a new element in the struggle for survival. The more adroit the choice, the more closely based on memories that reflected the requirements of the outer world, that is, the more it corresponded to the "truth", the greater the reward of these acts of "intelligence".

The next step permitted by macromolecular redundancy was the development of model responses. Instead of reacting by activating the motor nerves, retracings of memories, that is of images of actions, were evoked, as well as images of what (from memory-experience-learning) the consequences might be, followed, in the next loop, by a reviewing the more distant consequences. The result of this "thinking" process could then be compared with the results of motor action.

The practically unlimited scope of memory and thinking, based on the macromolecular system, enabled the organism to observe over and over again the sequence of cause and effect and, thus, to direct thinking into verifiable tracks. These included the development of tools, of thinking in terms of models for the reduction of trial and error approaches, and of forms of systematic thinking such as logic. A parallel development must have been the assignment of specific sounds to specific objects, actions, and happenings, i.e. the evolution of language. The correspondingly more accurate communication allowed a quantum leap in learning, as experiences were not only transmitted, but could be explained and coded. From this beginning technology was born, which extended Man's reach and grasp by many powers of ten.

A particular consequence of the development of technology was the growing recognition of the nature of macromolecular substances with which Man was familiar from his farthest past, i.e. fibres, wood, hair, hides, horns, clays, etc. This was followed by the synthesis and study of the new organic polymers which led to the full interpretation of macromolecular nature, and to the understanding of their scientific and technological potential.

The consequences of forming such concepts as: rubber extension and contraction, viscoelastic memory, the glassy state, the limits of material strength, large unsaturated, conjugated, and polyelectrolytic linear molecules, cross linking, copolymers, large rings and helixes, interactions of ternary and quartenary molecular structures, nematic fluids, viscosity modification, plasticization, the principles of heat and shock resistance, the intricacies of composites, etc., are yet far from being fully exploited. It is certain that the next decade will see great progress in our exploiting material behavior and taylor making of new materials, based on our molecular interpretation of microscopic and macroscopic observations.

In addition, macromolecular science has been abundently fertile in advancing molecular biology and paleontology, bio- and enzyme engineering, monoclonal antibody and gene technology, and our fledgling attempts to reconstruct the beginnings of life on Earth.

In an important departure, reshaping our lives and thinking in yet another way, our biological minds opened new intellectual territory by advancing solid state chemistry and physics, electronics, and computer technology. True, whatever efficiency and kind of activity we can build into our computers and robots reflects the ways of our thinking. Electrical or optical impulses are imprinted on the structure of the semiconductor chip in a digital or analog manner, modelled in accordance with our logic of language and counting, but conversely, the immense new potential of numerical and functional computations expands our horizons. The advantage of an inorganic brain, if ever independent of macromolecular organisms, would be a greater tolerance to the harsh conditions of the Universe outside the selfcreated, very special aquatic environment and the protective umbrella of the biosphere. With our ways of thinking geared to the manipulation of outer objects (tool making) and, until recently unaware of the so much more subtle and complicated machinery of perception and action developed by the living world, and further because of the vulnerability of our waterborn macromolecules, it was probably more than a historical accident that we used metals, semiconductors, and the corresponding network theory for designing our calculating and computer devices. This initiated a leap in the development of semiconductor technology on which our present efforts at building intelligent or thinking machines are based. It is too early to say whether this development will continue in the forseeable future, i.e., whether inorganic devices will ever develop their own mode of thinking and reach the sensitivity, variability and effectiveness of the macromolecular apparatus. It is on the other hand possible that, for instance, advances in ceramics will one day permit the development of different types of switching devices and networks. Another possibility is the development of molecular switches and mixed ionic and electronic conductance systems based on easily shifting charged states, or ion bindings, of conjugated and aromatic molecules. These are preferably incorporated in or attached to non-biological macromolecules. Such systems developed under the name of molecular computers, come much closer to the proteinic ones used by nature, and are much more robust in their chemistry than living macromolecules. Lastly, as we learn more and more about our own mechanisms, we may be able to advance these so as to become less dependent on the stringent terrestrial conditions for the survival of the present organisms, and better adapted to living beyond the biosphere.

Whether macromolecules or semiconductors will eventually overcome their present limits, the next decades will certainly see the development of spaceships with robotic instrumentation and guidance. Nothing would then stand in the way of building larger and more efficient spaceships, and eventually create small intelligent planetoids, an analog to L. Thomas' image of our Earth as a living supercell, or Bugliarello's superintelligence of globally connected computers. Eventually, it should be possible to collect matter from space and to assemble cosmic bodies capable of understanding and intent, which could then begin to re-order and re-shape the Cosmos.

Macromolecular chemistry, physics, and technology and their major offsprings, molecular biology and genetic and bioengineering, play ever-increasing roles in our daily lives, views, and plans, materially and intellectually. For a better understanding of what this means and to look ahead to what may yet come, the best course would seem to be to review what has happened, and is happening, as part of our planet's ongoing evolution which began some 4.5 billion years ago and has about as many years to continue. It has had its slower and faster evolutionary phases, during all of which macromolecules played a large role.

In the past, congregated macromolecules converted themselves into microbes and evolved into the human species. Now Man makes macromolecules faster and more varied than they could otherwise have arisen on this planet. Rephrasing the question [1], whether "selfish" nucleic acids make phenotypes so that phenotypes make nucleic acids, or vice versa, one might ask whether Man is just an instrument of, and for, macromolecules to boost their further development.

The first, inorganic macromolecules which appeared on this planet probably were silicates. As the temperature dropped and liquid water appeared, water-soluble carbon compounds evolved prebiotically from the gases of the primitive, weakly reducing atmosphere of CH_4, CO_2, NH_3, $(CN)_2$ and H_2O, plus dissolved salts and P and S compounds, acted upon by abundant sources of solar and vulcanic energy [2, 3]. Among the building blocks of life that appeared were oligomeric, or low polymeric condensates of carbohydrates, phosphates and sulfur compounds, amino acids and nucleotides, as well as oligomeric hydrocarbons and their derivatives. Laboratory experiments with model reactions, under conditions presumed plausible according to geological records, have indeed produced most of these compounds in their monomeric or low polymeric forms [2]. We may assume that all types of molecules necessary for the emergence of life, including nucleic acids, polypeptides capable of catalytic activity, phospholipids, and so on were available on a prebiotic basis [4, 5].

Polymers in mixed solutions tend to cluster, forming interpenetrating aggregates [Fig. 1(a)], coacervates, or micelles. We theorize that temperature, humidity and drying periods, intense radiation, and chemical energy from the degradation of other organics, fostered cyclic homo- and heteroautocatalytic processes within clusters dispersed in solution, or adsorbed on surfaces of suspended matter. This provided a basis for the syntheses of a large variety of polypeptides and polynucleotides and their interactions which further modified and developed their functions and roles [5]. We know the outcome of these interactions, but what went before is the first of several crucial unanswered questions.

The first concerns any indigenous ordering and complexing tendencies during the formation of macromolecules. Indeed, one of the basic features of co-

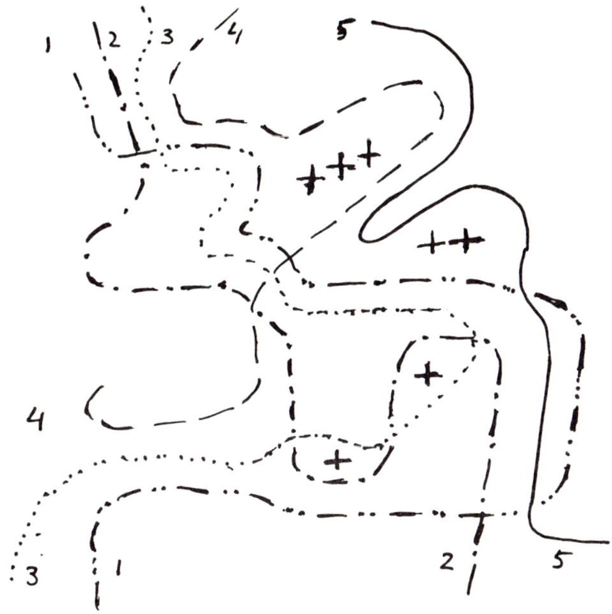

(a)

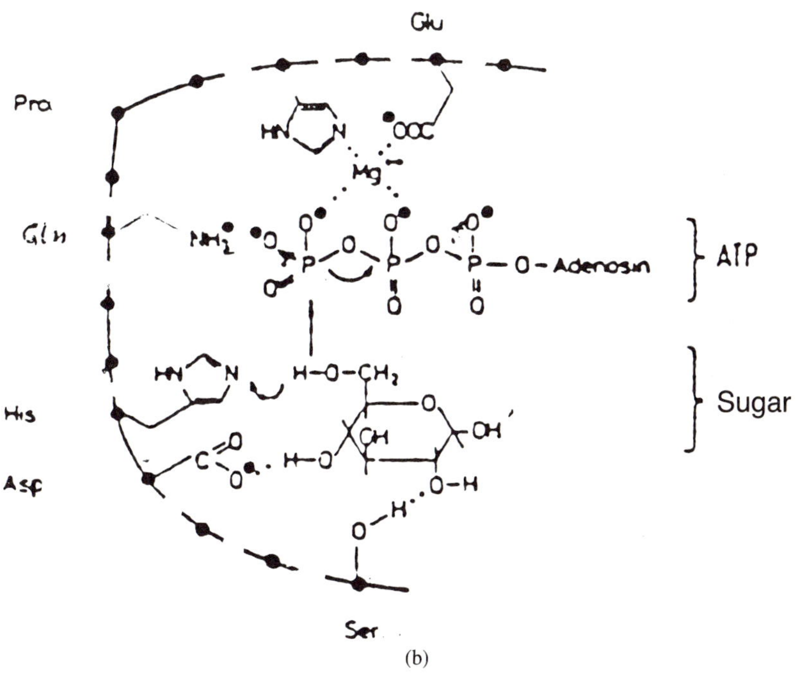

(b)

FIG. 1.(a) Two-dimensional schematic presentation of a cluster of five macromolecules that form pockets of potential catalytic activity. Sites for collecting substrates. See also Fig. 5(a). (b) A specific example of one such a complex, forming a pocket on polymerization that might help the phosphorylation of glucose by holding the two reagents in an advantageous reaction position [after Follmann, ref. 4]. The processes of collection, complex forming and polymerization would be enhanced, e.g. by clay.

polymerization is an ordering due to specificity of monomer addition which leads to averages of nonrandom compositions and sequences. Extensively studied for free radical, ionic, and ring-opening copolymerization, the addition process is affected by the substituent group as it determines enthalpies and reaction rates. In co-condensation involving acids, addition rates and orders are mostly dependent on the pK of the COOH group and all factors influencing it. During the copolymerization of RNA no ordering has so far been detected for individual short oligomers, though Takemoto reports some results during the copolymerization of nucleic-base substituted olefins [10j]. See also Challa [10k]. There is, of course, the well known strong ordering by base pairing, through which nucleotide chains act as templates.

During amino acid co-condensation, direct intersubstituent effects on the course of the addition are also weak [9c], while some selectivity via the substituent influence on the pK is observed. The ordering becomes stronger when the oligomers reach lengths that allow helix or β-sheet formation. As for specific results, evidence of formation of preferred amino acid sequences during the synthesis of oligo- and polypeptides has been observed [3b, 9a]. That is, among the countless numbers of possible amino acid sequences (see Fig. 4 below) among which the accidental formation of catalytically useful species would be vanishingly small, striking restrictions to a small number of peptides of some regularity have been seen during clay-enhanced aqueous polymerization. Thermal polymerization, within a helter-skelter mixture produces polypeptides of near biological bulk composition, some even showing catalytic activity [3b]. Thus, an innate self-alignment among amino acids joining to form polypeptides may have well shaped the first life-promoting macromolecules.

In general, macromolecules of different structures of low polarity, forced into a joint solution, will separate into individual phases as a result of the large space requirement for unhindered thermal motion and a low energy of mixing. On the other hand, charged or strongly polar macromolecules form intermolecular complexes by ion or dipole pairing. Though processes of this nature are of great technological importance, there has been little systematic work on well characterized biomacromolecules for the sake of studying the molecular mechanics. Polypeptides, unless negatively charged, may attach themselves to the polynucleotides, or vice versa. The associations thus created are extremely difficult to characterize but offer many constellations of potential catalytic activity. During biological evolution such aggregates led to the formation of ribosomes and mitochondria located in the cytoplasm, and to the chromatin fibers [5c].

The more readily ascertainable genetic template action of nucleic acids has been studied intensively. Probing into the evolution of RNA with the help of biological enzymes, it was indeed confirmed that those oligonucleotides that paired more extensively with the complementary mono- or oligonucleotides partners, and thus produced more templates for further reproductions, were selectively favored. Spiegelman [8], moreover, showed that even after removal of the original viral RNA while retaining the enzyme, the progeny evolved rapidly in vitro and could even adapt to disadvantageous conditions. Perhaps, notwithstanding their generally poor catalytic properties, polynucleotides may have evolved a genetically useful code on their own or with the help of simple peptides. The above experiments, though showing the need for polymerases, could have been interpreted equally well as pointing to a polypeptide evolution that preceded that of the nucleotides, a conclusion less widely held because evolution of polypeptides on their own toward improved polymerases has not yet been accomplished, and because polypeptides do not determine nucleic acid sequences in living systems.

We have only speculations on the origin of today's genetic code [6, 7]. The mechanism of its formation would tell us much about the transition from inanimate to living matter. DNA probably did not precede RNA but is itself a product of advancing evolution. We know even less about the origin of the earliest RNAs. The most likely mechanism should have been a coevolution of polypeptides and polynucleotides, but while thought of, modelling has not yet been undertaken. Lacey et al. [7] proposed a mutually reinforcing cyclical self-synthesis of polypeptides and nucleotides from their acylates (see Fig. 2). The proposed mechanism is based on the observed affinity between amino acids and their anticodonic ribonucleotides, which may lead to the formation of the anticodonic amino acid-nucleotide acylates as "monomers." When the two components undergo their individual polymerization, homogeneously or, better, on supporting surfaces such as clays [9], the order in which the amino acids enter the growing peptide chains should affect that of the nucleotides during the synthesis of their chains. The latter may then assemble their supplementary nucleotides which, when they polymerize, form strands capable of playing the role of a proto-m-RNA, that is, induce the replication of the primary RNA and its polypeptide (see Fig. 2). This reaction model, if it could be realized, could explain the origin of the genetic code and, based on amino acid loop-substrate affinities (Fig. 1b), possibly also the enhanced synthesis of enzymatically effective polypeptides. Syntheses of this kind, of a limited number of chains, could have occurred within a small niche or cranny [8], namely, within a microsystem [7], which could have helped the start of small, growing, self-synthesizing units. Eventually, some of the resulting cluster systems were able to replenish themselves faster than the rate of loss by chain degradation until they reached a state of maintaining and reproducing themselves through mechanisms which we may, or may never, be able to recreate. When an enveloping skin was acquired, consisting of a semipermeable membrane of peptides and phospholipids, by another mixed aggregation process [4, 13], such units became the ancestors of all life forms. From what we have learned, then, through polymer science and molecular biology, our Earth entered the life-bearing phase of planetary and biomacromolecular evolution [2, 4, 5] as soon as conditions permitted the blueprint of chemistry, inherent in the atomic properties, to become realized up to macromolecular dimension and then to evolve into self-sustaining systems. The subsequent development of a biosphere deeply affected the future of our planet's outer shell [4].

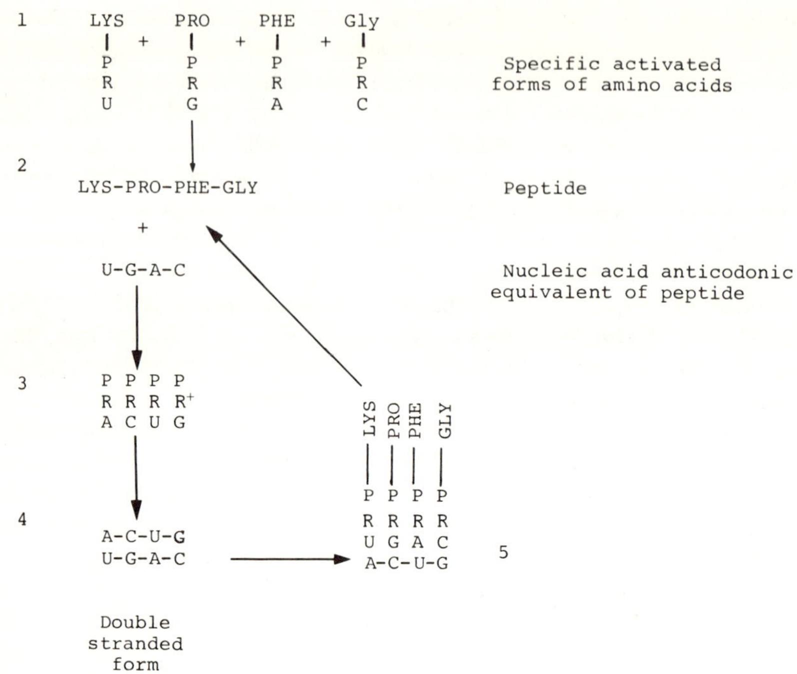

FIG. 2. Schematic of processes whereby a system could generate information *de novo* as a result of intermolecular affinities. (1) Amino acids are activated by reaction with specific nucleotides, e.g., ATP activates all hydrophobic amino acids, GTP activates amino acids of intermediate hydrophobicity, CTP activates all hydrophilic amino acids, and UTP activates charged hydrophilic amino acids. The proposed activated form is the aminoacyl nucleotidyl diphosphate anhydride in which the carboxyl group of the amino acid is attached to the terminal phosphate of the nucleotide. (2) The amino acids polymerize with each other yielding a peptide, while the nucleotides polymerize with each other, yielding a nucleic acid which is anticodonically equivalent to the peptide in a one-to-one relationship. (3) The nucleic acid portion can serve as a template for the synthesis of a complementary strand of nucleic acid which bears the same relationship to the peptide that a contemporary messenger RNA bears to its protein. (4) A double-stranded nucleic acid results from reaction no. 3. Disruption of the double strand gives two strands either of which can serve as a template to synthesize more double strands. (5) Also, the upper strand can now serve in a messenger capacity associating with specifically activated amino acids to generate the coded synthesis of more of the original peptide, as in reaction no. 2. Whereas the synthesis of the original peptide in step 2 produced a peptide simply by probability, natural selection processes allowed the system containing that peptide and its nucleic acid complement to persist. This persistence then allowed the other steps to take place, yielding more of all components of the system. NOTE: In line with RP, A, C, G, U stand for the bases; when connected, A–C–U–G stand for RNA's.

The emergence of life may be viewed as a unique case of self-ordering of matter [10]. Though the entropic tendency of our universe toward chaos has received much publicity, we are equally well acquainted with phenomena of the opposite tendency: crystallization, epitaxis, and phase separations that lead to uniform or ordered static states. Examples of dynamic functional (spatial or temporal) ordering extend from gravitational, cosmic phenomena to terrestrial tornados and eddies, the Benard phenomenon, and shock-wave fronts. Self-ordering chemical systems include reactions in structured fluxes such as the oxidation chains of steady flames, syntheses in streaming systems, the Zhabotinsky-Belusov reaction [10g], the emulsion polymerization in a monomer dispersion after reaching the critical micelle concentration of the emulsifier, interfacial polymerizations, or the polycondensation of activated amino acids on becoming adsorbed on bentonites in neutral dispersion [9].

These phenomena of self-ordering in nonequilibrium systems, particularly because of their relation to the emergence of life, have prompted theoretical treatments by mathematicians and physicists [10]. Self-ordering processes in open systems are postulated to be chance consequences of fluctuations within fluxes of chemical energy which grow to instabilities that lead to bifurcations of previously smooth functions. One of these branches may permit the emergence of sustained dissipative structures and chemical cycles [10c, d]. In a different context, it has been shown that reagents ("morphogens") within a mobile, ring-shaped reactive matrix may, by fluctuations of diffusion, produce geometric growth patterns and shapes that are found in animal or plant bodies [11].

For chemists it seems more appropriate to consider the occurrence of life as a special instance of self-ignition of cyclic chain reactions within self-ordering molecular aggregates of the micellar or coacervate type, which happened also to imbibe energy-supplying reagents. One could, for example, imagine that clusters of polypeptides and nucleotides, meeting through Brownian motion, became tied in matted structures as a result of hydrophobic and ionic bonds. Asymmetric molecular alignments, as they prevail at most interfaces, favored undirectional diffusion of nutrients like polyphosphates, glucose phosphate, amino acids, adenosine di-phosphate (ADP), and other RNAs into cluster pockets where they underwent reactions [Fig. 1(b)]. The sugars were anaerobically decomposed with the help of the polypeptides, yielding adenosine triphosphate (ATP), lactic acid, and water. The latter diffused out, away from the in-flow of nutrients. The ATP was used to form more peptide-nucleotide acylates and polymers. A first primitive coupling between activated diffusion and directional energy flow would have continued to drive the process of cluster maintenance and chain growth. Replication errors within this process then led to the accidental formation of more effective poly-peptides and nucleotides, thus initiating an evolutionary spiral.

The nature of a plausible mechanism of RNA evolution poses the next principal question. A major effort of its elucidation has been undertaken by Eigen [12] and co-workers, who studied the cycle of molecular variations and eliminations that lead to sharply selective survival, supported by an experimental replication machine (Fig. 3) [12b]. Though based on the activity of a viral enzyme as nucleotide polymerase, Eigen's approach offers a way to study the nature of the reproductive and evolutionary mechanisms, including that proposed by Lacey. For an *ab initio* beginning, various authors estimate that chains of about 20 to 30 amino acids were enough to form enzymatically ac-

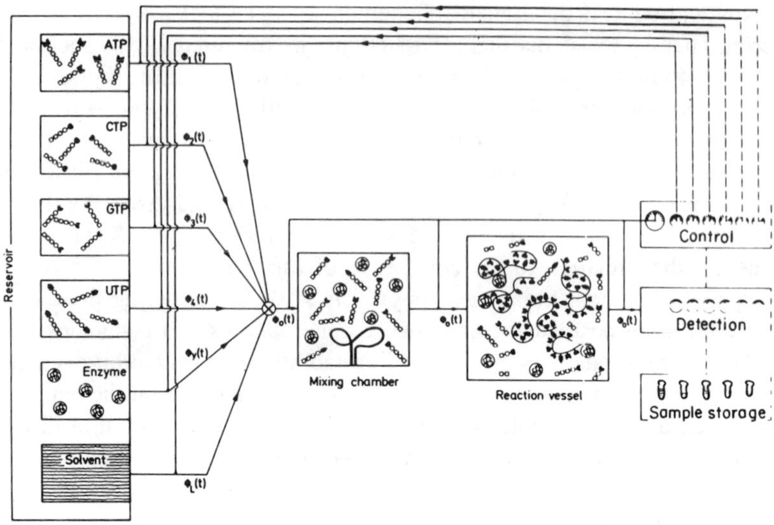

FIG. 3. Scheme of an evolution machine. The reactor is operated with an adjustable flow rate. The solvent entering the reaction vessel contains a replicase and four kinds of nucleotide monomers. The concentrations can be adjusted by varying the flow rates into the mixing chamber [12b].

tive, if only imperfect pockets. It does seem possible then that a few polypeptides of fewer than 100 residues and some RNA strands holding two to three times as many bases may have sufficed for the very first reproductive process [7]. Syntheses of peptides and nucleotides of such length are within today's synthetic capabilities. However, since the number of possible amino acids of DP 100 is still astronomical, say, 10^{100} for peptides containing the 10 primitive amino acids, one has first to develop guidelines for the synthesis of polypeptides that act as polymerases. Polypeptides of sequences able to complex with their metabolic or nucleic substrates to be [4] may be helpful indicators for the search [see Fig. 1(b)].

In all these attempts we are faced with the difficulty that, for our search for mechanisms that started life, or may start lifelike functions in the laboratory today, we have no parallels. If, as it is likely, the beginning of life stems from a small group of similar events, we are left to guess the conditions, the mechanisms, and the level of complexity at which peptides and nucleotides could begin to foster each others' synthesis. Even after a stable population was established, it may have been a long time before forms emerged that we would call "living" by comparison with today's simplest heterotrophic bacteria. Individual steps of evolution may have been relatively sharp, but there was probably no particular point at which one could have said that life started.

Once replication took hold, chemical evolution turned into Darwinian evolution and the organisms had to adapt progressively to conditions of need and supply. Environmental "signals" in the form of changes of nutrients, light, pH, or salinity, and collisions acting as stimuli affected the structure of the membranes, the loci for osmotic pressure regulation by mediated ion and energy exchanges, transport of food, water, and metabolic products. This poses the next major open question. How do physical chemical changes in the membrane, following "stimuli," lead to appropriate responses? There will have been changes in membrane permeability, strength, electrical potential, and by causing dilations or contractions, membrane undulations that led to locomotion; but which factors guided the cell to utilize the survival-promoting effects and suppressed the deleterious one? There is a crucial difference between passive acceptance of imposed changes and reactive, purposeful actions. The latter require prescient instructions or responsiveness, the very attributes of life which our automatons had yet to develop.

In this dilemma we may resort to v. Neumann's thesis. The latter demonstrates [3c] that a self-propagating automaton must be composed of four components: an automated factory, a set of written instructions that specify all needed actions, an automated duplicator for making copies of the instructions, and a controller that orders the duplicator to prepare copies and decides when they are given to the factory to build the four components.

The findings of molecular biology of the roles of the genetic material, the nucleic acids as instructions, the proteins that help to duplicate them, the ribosomes as factories, and the inducer and repressor molecules as controllers confirmed v. Neumann's postulate; however, the question of the origin of the four components and therefore of the origin of life and its evolution is left wide open. Where did factories and instructions come from if they were not the sooner or later incidental results of physicochemical processes? The chemical evolution of macromolecules discussed above goes part of the way toward answering the development of instructions, duplications, and factories, but for automatons to not merely exist but improve, it must be possible to vary the instructions. This could have occurred by feedback mechanisms which may have existed for the coevolution of polypeptides and polynucleotides [7], but were not reinvented for the adaptation of automata to their environment. Ours is the worst of all possible worlds, not based on any principle higher than on "getting by," that is, survival of the individuals until they are reproduced. Thus the simplest, if costliest system, incomplete fidelity of reproduction due to scrambling, and/or errors in the instructions followed by Darwinian evolution, prevails till today.

When the early membranes responded to stimuli, among the millions of passive ones must have been some more sensitive to temperature extremes and other environmental insults. When such cells happened to respond by membrane tightening and constrictions on the danger side, and extension and bulges in the safer directions, such undulations afforded escape. For opposite signals, there resulted motion toward food sources. Cells into which these responses had been built by chance alterations of the instruction, survived and their factories built more of them, including the altered instructions. This Darwinian selection by indirect feedback through survival is the current answer to the thorny question into the basis for adaptation. It has the advantage that built-in life-serving responses may also be seen as the beginning of behavior that in higher animals manifests itself as reflective, instinctive, or purposeful.

The polymeric systems that became responsive and survived were surely much simpler than today's bioconstituents, but even then macromolecular complexity and versatility were needed to fulfill the many exacting structural and conformational requirements of catalysis and regulation on which the living process is based. Dr. H. F. Mark, many years ago in class, pointed out the unimaginably large number of different molecules that exist in any batch of synthetic polymer, counting those due to molecular weight (MW) distribution and isomers, tacticity, optical isomerisms, branched and/or head-tail sequences even for homo-, and more so for copolymers. It is most unlikely that in all of polymer manufacture ever two alike molecules were produced. Thus, when talking of macromolecules one should be aware that no two are ever identical but that they constitute unusual populations of unique members where the vast majority of possible individuals is missing. Consider a polypeptide with 20 L-amino acids as components and say, of, 300 amino acid residues in a chain. If all the possible isomers were realized, there would be $20^{300} \sim 10^{400}$ different molecules whose mass would greatly exceed the mass of the universe (Fig. 4). Moreover, each of these macromolecules is capable of occurring in large numbers of three-dimensional conformations. Assuming again a chain of 300 amino acids with perhaps 100 rotating links and 2 choices (gauche-droit), the number

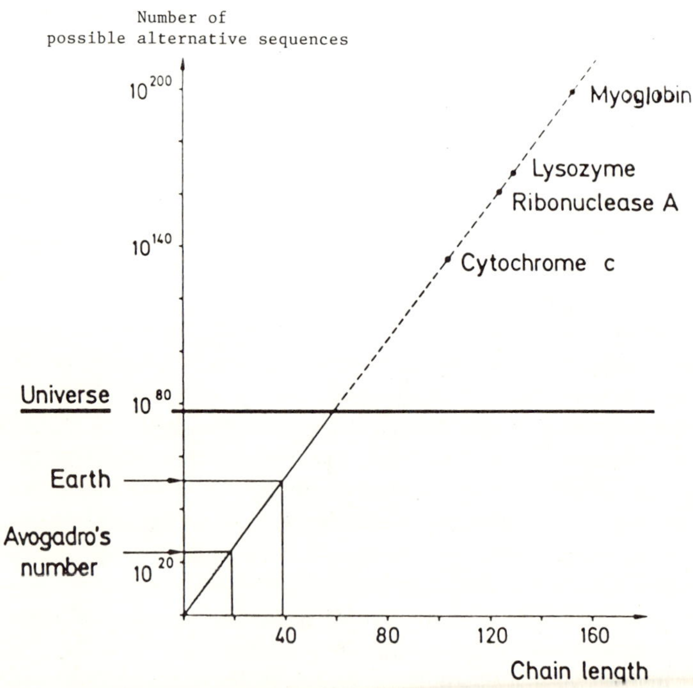

FIG. 4. The possible number of protein structures. The electron-transport enzyme cytochrome with 104 amino acids in its chain, is one of the smallest naturally-occurring proteins with catal function. Even so, a protein of this length has 10^{135} possible alternative sequences (of which o the active ones occur). In comparison, the mass of the universe is 10^{78} times that of a hydro atom, and that of the Earth a mere 10^{51} times this unit [12b, e].

of conformations that may be taken up by each of these 10^{400} molecules mostly with minor changes in the energy per chains, is of the order of 2^{100}. It follows that the tertiary, three-dimensional conformations in which peptide chains fold as a result of residue sequence and rotation, constitute extreme cases of ordering due to instructions that are as precise as their origin is lost in antiquity. We are inclined to assume that they evolved under constraints [7] within complexes of substrates and nucleotides [see again Fig. 1(b)]. Proteins structured to act as effective catalysts helped their organisms, and the instructions, to survive.

Related proteins are often very similar for similar functions. The chains of hemoglobins for example have an extensive sequence and structure homology with the also oxygen-carrying myoglobins [16] [see Fig. 5(a)]. The peptide-splitting serin proteases, too, have very similar structures and reactive pockets. This suggests that these enzymes are the progenies of divergent evolution from a common ancestor. Convergent evolutions of polypeptides, when different molecules became adapted to fulfill similar functions, are also known. In any event, even though many structural features must remain unchanged to maintain a given catalytic action, the remaining number of possible sequences and conformations for molecules of the same function is still so large that every species, dead or alive, has its own characteristic sequences.

The broad scope of life is based on the underlying diversity of instruction (= information) that macromolecules are capable of containing. The information content, I, of a system was formulated by Shannon [27a] in line with the formulation of entropy, since "information vs. noise" is analogous to "order vs. disorder": $I = k \log(W_p/W_r)$. Here, W_p is the number of possible chains of N nucleotide links (i.e., 4^N), and W_r is the number of realized chains following stringent chemical and Darwinian selection. Since W_r is large, W_p has to be much larger still, so large that only macromolecules are capable of the information content on which life on this planet is based. In a study of the principles that may guide the transition from the structural-chemical to biological information, Anderson et al. [10a, i] developed a system that is more general than Eigen's in that it is not dependent on specific chemical details. His results agree well with the findings of Eigen's work on nucleotides [12d].

The sequencing of the amino acids which occur in all organisms and were presumably essential from the beginning of life, revealed remarkable similarities of composition for related animals and plants [Fig. 5(b)] [14, 16]. One observes a convergence of composition, as one goes to lower and more primitive creatures, that permits tentative extrapolations back in time to the original sequences of, say, the group of cytochromes [12], or of the ferredoxins. Conversely, the divergence from chemical "ur"-forms toward today's structures permits us to follow the courses of evolution by this molecular clock with greater certainty than by comparing the taxonomies of fossils. The branches of the trees show impressively how, for all their individual changes with time, the proteins have preserved throughout all the kingdoms their "essential" composition and structure, that is, the sections of the chains immediately involved in enzyme actions. This fundamental internal relationship confirms the all-embracing brotherhood of life, and also the universal adherence and adaptation to function as the basic ordering evolutionary principle.

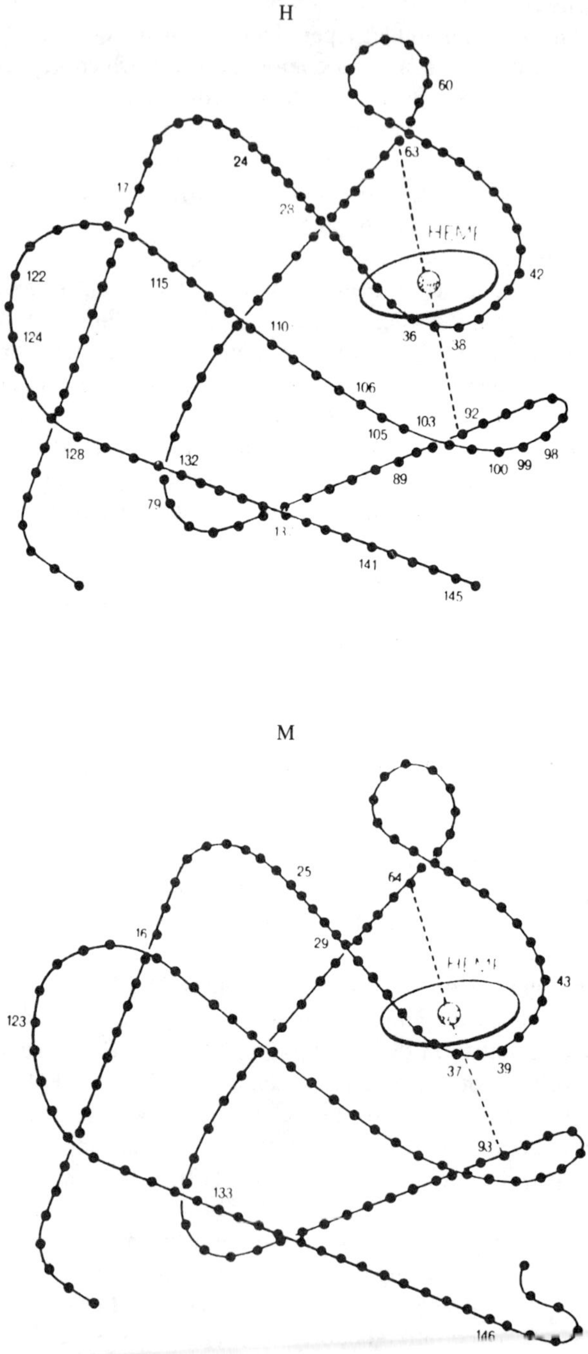
H
60
63
24
17
28
HEME
122
115
42
110
124
36
38
106
103
92
105
128
132
100
99
98
89
79
141
145
M
25
64
16
29
43
123
37
39
93
133
146

FIG. 5.(a)

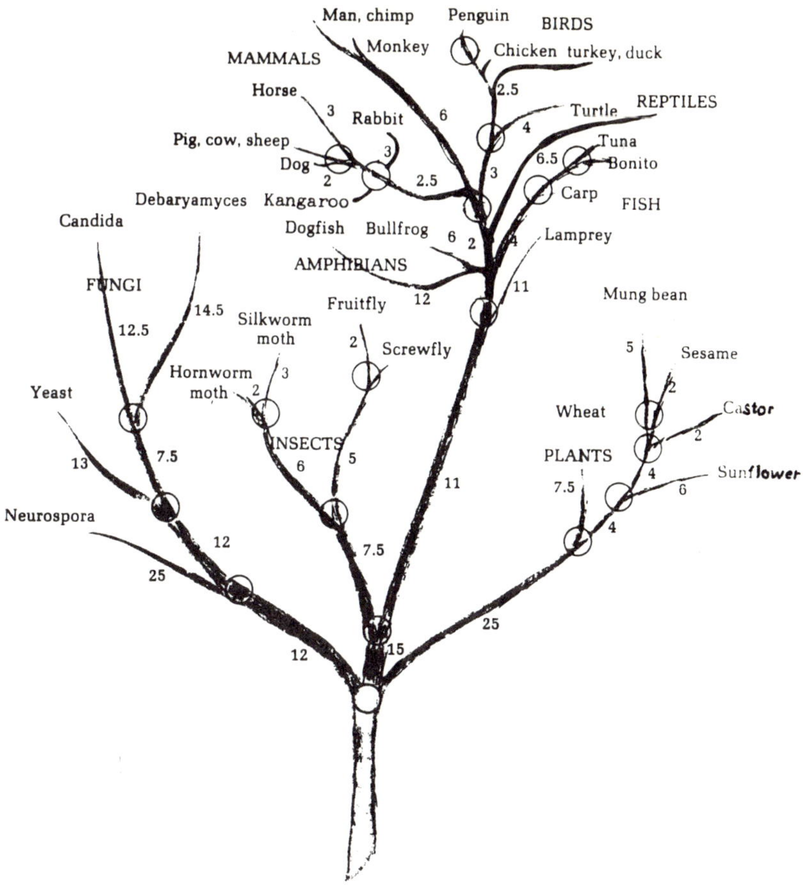

5(b)

FIG. 5.(a) Invariant sites are identified on knotlike shapes representing the three-dimensional structure of the polypeptide chains of hemoglobin H and myoglobin M. The 26 numbered sites at left are occupied by the same residues in all known hemoglobin chains. Eleven of these same sites (assigned slightly different numbers) are occupied by the same residues in all known chains of hemoglobin and myoglobin. Presumably the invariant sites are important in establishing the structure and function of these polypeptides [16b].

5(b) Phylogenetic family tree from comparative sequence analyses of the protein cytochrome C [16a]. The numbers on the branches are the minimum numbers of *nucleotide* substitutions in the DNA of the genes that could have given rise to the observed differences in the amino acid sequences. The overall relations agree fairly well with those inferred from the fossil record and other sources. There are only a few deviations of the cytochrome C phylogeny from the traditional one. Thus the chicken appears to be related more closely to the penquin than to ducks and pigeons; the turtle, a reptile, appears to be related more closely to birds than to the rattlesnake, and man and monkeys diverge from the mammals before the marsupial kangaroo separates from the placental mammals. Despite these differences the amino acid sequence of cytochrome C reflects astonishingly well the evolutionary kinship of organisms as observed on the macroscopic level. (From Ayala, 1978 [16c], and Küppers, 1983 [12b].)

Concerning yet another aspect of macromolecular multiplicity, biological evolution was able to elect the random path of generational chemical changes, the chain of life based on the death of individuals, only because macromolecules delivered the needed enormous number of slightly different molecules of the same general structure and function. Otherwise, life may have evolved by feedback and the gradual perfecting of the instructions during individual life spans of comic duration.

The earliest truly living cells, the "progenotes" [14], were presumably not exactly alike and were also able to absorb, or to exchange, subcellular units to evolve on the intracellular level [15]. Based on their range of properties they became the ancestors of today's three main branches of living cell lineages, of the archebacteria, eubacteria, and the eucaryotes [14]. During the long series of Darwinian evolutionary steps (Fig. 6), according to microfossil and microbiological evidence [5, 17], the first heterotrophic forms were rather soon succeeded, about $3,5.10^9$ years ago, by the autotrophic sulfur, purple, and sulfate bacteria, with their anaerobic photosynthesis. About 10^9 years later, the cyanobacteria developed the water-based photosynthesis and with it the beginnings of a oxygen richer atmosphere. Some lines remained anaerobic or retained facultative anaerobism, while others converted fully to the energetically much more efficient aerobic metabolism and produced our atmosphere. Their evolution included adaptation to practically all terrestrial environments, and the development of sophisticated, functional as well as defensive, multilayer outer membranes. Also perfected was an orderly arrangement of the internal working groups of macromolecules within the cytoplasm, a highly complex yet flexible machinery for metabolism, water and ion maintenance (osmotic pressure), responses to signals from the membrane, and capable of macromolecular defenses against invaders [15].

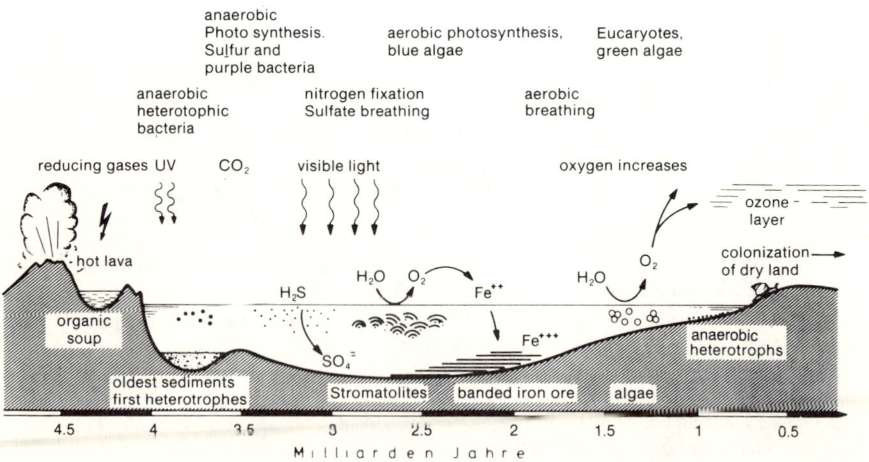

FIG. 6. Schematic representation of prebiotic and early biological Evolution [after Follmann, Ref. 4].

For the parallel developing eucaryotic ancestors, the principal progress was the separate packaging of their genetic material in the nucleus, followed by the symbiotic imbibing of purple bacteria which, as mitochondria, became the future power centers [14, 15]. The further incorporation of chloroplasts from cyanobacteria led to future plant life, while adopting and developing the endoplasmatic reticula of unknown heritage became characteristic for animal cells. Thus equipped, the eucaryotic cells retained their basic molecular and cellular structure until today. The animal cells did not acquire photosynthesis, or lost it again, learned to live in symbiosis or to feed on their autotrophic relatives, an advantage that permitted them to develop into much more complex, mobile higher organisms. The price paid was that they became predators, hosts, or parasites, dependent on plant life, and subject to an eat-and-be-eaten order.

The bacteria remained small, simple, with some of their genetic material in the cytoplasm attached to the inside of the membrane. Budding or involution of membrane-RNA(DNA) pieces created wandering bits of genetic material that could be shared widely within the bacterial population. The process that survives as plasmid transfer is still today a powerful source of genetic adaptability. It is similar to that of micellar interchanges, well known to inanimate colloid chemistry. The bacteria also became the beneficiaries of the high rate of biomass accumulation which the sophistication of the eucaryotes permitted, and developed their own version of predators and parasites, the phages, and viruses. Considering microbial viability one might well conclude that the appearance and existence of the "higher" organisms represents just an episode of the story of life on Earth.

Looking at the relative histories, the way from the progenotes to eucaryotes took an exceedingly long time, many times longer than from macromolecules to the first life forms, and also much longer than the evolution of all multicellular creatures, or than the development from simian brains to those capable of colonizing outer space. Probably, the long time span needed for unicellular perfection reflects the syntheses and elimination of billions after billions of macromolecular compositions and conformations engaged in the complexing with substrates to be [see again Figs. 1(a, b)] before the needed metabolic reactions were optimized by Darwinian selection.

Among the evolving enzymes were those that had to obtain the ability to reduce RNA to DNA which is more precisely specific for the collection, alignment, and polymerization of nucleotides, and thus for the regulation of their own reproduction. The rapid growth in the amount of haploid DNA per cell weight coincident with the evolution of multicellular organisms must have been enhanced by the advances of all metabolic processes by their specialization and, in turn would have been one of the reasons for the accelerated evolution. It is striking that the amounts of DNA increase along the evolutionary ladder much more than the number of genes necessary to express the higher phenotypes. In other words, the higher the organism, the greater the amount of "introns," the noncoding sections of the DNA chains, relative to the "exons" which code for the needed syntheses. The suggestion has been made [18] that those spare sections increase with complexity of the organism to allow mutations to add to the flexibility and speed of evolution [14, 18], or that the introns serve other control functions. The bacteria, which either never had, or lost spare DNA, have the simpler reproduction due to an all-functional DNA.

Precision became of even greater importance when the cells began to aggregate to colonies, or grew into groups of adhering daughter cells. The records of microfossils indicate that the cells aggregated early to form fibrous strands and mats much as the stromatolites and certain cyanobacteria do today [17]. As long as they stayed joined to single file filamentous, sheetlike, or hollow cylinder groups, the cells retained their direct contact with the environment. More complicated groupings required specialization of function, or the symbiotic colony formation of eucaryotes which had already become specialized through their ancestry. This formation of organs required material transport and communication. The inner membranes, in addition to their earlier functions, had to become the carriers of all organizational needs. We conjecture that the lipid-protein bilayers retained their general structure and had [3, 5], or acquired enzymatic and receptor qualities through additional embedded or attached polypeptides. Carbohydrate markers at the end of the peptide chains served intercellular and individual recognition, while extracellular helical peptide threads formed the tissue that held the animal cells together. The reinforcement for plant cells was constructed from exuded sugars that polymerized into macromolecular cellulose. Together with lignin they are the preponderant part of today's biomass. The exoskeletons of mollusks, insects, crustaceans, and so on, are also based on macromolecules, per se, or as reinforcing matrices.

The parallel development of aerobic metabolism and of multicellular organisms (Fig. 6) was most likely no accident, as the latter required the efficiency of the former. The transition from individuals to communities and then to an orchestra of interacting units whose whole is much more than the sum of its parts, present difficult problems of organization [19] including developments of the most functional shapes [11]. It has been shown that features of the cell surface, as well of the endoplasm, decide the rearrangements of the cells, whereby patterns of geometric symmetries and enlargement are decisive [4, 11] as well as the principles of assembly. A society of units may have abilities which the units do not possess.

Cell division itself, triggered by molecular signals from the cytoplasm when growth reaches a critical state, utilizes general mechanical patterns of membrane stability and deformation. When spherical membranes grow and extend to ellipsoidal shapes, the inherent instability of expanding ellipsoid bodies of constant shell thickness causes them to develop waists followed by membrane division along the waist. Another variety of division is budding, the local growth of daughter cells by protuberances that form a tightening waist at their base until they part, a process well known also to nonbiological surface chemistry. The differentiation of the cells during morphogenesis follows a sequential growth and retardation pattern regulated by gradients of synthesis-promoting and retarding proteins, the morphogens, and the residence time of a cell in this gradient [14, 15, 19].

Evolutionary differences notwithstanding, the appearance of the nervous systems in very primitive animals [20] indicates the heightened importance of sensation, response, and communication for the next phase of evolution. Living matter does not allow for electronic conduction, while ionic conduction is too slow. Thus, the principle of a traveling electrical wave analogous to that of transversal wave pulses was utilized, in that changes in membrane potential which require electron or ionic transport merely across the thickness of the membrane, travel along the membrane in domino fashion.

The electrical potential difference between cell interior and exterior is determined by ionized groups of the lipid and polypeptide molecules of the membranes. Active diffusion maintains laterally coupled charge differences across the membrane and vice versa. Any disturbance or stimulus spreads rapidly sideways as a tangential soliton wave of change of potential by a most remarkable cooperation between millions of molecules of Na-K-ATPase and of other membrane enzymes. This stimulus-impulse transport is so efficient that even in today's complex organisms the peripheral receptors are not separate, but are located on long extensions of the nerve cells that lead to the central part of the nerve cells, the somas. From there, the pulse speeds again via axons and dendrites to processing nerve centers, the ganglions which are connected to a center of efferent nerves from where actions are initiated [21].

The most important advance following the development of nerve functions was the evolution of memory, the molecular mechanism of which is still poorly understood. According to the available evidence, the sequence of action potentials, a pulse, which results from the stimulation of the nerve endings is transmitted to the soma. If larger than a threshold value, the pulse will be conducted with the same amplitudes and frequencies from the soma along other axons to dendrites of connected cells, from there to their nerve bodies and so on (Fig. 7). The transmission across the synapses involves the release, upon arrival of the electrical pulse, of chemical messengers from storage vesicles triggered by calcium-ions. The messengers, a class of amines called neurotransmitters travel across the synaptic cleft and by binding to receptors create corresponding potentials in the membrane of the opposite dendrite (Fig. 8) [23]. There, after reconversion into electrical pulses, spikes travel toward their soma initiating a chain of responses [21]. Memory, or habituation, follows pulse arrivals in the soma and at synapses, and is based on molecular changes as records of the event [22]. When a second pulse of the same characteristics follows at some later time, the neuronal consequences are no longer exactly the same as for the first passage. The neuron "remembers," it has "learned," thus enabling the organisms to respond faster and more effectively to repeated stimuli, i.e. to react purposefully.

The nature of the traces that modify subsequent responses to stimuli is much debated. Mechanisms proposed range from reverberating electrical circuits or oscillations in the synapse, to ion motions and binding, polarizations, membrane changes, and syntheses of RNA and protein whose structures reflect the nature of the pulse. Proteins are certainly involved in some ways [22, 23].

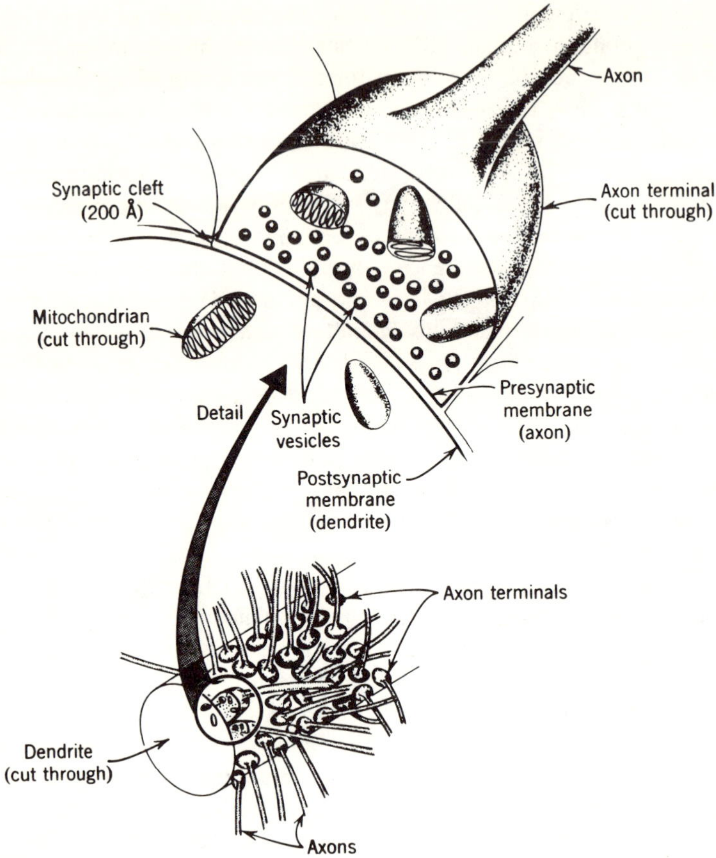

FIG. 7. Fine structure (schematic) of a synapse. The pulse would proceed from upper right to the lower left [after Stevens, Ref. 21].

The evidence is impressive that the learning process is accompanied by an increased RNA and protein content in the neurons. Reports that polypeptides, extracted after an intense learning period and injected into, or fed to untrained specimens, shortened the latters' learning time are, however, treated with scepticism.

Apart from any objective evidence, the practically infinite suitability of macromolecules to function as unique information carriers, mentioned earlier, the speed and the very low energy required for conformational changes make polymers obvious candidates for their participation in the formation of memory traces. The rationale for the proposed two mechanisms of ionic short-term and conformational long-term storage can also be easily seen. The first would be based on the patterns of Ca-ion induced charge distribution of the polyelectrolytic state, the other on tertiary or even quaternary structures of the polypeptides, imprinted by, and retained after, the passage of a particular pulse. Memory recall would involve the synthesis of polypeptides, patterned according to those imprinted by the original pulse, brought up to the synapse to create

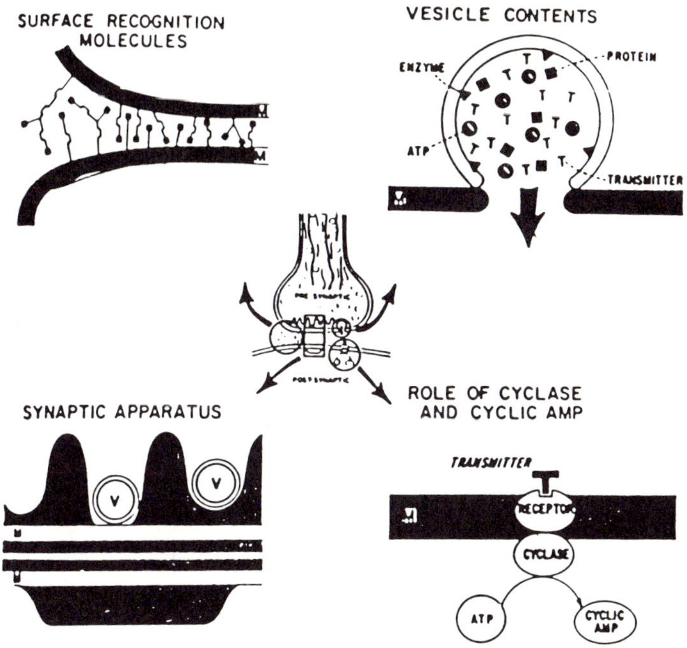

FIG. 8. Models of presynaptic and postsynaptic membrane interfaces showing the synaptic apparatus and various biochemicals involved in synaptic transmission. (From Schmitt & Samson, 1969.) "Brain Cell Microenvironment" *Neurosciences Research Progress Bulletin,* 1969 [22b].

the release of the neurotransmitters that marked the "remembered" event. The signature of the pulse could thus be preserved as long-time memory, allowing unlimited numbers of prints to be made from the first mold. After use, the prints could be depolymerized and the amino acids reused for a later synthesis. Even if trillions of creatures with billions of neurons receive trillions of stimuli involving millions of molecules for billions of generations, a number of say, 10^{50} polypeptide structures and conformations, this is still vanishingly small compared with the number of possible ones.

The molecular mechanism that would support this course of events raises once again large questions. Does the electrical pulse tie waiting proteins directly into the ordered, imprinted conformations? Where do the proteins come from and are they "naive," in other words, do they have a readily alterable structure waiting to be shaped? If so, how could copies of the imprinted protein be prepared, since feedback from protein to nucleic acid and from protein to protein is not known? Are there special memory cells?

The likely answers come from a number of data and theories, more or less confirmed, of cellular mechanisms. Briefly, for lack of a direct feedback of cellular experiences and changes into the structure of DNA (an absence that necessitated Darwinian selection as the source of structural changes before) another feedback mechanism was developed at this later stage that acts on the function of DNA rather than on its structure. This allows adaptive, though not hereditary changes in the chemistry of the individual response mechanism.

DNA carries the instructions for all proteins for all life-sustaining functions, but not all proteins and enzymes are needed all the time. Sections of DNA that code for proteins that are unwanted at a given time are blocked by a group of regulatory cell constituents, again proteins, called repressors. At the appropriate time, dictated by the needs of the cellular cytoplasm that now assumes the role of the controller, the repressors are removed by other molecules called inducers, so that the freed DNA section may now transcribe and initiate the syntheses of the protein required for the moment.

This well substantiated series of actions [24, 25] has been formulated as the "derepressor hypothesis" to explain, at least part of the memory processes (see Fig. 9). The arriving electrical pulse changes the activity of the endoplasm of axon and soma, freeing a DNA section (an operon) to synthesize a special m-RNA that in turn gives rise to a special protein, both specific for the impulse. The m-RNA and protein, new components of the endoplasm, ready the cell for modified activity: to "remember." Should the same, or a related im-

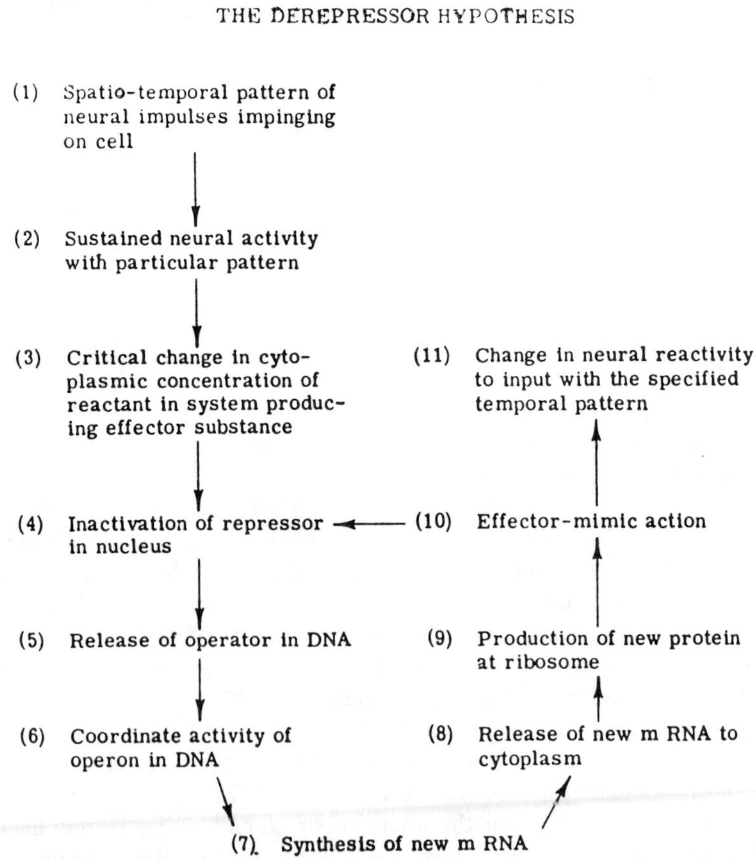

FIG. 9. Schematic representation of the memory-forming steps according to the derepressor hypothesis.

pulse arrive again, a revised response is transmitted, in the simplest case to the center (ganglion) of the efferent neurons from where a modified action may be signalled to the motor nerves. The imprint on the new endoplasmic proteins represents the memory trace, the modified output corresponds to the recall, and the ability to modify the outgoing signal is equivalent to acquired knowledge, to having "learned." Even this barest outline, neglecting a multitude of essential steps and factors, shows a process of staggering complexity made possible only by the all but infinite number of possible protein structures, instructed by the nucleotide code and adapted by cytological processes. Up to this point, learning is all automatic; indeed, memory feats can be performed by beheaded insects [25]. Also, all neurons should be able to retain some memory. This actually fits the observation that memory is not the prerogative of any special section of the brain, though some areas are more involved than others.

The main difficulties raised by the derepressor theory are connected with several facts. Inhibition of RNA synthesis does not seem to interfere with memory. The removal of the repressor would open an indigenous section of the DNA that is in no way touched by the pulse. The same is then true for the transcribing m-RNA and the protein. How can naive protein have been imprinted when it has been synthesized an appreciable time after the arrival of the pulse? Thus, one needs to make the additional assumption that the pulse was lingering sufficiently long at the synapse to be still active when the naive protein arrived.

Actually this would fit another set of assumptions [26], which postulate that memory processes consist of three kinds (see Fig. 10). The first is supposed to

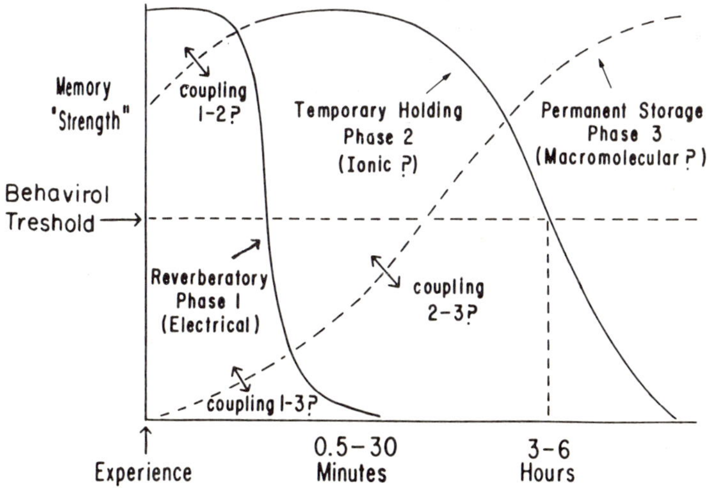

FIG. 10. The multiple trace theory of consolidation of memory envisages three phases in the storage of information: (1) a short-term process, probably consisting of reverberatory neutral activity, representing the immediate effects of an experience upon the neurons; (2) an immediate holding process, perhaps mediated by the binding of extracellular calcium, potassium or sodium responsible for temporary conformations and retrieval; and (3) a permanent storage process, probably of RNA-mediated changes in the synthesis of proteins or catecholamines, by affected reactions initiated during the reverbatory activity (John, 1970).

be strictly ionic. The arriving pulse sets responding ionic reverberations in motion which continue for a short period, seconds or minutes, and which constitute the short-term memory. This is followed by a holding period. Its duration allows time for the arrival of the newly synthesized protein to become imprinted by the vibrations of trains of ions or of sections of neuronal fibers or membranes, before the reverberations die down. The imprinted proteins constitute the permanent memory. Copies are being made without the help of m-RNA from additional naive protein by tying it into conformations expressive for the original pulse by an electrostatic alignment of the new protein against the permanent record.

Whether it is this mechanism, or any other that may pertain, the story so far refers to single neurons and their synapses. The complications increase greatly when all the many axons and synapses are considered that are sequentially active in a single event, and the large societies of neurons that may have to be excited in parallel to form a memory. Further complications arise from the fact that the reactions to the first stimulus must also be remembered for a proper learning process. Future responses must rest on an integration and judging of all that went before. Here we enter the field of neurophysiology and behaviorism, where the reduction of the observations to macromolecular mechanisms becomes, at this time, impossible. However, the influence of memory in all its stages and manifestations on the further progress of evolution is too intriguing not to attempt some extrapolations, even if they may border on science fiction.

Memory, the first source of instruction evolved since the perfection of the genetic code, is not as compelling as the latter, but in some ways more powerful, since it is, or can be, continuously improved by feedback of immediate relevance. Thus, memory is not only the basis of all learning, but also that for the first lesson, that of cause and effect, and for the second lesson, that there can be actions for a purpose, that will answer needs.

Memory, as the imprint of earlier experiences and to the extent that it is reliable, becomes the basis for realistic action. It becomes even more powerful through transmission, or sharing, executed and received as communication expressed in the form of a language, be it through behavior, specific signs, sounds, or other means. The imprint of memory can be scanned for pertinent information. Fragments can be combined to new useful combinations in contrast to the blind shuffling of genes. The combination of memories allowed also for the forming of previously unseen images, as in creativity.

The beginning of memory constructs already exists in insects, supposed to act largely as automatons. The dance of homecoming bees expresses the memorized location of food sources in coded body language. Even more impressive is the use of memory of those bees who have learned of the location and now go out to find it. When v. Frisch set out an attractive honey source some distance from the hive, and then increased the distance every day by a certain amount, by the fourth day the bees had extrapolated direction and distance increments and were before him at the projected location.

The existence and the consequences of memory lead directly to topics central to human communications, linguistics, and computer science. In fact, information theory stands to learn much from the structure of memory, from the information content of the tertiary structure of proteins and the linearly coded information held by the genetic code and expressed in translation and transcription. It is interesting that the biological coding system for impression and expression, because of its longer "words," is more complicated than human and computer languages. All information and communication systems however, are subject to the same laws of grammar, content, redundancy, expedience, and noise, that is, subject to intended structure and order of a message as against the unintentional randomness or noise, the opposite of message and order [27]. As mentioned, the formalism of the statistical mechanics of entropy has been found applicable to all forms of communication [27].

The availability of a memory had many other profound effects. In response to a repeated stimulus of a given kind, memory permits as we have seen a much faster and more efficient response, and a wealth of combinations with earlier related responses. The acquisition of further memory, and of the memory of past utilizations makes choices available, a new element in the struggle for survival. The more adroit the choice, the more closely based on imprints that reflected the requirements of the outer world, that is, the more the memory corresponded to "reality," or "truth," the greater the reward of these acts of "intelligence."

Perhaps the most important step permitted by memory and the available pool of individual macromolecules was the development of model responses. Instead of reacting by activating the motor nerves, retracings of memories could be evoked, as well as images of what, from memory-experience-learning, the consequences of actions might be. Thinking became possible. The result of this first analytical and then integrating process could be compared with the results of motor action and help to arrive at effective decisions. Though purposiveness increased over billions of years, it was only in the last 3-4 million years that self-awareness arose and increased in the human species. It allows an individual to see itself as an actor on a field of possibilities, visualizing the consequences of his acts, to plan ahead and develop the concept of his position in his environment.

The practically unlimited scope of memory and thinking, based on the macromolecular recording and recall systems, enabled the more advanced organisms to observe over and over again the sequence of cause and effect and, thus, to direct thinking into verifiable tracks. These included the development of tools, of thinking in terms of models for the reduction of trial and error approaches, and such forms of systematic thinking as logic. A parallel development must have been the assignment of specific sounds to specific objects, actions, and happenings (i.e., the evolution of a wide range of language). The correspondingly more accurate communication especially in writing allowed a quantum leap in learning, as experiences were not only transmitted by signals, but could be voiced, explained, and shared precisely. From this beginning, technology and science were born which extended Man's reach and grasp by many powers of ten.

A particular consequence of the development of technology was the growing recognition of the nature of macromolecular substances with which Man was familiar from his farthest past (i.e., fibers, wood, hair, hides, horns, clays, etc.). This was followed by the synthesis and study of the new organic polymers which led to the full interpretation of the nature of macromolecules and to the understanding of their scientific and practical potential.

The consequences of forming such concepts as: rubber extension and contraction, viscoelastic memory, the glassy state, the limits of material strength, large unsaturated, conjugated, and polyelectrolytic linear molecules, cross-linking, copolymers, large rings and helices, interactions of ternary and quatenary molecular structure, nematic fluids, viscosity modification, plasticization, the principles of heat and shock resistance, the intricacies of composites, etc., are yet far from being fully realized. It is certain that the next decade will see great progress in adjusting material behavior and tailor-making new materials, based on our molecular interpretation of microscopic and macroscopic observations on macromolecular substances [29].

In addition, macromolecular science has been abundantly fertile in advancing molecular biology and paleontology, bio- and enzyme engineering, work with monoclonal antibodies, and gene technology. Biology in all its aspects allows us to develop an understanding of ourselves in terms of living units, including our fledgling attempts to reconstruct the beginnings of life on Earth [1–10, 28].

In an important departure, reshaping our lives and thinking in yet another way, our biological minds opened new intellectual territory by engaging in solid-state chemistry and physics, electronics, and computer technology. So far, whatever efficiency and kind of activity we can build into our computers and robots reflects the ways of our thinking. Electrical or optical impulses are imprinted on the structure of the semiconductor chip in a digital or analog manner, modelled in accordance with our logic of language and counting. Conversely, the immense new potential of numerical and functional computations expands our horizons. The advantage of an inorganic brain, if ever independent of macromolecular organisms [29] and its need for temperature control, should lie mostly in a greater tolerance of the harsh conditions of the universe outside the self-created, very special aquatic environment and the protective umbrella of the biosphere.

Our way of thinking was until recently geared entirely to the manipulation of objects (tools, i.e., extensions of ourselves), mostly unaware of the so much more subtle and complicated machinery of perception and action developed by the living world. Adding to this the vulnerability of our waterborne macromolecules, it was probably more than a historical accident that we used metals, semiconductors, and network theory for designing our calculating and computer devices on which our present efforts to build thinking machines are based. It is too early to say whether this development will continue in the foreseeable future, in other words, whether man-made machinery [27], including inorganic devices, will ever develop their own mode and ways of reproduction and of thinking, and reach the sensitivity, variability, and effectiveness of the macromolecular apparatus. It is though, also possible that other advances, for instance in ceramics will one day permit the development of different types of switching devices and networks. Another possibility is the development of molecular switches and mixed ionic and electronic conductance systems based on easily shifting charged states, or ion binding to conjugated and aromatic molecules. Incorporated in or attached to nonbiological macromolecules, such systems are being developed under the name of molecular computers. They may come much closer to the proteinic ones used by living nature of our kind, and are much more robust in their chemistry than living macromolecules, and even silicon chips. Lastly, as we learn more and more about our own mechanisms, we may be able to advance these to become less dependent on the stringent terrestrial conditions and better adapted to living beyond the earth.

Whether macromolecules, semiconductors, or other devices will eventually overcome their present limits, the next decades will certainly see the development of spaceships with robotic instrumentation and guidance. Nothing would then stand in the way of building larger and more efficient space vehicles and habitats and eventually creating small intelligent planetoids, an analog to L. Thomas' image of our Earth as a living supercell, or Bugliarello's superintelligence of globally connected computers. Eventually, it should be possible to collect matter from space, assemble cosmic bodies, and render them capable of understanding and intent, which could then begin to reorder and reshape the cosmos, a takeover in line with the intelligent principles and their ways of thinking of whichever creatures macromolecules will have developed into at that time.

In the footnote to the title, I stated that this was a personal attempt as a chemist to develop a coherent understanding of our past evolution. I indulged further, in this lecture, in an extrapolation that was optimistic in that it pointed toward a future for life, perhaps the only one in this universe. Past and future life were seen as being based on self-ordered macromolecular systems advancing themselves through ordered reaction cycles. As an expression of potential in a world permissive, but not supportive, of order, living systems are malleable but also vulnerable. Life as a whole may easily expire. The "wish to live" is just an expression of a dynamic balance. The awareness of threatening demise is a stimulus to hold each system near its steady state, and at the same time is a destabilizing factor. Hence, the search for a yet-supportive order in the universe is an inevitable need of any self-aware existence. The development of tools leading to the discipline of science and to operational truths also opened the door to the study of evolution. This study too, is an example of the search for a life-friendly ordering principle, since it permits extrapolation of existing ordered macromolecular states into the future. Yet, in terms of Monod's [30] famous discussion, all forms of ordering and their ongoing existence are events with chances of occurrence that border on inevitability, but still remain chance.

REFERENCES AND READINGS

[1] L. E. Orgel and F. H. Crick, "Selfish DNA," *Nature, 289*, 604 (1980). See also W. T. Doolittle and C. Sapienca, *Nature, 289*, 601.

[2] (a) K. Dose and H. Rauchfuss, *Chemische Evolution and der Ursprung Lebender Systeme*, Wissenschaftlicher Verlag, Stuttgart, 1975. (b) S. Miller and L. Orgel, *The Origins of Life on*

Earth, Prentice Hall, Englewood Cliffs, NJ, 1974. (c) J. D. Bernal, *Origins of Life,* Weidenfeld and Nicolson, London, 1974.

[3] (a) H. Hartman, J. Lawless, and P. Morrison, *The Search for the Universal Ancestor,* NASA, Publ., S-P. 477. (b) S. Fox and K. Dose, *Molecular Evolution and the Origin of Life,* Dekker, New York, 1977. (c) J. V. Neumann, "Theory of Self-Reproducing Automata," U. of Illinois Press, 1966.

[4] H. Follmann, *Chemie und Biochemie der Evolution,* Quelle and Meyer, Heidelberg, 1981.

[5] (a) M. Calvin, *Chemical Evolution,* Oxford, London 1969. (b) D. L. Rohlfing and A. I. Oparin, (Eds.), *Molecular Evolution,* Plenum Press, New York, 1967. (c) A. Lehninger, *Biochemistry,* 2nd Ed., Worth, New York, 1975. (d) A. I. Oparin, *Life, Its Nature, Origin and Development,* Academic, New York, 1964.

[6] (a) F. H. Crick and L. E. Orgel, *Directed Panspermia,* Icarus, *49,* 341 (1973). (b) F. H. Crick, "The Origin of the Genetic Code," *J. Mol. Biol., 38,* 367 (1963). (c) E. Chargaff, *The Inexplicable Riddle,* Klett-Cotte, Stuttgart, (1980). (d) T. H. Jukes, *Molecules and Evolution,* Columbia University Press, New York, 1966. (e) H. R. Pagels, *The Cosmic Code,* Simon and Schuster, New York, 1982.

[7] (a) J. C. Lacey, Jr., A. L. Weber, and K. M. Pruitt, "The Edge of Evolution," in *The Encyclopedia of Ignorance,* R. Duncan and M. Weston-Smith, Eds., Pergamon, New York, 1977. (b) J. C. Lacey, Jr. and A. L. Weber, "The Origin of the Genetic Code," in *Protein Structure and Evolution,* J. L. Fox, Z. Doyl, and A. Blazej, Eds. Dekker, New York, 1976. (c) P. W. Anderson, D. L. Stein, and F. E. Gates, Eds., *Self-organizing Systems: The Emergence of Order,* Plenum, New York, in press. (d) J. J. Hopfield, "Origin of the Genetic Code," *Proc. Natl. Acad. Sci. U.S.A., 75,* 4334 (1978). (e) A. Fersht, *Enzyme Structure and Mechanism,* 2nd ed., Freeman & Co., New York, 1985. (f) J. C. Lacey, Jr. and D. W. Mullins, Jr., "Isolated Microsystems in Evolution," in *The Origin of Life and Evolutionary Biochemistry,* K. Dose, S. Fox, G. A. Deborin, and T. E. Pavlowskaya, Eds., Plenum, New York, 1974. (g) M. Kasha and B. Pullman, Eds., *Enzyme Structure and Mechanisms,* Academic, New York, 1962.

[8] (a) H. Kuhn, "Model Considerations for the Origin of Life," *Naturwissensch., 63,* 68 (1976). (b) H. F. Blum, "On the Origin and Evolution of Living Machines," *Am. Sci., 49,* 474 (1962). (c) S. Spiegelman, *Am. Sci., 55,* 221 (1967).

[9] (a) M. Paecht-Horowitz, J. C. Berger, and A. Katchalsky, *Nature, 228,* 638 (1970). (b) F. Eirich, in *Origin of Life,* Y. Wolman, Ed., D. Reidel, Dordrecht, 1981. (c) F. Eirich, in "Evolution of Protein Molecules," H. Matsubara and T. Yamanaka, Eds., Japan Scientific Pr., Tokyo, 1978, p. 121.

[10] (a) P. W. Anderson, "A Model for Prebiotic Evolution: The Use of Chaos," *Natl. Acad. Sci. USA, 80,* 3386 (1983). (b) A. M. Turing, "The Chemical Basis of Morphogenesis," *Phil. Trans. Royal Soc., London, B273,* 37 (1952). (c) G. Nicolis and I. Prigogine, *Self-Organization in Non-Equilibrium Systems,* Wiley-Interscience, New York, 1977. (d) I. Prigogine, *From Being to Becoming,* Piper, Muenchen-Zuerich, 1982. (e) B. Hess, E. M. Chance, A. R. Curtis, and A. Boitex, "Complex Dynamic Stuctures," in *Non-Linear Phenomena in Chemical Dynamics,* C. Vidal and A. Pacault, Eds., Springer, Berlin (1981). (f) H. Othmer, "Mathematical Aspects of Temporal Oscillations in Reacting Systems," in *Oscillations and Travelling Waves in Chemical Systems,* R. J. Field and M. Burger, Eds., Wiley-Interscience, New York, 1974. (g) A. T. Winfree, "Rotating Chemical Reactions," *Sci. Am., 230,* 82 (1984). (h) M. Eigen, "Self-Organization of Matter and the Evolution of Biological Molecules," *Naturwissensch., 58,* 465 (1971). (i) D. L. Stein, "A Mathematical Model of the Origin of Genetic Information," in *Molecular Theories of Cell Life and Death,* Internat. Colloquium, Rutgers University, New Brunswick, NJ, 1986. (j) K. Takemoto, in "Polymeric Drugs," L. G. Donaruma and O. Vogl, Eds., Academic, New York, 1978, p. 103. (k) For ordering in macromolecular Complexing, see D. G. Challa, in "Macromolecules 86," Symp. on Functional and Bio-polymers, Oxford, Sept. 1986.

[11] (a) M. Ghyka, *The Geometry of Art and Life,* Dover, New York, 1977. (b) D'Arcy Thompson, *Growth and Form,* Ltd. Ed., T. Bonner, Ed., Cambridge Univ. Press, London, 1961.

[12] (a) M. Eigen and P. Schuster, *The Hypercycle,* Springer, Berlin, 1979. (b) B-O. Kuepers, *The Molecular Theorie der Evolution,* Springer-Verlag, Berlin, 1983. (c) F. K. Biebricher, M. Eigen, and R. Luce, "Product Analysis of RNA Generated de novo by Qβ-Replicase," *J. Mol. Biol., 143,* 369,391 (1981). (d) M. Eigen and R. Winkler, Oswatitsch, *Naturwissensch., 68,* 217, 228 (1981). (e) H. P. Yockey, *J. Theoret. Biol., 67,* 345, 377 (1977).

[13] (a) M. L. Fishman and F. R. Eirich (1970, 1975). *J. Phys. Chem., 75,* 335; *79,* 2740. (b) See also Reference 3, p. 2.

[14] K. Woese, "The Primary Lines of Descent and the Universal Ancestors," in *Evolution from Molecules to Man,* D. S. Bendall, Ed., Cambridge University Press, London, 1983.

[15] (a) L. Margulis, *Origins of the Eucaryotic Cells,* Yale University New Haven, 1970. (b) K. R. Porter, "The Cytomatrix and the Integration of Cell Function," in *Molecular Theories of Cell Life and Death,* Internat. Colloquium, Rutgers University, New Brunswick, NJ, 1986.

[16] (a) M. O. Dayhoff, P. J. McLaughlin, W. C. Barker, and L. T. Hunt, "Evolution of Sequences within Protein Superfamilies," *Naturwissensch., 62,* 154 (1975). (b) E. Zuckerkandl, "The Evolution of Hemoglobin," in The Scientific American, Freeman and Co., San Francisco and London, 1968. (c) F. Ayala, *The Scientific American, 234,* 48 (1978).

[17] J. W. Schopf, "Microflora of the Bitter Springs Formation," *J. Paleontology, 42,* 651; also in C. Ponnamperuma, Ed., *Exobiology,* North Holland, Amsterdam, 1972.

[18] W. Gilbert, "Why Genes in Pieces?," *Nature, 271,* 501 (1978).

[19] (a) C. H. Waddington, *Principles of Development and Differentiation,* Macmillan, New York, 1966. (b) L. Wolpert, "Pattern Formation in Biological Development," *Sci. Am., 239,* 154 (1978). (c) G. R. Welch, "Cytosociology," in *Molecular Theories of Cell Life and Death,* International Coll., Rutgers University, New Brunswick, NJ, 1986.

[20] (a) E. R. Kandel, "Cellular Aspects of Learning," in *Brain Mechanisms in Memory and Learning,* M. A. B. Brazier, Ed., Raven, New York, 1979. (b) D. L. Alcon, "A Biophysical Basis for Molluskan Associative Learning," in *Conditioning,* C. D. Woody, Ed., Plenum, New York, 1982.

[21] C. F. Stevens, *Neurophysiology,* Wiley, New York, 1966.

[22] (a) S. Bogoch, *The Biochemistry of Memory,* (1968). Oxford University, London, 1968. (b) Schmitt and Samson, "Brain Cell Microenvironment," Neurosciences Res. Progress Bull., 1969.

[23] R. W. Thatcher and E. R. John, *Foundations of Cognitive Processes,* Lawrence Erlbaum Assoc., Hillside, NJ, 1977.

[24] F. Jacob and J. Monod, "Genetic Regulatory Mechanisms in the Synthesis of Proteins, *J. Mol. Biol., 3,* 318 (1961).

[25] E. R. John, *Mechanisms of Memory,* Academic, New York, 1967.

[26] E. R. John, "Summary, Symp. on Memory Transfer," in *Molecular Approaches to Learning and Memory,* W. L. Byrne, Ed., Academic, New York, 1970.

[27] (a) C. E. Shannon and W. Weaver, *The Mathematical Theory of Communication,* University of Illinois Press, Urbana, Illinois, 1949. (b) L. Gatlin, *Information Theory and the Living Systems,* Columbia University, New York, 1972. (c) J. Campbell, *Grammatical Man,* Simon and Schuster, New York, 1982.

[28] L. Margulis and D. Sagan, "Strange Fruit of the Tree of Life," *The Sciences, 238,* 38 (1986).

[29] H. F. Mark, "From Revolution to Evolution," *Chem. Tech.,* in press.

[30] J. Monod, *Chance and Necessity,* Piper, Muenchen, 1971.

FREDERICK R. EIRICH

ACADEMIC DEGREES

PhD., University of Vienna, 1929.
Dr. Phil. Habil., University of Vienna, 1938.
M.A. Hon Stat., University of Cambridge, 1939.

PROFESSIONAL AND TEACHING EXPERIENCE

Post. Doc. Fell., Instructor and Research Supervisor, U. Vienna 1929–1938.
Instructor U. Cambridge 1938–1940.
Senior Research officer, Univ. of Melbourne, 1941–1943.
Lecturer, Physical Chemistry, University of Cambridge, 1943–1947.
Assoc. Prof. Polytech. Inst. 1947–1951; Professor 1951–.

HONORS, OFFICES, AND FELLOWSHIPS

Distinguished Professor, Polytechnic Institute,
President, Society of Rheology 1972–1973; Chairman, AAA, Gordon Research Conferences (Polymers, 1959, Adhesion, 1961; Interfaces, 1967), Councillor 1961–1966;
Gov. Bd., AIP '72–'74
Sigma Xi Distinguish Service Award, 1970.
Chrmn, Div. Colloid and Surface Chemistry, ACS
Chrmn, Division of Chemistry, N.Y. Academy of Sciences
Fellow N.Y. Acad. Sci., Faraday Society
Visiting Prof. Polymer Science, Uppsala, 1950.
Visit. Unilever Prof., Dept. of Chemistry, Univ. of Bristol, 1964–1965.
Dean of Research, Polytech, 1967–1970: Lecturer, Robert A. Welch Foundation, 1973.
Hon. Member, Intern. Fract. Inst., 1974.
Merit Award Colloid Division, 1975.
Merit Award, Rubber Div., ACS., 1978.
Humboldt Prize, 1981.
Bingham Medal, Soc. of Rheol., 1983.
Maurice Huggins Award, 1985.

CURRENT RESEARCH OR OTHER PROFESSIONAL ACTIVITIES

Chemical Evolution, Structure and Strength of Polymeric Materials, Adsorption and Adhesion of Macromolecules, Colloid Chemistry of Calcification, Rheology, Polyelectrolytes, Biopolymers.

PUBLICATIONS

About 150 scientific papers and articles in professional journals;
Editor of "Rheology," "Science and Technology of Rubber"
Co-editor, "Colloid and Surface Science," "High Speed Testing"

STUDIES OF POLYMERS CARRYING FLUORESCENT LABELS*

HERBERT MORAWETZ
Department of Chemistry, Polytechnic institute of New York, Brooklyn, New York 11201

SYNOPSIS

Excimer fluorescence, nonradiative energy transfer and the sensitivity of some fluorophores to the nature of the solvent medium observed in systems containing polymers carrying fluorescent labels may be used to investigate a variety of polymer problems. This approach is illustrated on studies of the dynamics of conformational transitions of chain molecules in dilute solution, polymer miscibility, the degree of chain entanglements in dilute polymer solutions as a function of the polymer concentration, self-diffusion in polymer melts, and the kinetics of polymer complex formation and dissociation.

INTRODUCTION

The use of fluorescent labels for the study of the physical chemistry of macromolecules was first applied by Weber for the estimation of the rotational diffusion constant of globular proteins [1]. The use of fluorescence for the characterization of synthetic polymers was pioneered by Nishijima et al. [2], who showed that it could be used to investigate the dynamics of flexible chains in dilute solution, and by Oster and Nishijima [3] who used it for the study of the rigidity of systems containing polymers. Over the past ten years, our research group has shown that fluorescent labels attached to flexible chain polymers can furnish information concerning a surprising variety of problems. In this work we used three properties of such labels, (a) the so-called excimer emission observed from sandwich complexes of an excited and a ground-state aromatic chromophore, (b) the nonradiative energy transfer from a "donor" fluorophore whose emission spectrum overlaps the absorption spectrum of an "acceptor" fluorophore, and (c) the sensitivity of the emission characteristics of the "dansyl" label on its contact with water molecules.

THE NATURE OF THE CONFORMATIONAL TRANSITIONS OF FLEXIBLE CHAIN MOLECULES IN DILUTE SOLUTION

Almost forty years ago, Kuhn and Kuhn [4] considered the dynamics of hindered rotations in the backbone of long chain molecules in dilute solutions and

*This paper is affectionately dedicated to Professor H. Mark whose enthusiasm for polymer research has been a constant inspiration and whose kindness has been one of the great pleasures through all the years of our association.

pointed out that such a rotation around a single bond would require a long chain segment to be dragged through the viscous medium with a very large expenditure of energy. They suggested that this difficulty could be circumvented by a correlated rotation around two bonds, such that only a short loop of the chain would have to move through the solvent (see Fig. 1). This type of "crankshaft-like motion" was later proposed as a mechanism for conformational transitions of polymers both in bulk [5–7] and in dilute solution [6, 7]. However, it was noted that a strict simultaneity of the passage over two energy barriers should double the activation energy of the process [8, 9]. Even if the energy barrier for a hindered rotation is only about 3 kcal/mole, this would mean that at ambient temperatures such rotations would be about a hundred times slower in the backbones of long molecular chains than in analogous small molecules in which only a single energy barrier has to be surmounted. Relaxing the requirement for simultaneity of the two rotations in the crankshaft-like motion would reduce the effective activation energy, but conformational transitions would still be expected to be significantly slower in polymer backbones than in their low molecular weight analogs.

In view of the above considerations it was desirable to compare the rates of hindered rotations in the backbone of a polymer and a small analogous

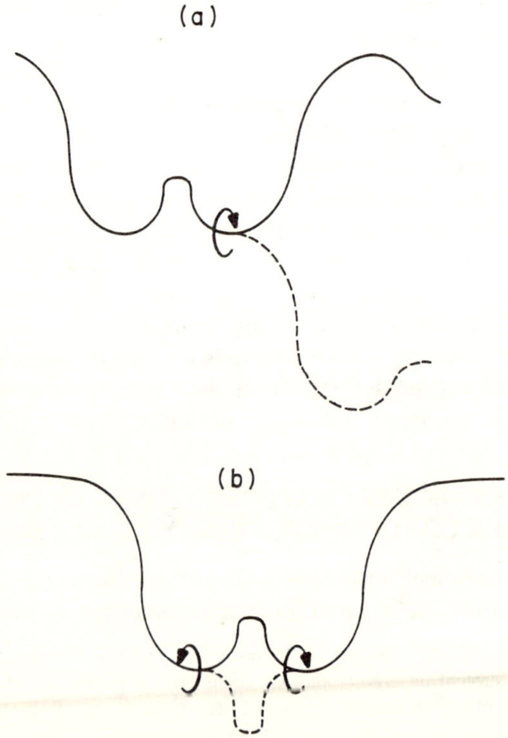

FIG. 1. Schematic representation of a conformational transition in the backbone of a polymer chain. (a) Rotation around a single bond. (b) Two correlated rotations in a crankshaft-like motion.

molecule. For this purpose, the phenomenon of excimer fluorescence provides a powerful experimental tool. In 1954, Förster and Kasper discovered that aromatic fluorophores may exhibit, above a critical concentration, an emission which is red-shifted from that observed in dilute solutions [10]. This phenomenon was eventually shown to be due to the formation of a sandwich complex between the excited and the ground-state molecule which became known as an "excimer." Somewhat later, Hirayama studied a series of α, ω-diphenylalkanes and found that their emission spectrum was similar to that of toluene except for the case of 1, 3-diphenylpropane which exhibited in addition to the normal emission a typical excimer fluorescence [11]. This was independent of solution concentration and, therefore, due to an intramolecular interaction of an excited and a ground-state phenyl group. Goldenberg et al. [12] showed that dibenzyl ether behaves in a similar manner and that excimer formation is particularly pronounced in acylated dibenzylamine. Since the g^+g^- conformation, in which the two phenyl groups lie face-to-face to each other, is unstable in the ground state, a $tg \rightarrow g^+g^-$ conformational transition has to take place during the lifetime of the excited phenyl group if excimer emission is to be observed:

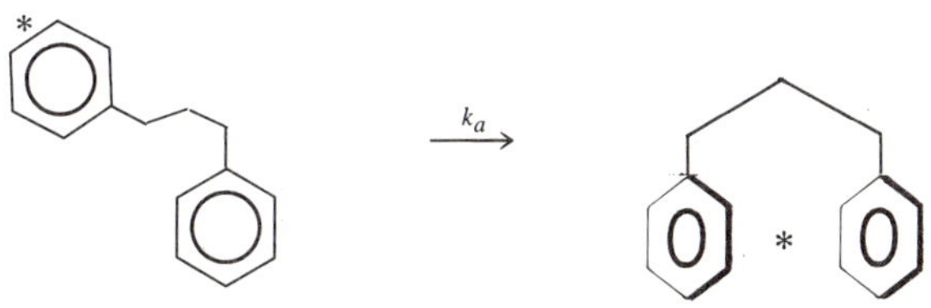

The ratio of the excimer and "monomer" emission intensity, I_e/I_m, should then be proportional to the product of the lifetime of the excited phenyl group and the rate constant k_a for the hindered rotation by which the excimer is formed.

Using this principle, we compared [13] a series of compounds of the type

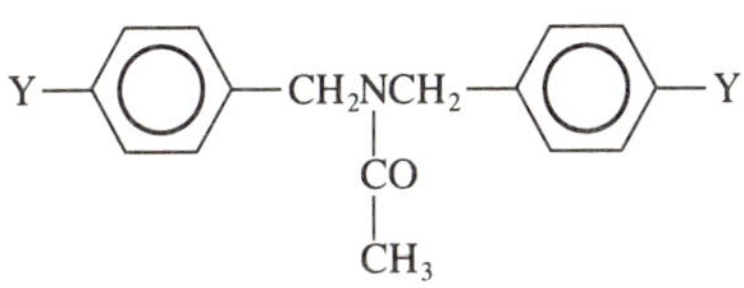

and found that I_e/I_m was reduced only by 27% when Y = CH_3 was substituted by a polyoxyethylene chain with a molecular weight of 3,000. Moreover, when the excimer-forming grouping was incorporated into the center of the polymer backbone, the temperature dependence of I_e/I_m was indistinguishable from that observed in an analogous small molecule. We concluded that the results are incompatible with the concept of a crankshaft-like motion and suggested that the stress introduced into the chain by one hindered rotation is relieved by many small distortions of internal angles of rotation. Computer simulation studies [14, 15] confirm our conclusion that the activation energy for conformational transitions in the backbone of long chain molecules is similar to that expected for a single hindered rotation, although the authors of these studies advance different interpretations of this result than proposed by us.

POLYMER COMPATIBILITY

Whereas low molecular weight organic molecules are almost universally miscible with each other, with polymers miscibility is much less common. The cause for this difference in behavior was first explained by Gee [16] who pointed out that the entropy of mixing, which is a powerful driving force with small molecules, is negligible in systems composed of polymers. Thus, mixing will be resisted if the process is endothermic as will almost always be the case unless specific group interactions of the acid-base or hydrogen-bonding type provide an energetic driving force.

A wide variety of methods have been developed in recent years for the study of polymer compatibility [17]. The method which we have introduced depends on the energy transfer between one excited fluorophore (the "donor") and a second fluorescent species (the "acceptor") so that irradiation of the donor leads to emission from the acceptor. The theory of this phenomenon was formulated by Förster [18] who showed that the sixth power of a characteristic distance R_o, at which half of the absorbed energy is transferred, is proportional to the "overlap integral" of the emission spectrum of the donor and the absorption spectrum of the acceptor. With an appropriate choice of donor-acceptor pairs, R_o values of 2–5 nm can be obtained.

When donor-labeled and acceptor-labeled polymers are blended, the distance between the donor and acceptor will be much smaller if the two polymers are miscible with each other than when they are segregated into separate phases. Thus, the ratio of the emission intensity from the acceptor and the donor, respectively, will increase with the extent of interpenetration of the chain molecules of the two species. Figure 2 shows the result of experiments in which this principle was used to study the compatibility of poly(methyl methacrylate) with methyl methacrylate-ethyl methacrylate or methyl methacrylate-butyl

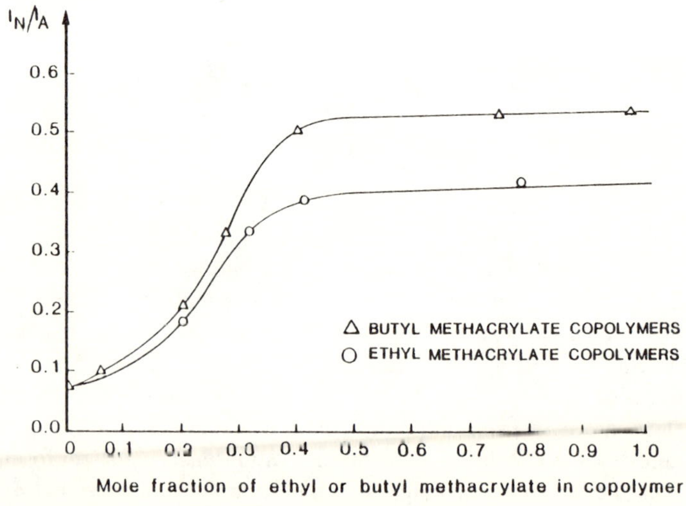

FIG. 2. Ratio of naphthyl and anthryl fluorescence intensities, I_N/I_A, from a film containing equal weights of anthryl-labeled poly(methyl methacrylate) and naphthyl-labeled methyl methacrylate copolymers.

methacrylate copolymers [19]. The extent of energy transfer first decreases continuously with an increasing difference in the composition of the two polymeric species, to attain eventually a limiting constant value. This suggests that there is no sharp transition from compatibility to incompatibility. In fact, Helfand and Tagami [20] have shown that for two polymeric species which are not too far from compatibility each polymer will have its chains penetrate, to a depth inversely proportional to the square root of the polymer-polymer interaction parameter, into the other polymer phase. The "fuzzy" phase boundary produced in this way has also been observed by electron microscopy [21].

These observations suggest that use of the energy transfer between fluorophore-labeled polymers reveals details of phase relationships in polymer blends which cannot be deduced from the appearance of glass transitions in DSC (differential scanning calorimetry) measurements—the most common approach to the characterization of polymer compatibility. In fact, this was the conclusion of two studies in Teyssié's laboratory [22, 23] in which the two techniques were carefully compared. For instance, in blends of poly(vinyl chloride) with syndiotactic poly(methyl methacrylate) (s-PMMA), DSC studies revealed a single T_g for 0–60 wt% of s-PMMA while the extent of energy transfer between fluorescent labels attached to the two polymers decreased continuously with an increasing s-PMMA content. Since the energy transfer was still much more efficient than expected for a system in which the dimensions of phase domains are much larger than R_o, the decreasing efficiency of energy transfer is perhaps more realistically thought of as due to increasing composition fluctuations rather than indicative of two phases with very small phase domains.

An interesting application of the energy transfer technique involved a study of blends of a block copolymer of styrene and *p-tert*-butylstyrene with the homopolymers of styrene and *p-tert*-butylstyrene [24]. Since the block copolymer can be labeled selectively in the polystyrene block, the mixing of the homopolymers with this block can be distinguished. As expected, this portion of the block copolymer mixes freely with polystyrene but excludes poly(*p-tert*-butylstyrene).

POLYMER CHAIN INTERPENETRATION IN DILUTE SOLUTION

It is well known that a strong entropic resistance opposes the interpenetration of flexible polymer coils in dilute solution. However, since these coils occupy a much larger volume than the "dry" polymer chains, a concentration is reached at relatively high dilution at which the total space is insufficient to allow the polymer coils to avoid entanglement. The significance of this "critical concentration" c* was first discussed by Weissberg et al. [25] who pointed out that c* should be proportional to the inverse of the intrinsic viscosity, $1/[\eta]$. In recent years, a paper by Daoud et al. [26] stimulated numerous experimental and theoretical studies of c* in polymer solutions.

While a variety of criteria allow an estimation of c*, it is obvious that the characterization of chain interpenetration should ideally do more than specify a "critical concentration," but describe the extent of such interpenetration as a function of polymer concentration. I believe that we have developed a technique which furnishes important insights into this problem: A solution of a polymer, of which half was labeled with a donor and half with an acceptor fluorophore, is rapidly frozen so that it can be safely assumed that the extent of chain interpenetration is not altered during the process. If the frozen solvent is now sublimed, the chain molecules in the resulting polymer sample retain a memory of the entanglement which existed in the solution from which they were derived. This will determine the energy transfer characterizing the reflectance fluorescence spectrum of a pellet pressed from the freeze-dried material. The feasibility of this approach was first demonstrated using polydisperse poly(ethyl methacrylate) [27] and more recently studied in greater detail with the use of monodisperse polystyrene [28]. Figure 3 shows the results obtained with polystyrene solutions in benzene and dioxane freeze-dried from an initial temperature of 25°C and cyclohexane solutions freeze-dried from an initial temperature of 40°C. In each case the ratio of the emission intensities of the anthracene acceptor and the carbazole donor, I_A/I_C, is compared with this ratio,

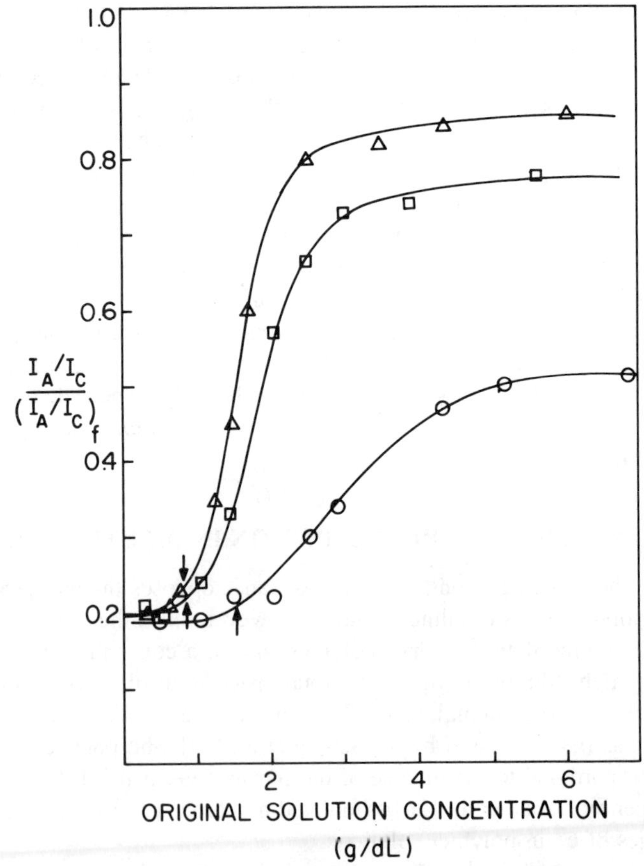

FIG. 3. Ratio of anthracene and carbazole emission intenities (at 450 nm and 365 nm, respectively) from freeze-dried mixtures of polystyrene (M = 410,000) carrying 9.2 mol/kg of the labels, compared to I_A/I_C in a film cast from these mixtures. Original solution in benzene at 25°C (Δ); dioxane at 25°C (□); cyclohexane at 40°C (○). Arrows indicate $1/[\eta]$ in the various solvents.

$(I_A/I_C)_f$, in a film cast from a mixture of the acceptor and donor-labeled polymer, where the two polystyrenes may be assumed to be randomly intertwined. The following features of the results may be noted: (a) When freeze-drying was carried out from high dilution, $(I_A/I_C)/(I_A/I_C)_f$ is independent of the solvent. In this range, there are no chain entanglements in solution, so that after freeze-drying each polymer molecule collapses into a spherical particle. (Such globules have been observed by electron microscopy [29, 30]). Energy transfer involves then only labels close to the surface of these globules. (b) Chain entanglements, as reflected by an increase in the I_A/I_C ratio, become significant above a concentration given approximately by $1/[\eta]$, i.e., at a concentration which increases with the solvent in the order benzene $>$ dioxane $>$ cyclohexane, the order of decreasing solvation of the polymer. (c) For $c > c^*$, I_A/I_C increases with an increase of the concentration from which the polymer was freeze-dried, but levels off at about $3/[\eta]$ to a value which decreases with a decreasing solvent power of the original solvent.

The leveling off of the energy transfer, indicates that beyond a certain polymer concentration the cores of the polymer coils are highly resistant to interpenetration. This observation is particularly important in the case of the cyclohexane solution, since a polystyrene solution in cyclohexane at 40°C is close to the theta point of 34.4°C [31]. In the Flory-Krigbaum theory of the second virial coefficient, A_2, of flexible chain molecules [32] the polymer chain was represented by a cloud of disconnected chain segments and A_2 vanished when the excluded volume vanished for a single chain segment. This led to the suggestion, implied by Orofino and Flory [33] and stated explicitly by Yamakawa [34], that polymer coils are freely interpenetrable in theta solvents. Yet, topological constraints render this conclusion highly improbable, lending credance to the result obtained by Olaj and Pelinka [35]. Their computer simulation indicated that in a theta medium the free energy of two polymer coils depends on the spacing of their centers of gravity, a moderate interpenetration being favored, while a very deep interpenetration is highly improbable; thus, the excluded volume vanishes because of a compensation of a positive and a negative term. Our results provide strong support for this conclusion.

SELF-DIFFUSION IN POLYMER MELTS

Self-diffusion of long chain molecules in polymer melts is generally assumed, as proposed by de Gennes [36], to take place by a reptation mechanism leading to a diffusion coefficient inversely proportional to the square of the chain length. This means that the self-diffusion coefficient D assumes extremely low values for long polymer chains, so that its experimental determination requires the use of unconventional techniques. The mean square displacement of a particle in a specified direction, $\overline{x}^2$, depends on time t as $\overline{x}^2 = 2Dt$, so that the smallest value of D which can be characterized within a reasonable time depends on the spatial resolution obtainable in a given experimental method.

The first attempt to check de Gennes' theory of the chain length dependence of D was carried out by Klein and Briscoe [37] who used IR (infrared) spectroscopy to follow the diffusion between two polyethylene sheets, one of which was deuterated. The method had a spatial resolution of 0.01 cm, which allowed over an experimental time of one day (10^5s) the characterization of D values of the order of 10^{-9} cm^2 − s^{-1}, namely, those of polyethylenes with molecular weights up to 23,000 at 176°C. Clearly, the development of methods with a much higher spatial resolution is needed to extend measurements to polymers of much higher molecular weight and to temperatures relatively close to the glass transition.

We have approached this problem [38] by using as a specimen a pellet pressed from the material obtained by freeze-drying from a highly dilute solution a mixture of donor and acceptor labeled polymer. As stated in the preceding section, this leads to the collapse of each chain molecule into a globule and energy transfer, as evidenced from the emission spectrum, will be small if the characteristic transfer distance R_o is small compared to the dimension of the globule. On heating the pellet above T_g, interdiffusion of the polymer will lead to an increasing energy transfer, which will tend towards an equilibrium value observed for films cast from a mixture of the donor- and acceptor-labeled polymer. Using the diameter d of the globules as a characteristic length and the time $t_{1/2}$ required for the ratio of acceptor and donor emission intensity to change half-way to its equilibrium value, we proposed that D should be of the order of $d^2/t_{1/2}$. Thus, with d of the order of 10^{-6} cm, one should be able to characterize D as small as 10^{-16} cm^2 − s^{-1} within a few hours. (For such small D, it would take a particle 10 million years to diffuse 1 mm!). Our published results [38] give data for poly(methyl methacrylate) with viscosity average molecular weights up to 5×10^5. Obviously, the significance of such data is uncertain if obtained on polydisperse samples and we are now trying to study by the same technique the behavior of monodisperse polystyrene.

It should be pointed out that the process we are observing involves the "unfolding" of the chain from a compact globular state to its equilibrium expansion and such a process is different from diffusion. It is possible to conjecture that the two processes should be closely related, but this is hard to prove. A rigorous method with the ability to characterize similarly small diffusion coefficients has been developed in Sillescu's laboratory [39]. Our method recommends itself, perhaps, by its experimental simplicity.

EQUILIBRIA AND KINETICS OF POLYMER COMPLEXATION

The mixing of two uncharged water-soluble polymers may lead under certain conditions to the formation of a water-insoluble complex. Such a phenomenon was first described by Bailey et al. [40] for the association of poly(acrylic acid) with polyoxyethylene (POE). This and other analogous systems have been studied in great detail in Kabanov's laboratory, leading to the conclusions that the process is highly cooperative [41], that hydrogen bonding provides the main driving force [42] and that complexation involves uninterrupted sequences of hydrogen bonds between the monomer residues of the two chains [43].

We have used a novel approach to the study of such complexes which lends itself both to the study of complexation equilibria and to the measurement of the rates at which complexes form and dissociate. In this procedure one of the interacting polymers carries a fluorescent label whose emission intensity is sharply reduced by contact with water molecules. Weber and Laurence [44] found in 1954 that the "dansyl" residuc has this property and this group has since found wide application as a "reporter" of the microenvironment of spccific points in biological macromolecules.

When a water solution of a dansyl-labeled poly(acrylic acid) (dan-PAA) is mixed with a water solution of POE, the formation of the complex displaces water from the vicinity of the dansyl label and the fluorescence intensity increases sharply [45]. Our first experiments [45] indicated that complexation is favored by an increasing chainlength of POE, as would be expected from the cooperativity of the process. However, a later study [46] led to the unexpected finding that the fluorescence intensity increases much less if POE with a molecular weight above 10^5, rather than a molecular weight of 20,000, participates in the complex formation. Moreover, while the intensification of the emission intensity of the dansyl label on transfer from water to an organic solvent is normally accompanied by a red shift of the emission maximum, a shift in the opposite direction is observed in this case. This has been tentatively interpreted as suggesting that the dan-PAA is complexed with the POE at widely separated regions along the contour of the chain molecule, while the intervening parts of the chain are highly extended, so that the dansyl label is even in a more intimate contact with water than in the absence of POE.

Using a stopped-flow apparatus with fluorescence detection, the rate of complexation of dan-PAA with POE may be followed [47]. Since the complex dissociates when the carboxyl groups are ionized, the dissociation of the complex after a pH jump may also be followed kinetically. Both complex formation and complex dissociation have very low activation energies, such as characterize diffusion-controlled processes. Their relative slowness, which makes it possible to follow them by a technique with a resolution in the millisecond range, is the result of an unusually high negative entropy of activation. This may be understood as follows: A carboxyl and an ether group diffusing toward each other will form the first hydrogen bond between the dan-PAA and the POE. However, to obtain a stable complex, this must be followed by extensive conformational transitions of the two polymers, to allow a sufficient number of interacting groups to form additional hydrogen bonds and this must lead to a large decrease in entropy.

KINETICS OF THE CONFORMATIONAL TRANSITION OF POLY(METHACRYLIC ACID) AFTER A pH JUMP.

It is well known that the chain molecules of poly(methacrylic acid) are highly contracted at low degrees of ionization but expand rather rapidly at a critical charge density [48]. Dansyl labeled copolymers of maleic acid and vinyl ethers have been used by Strauss and his students [49, 50] to monitor the transitions from a contracted state, in which the label is shielded from contact with water molecules, to the expanded state where water quenches the dansyl fluorescence and it is not surprising that dansyl labeled poly(methacrylic acid) should behave in an analogous manner. This induced us to try to use our stopped-flow apparatus to follow the kinetics of the expansion and the collapse of dansyl-labeled poly(methacrylic acid) molecules after a sudden increase or decrease of the pH [51]. We found that part of the fluorescence change was too rapid to be followed kinetically by this procedure. The remainder followed biphasic kinetics. The rate of the final approach to conformational equilibrium increased with increasing pH as would be expected, since in the more contracted molecules conformational transitions should be hindered by chain entanglements, hydrogen bonds between the carboxyls and hydrophobic interactions between the methyl groups. However, even at pH 6.83, the ralaxation time for this final equilibration at 25°C was 28 ms, orders of magnitude larger than one might have expected.

We are grateful for financial support of this research by the National Science Foundation Grant DMR-77-05210, Polymers Program.

REFERENCES

[1] G. Weber, *Biochem. J., 51,* 145, 155 (1951).
[2] Y. Nishijima, A. Teramoto, M. Yamamoto, and S. Hiratsuke, *J. Polym. Sci., A-2, 5,* 23 (1967).
[3] G. Oster and Y. Nishijima, *Fortschr. Hochpolym. Forsch., 3,* 313 (1964).
[4] W. Kuhn and H. Kuhn, *Helv. Chim. Acta, 29,* 857 (1946).
[5] T. F. Schatzki, *J. Polym. Sci., 57,* 496 (1962).
[6] W. Pechold, S. Blasenbrey, and S. Woerner, *Kolloid Z., Z. Polym., 189,* 14 (1963).
[7] W. H. Stockmayer, *Pure Appl. Chem., 15,* 539 (1967).
[8] S. Mashimo, *Macromolecules, 9,* 91 (1976).
[9] G. Allegra, *J. Chem. Phys. 61,* 4910 (1974).
[10] T. Förster and K. Kasper, *Z. Phys. Chem. (Frankfurt), 101,* 275 (1954).
[11] F. Hirayama, *J. Chem. Phys., 42,* 3163 (1965).
[12] M. Goldenberg, J. Emert, and H. Morawetz, *J. Am. Chem. Soc., 100,* 7171 (1978).
[13] T. P. Liao and H. Morawetz, *Macromolecules, 13,* 1228 (1980).
[14] E. Helfand, Z. R. Wasserman, and T. A. Weber, *Macromolecules, 13,* 526 (1980).
[15] D. Perchak and J. H. Weiner, *Macromolecules, 14,* 785 (1981).
[16] G. Gee, *Q. Rev. Chem. Soc., 1,* 265 (1947).
[17] O. Olabisi, L. M. Robeson, and M. T. Shaw, *Polymer-Polymer Miscibility,* Academic, New York, 1979.
[18] T. Förster, *Disc. Faraday Soc., 27,* 7 (1959).
[19] F. Amrani, J.-M. Hung, and H. Morawetz, *Macromolecules, 13,* 649 (1980).
[20] E. Helfand and Y. Tagami, *J. Polym. Sci., B, 9,* 741 (1971).
[21] A. Voyutskii, N. Kamenskii, and N. M. Fodiman, *Kolloid-Z., 215,* 36 (1966).
[22] B. Albert, R. Jerome, P. Teyssié, G. Smyth, N. G. Boyle, and V. J. McBrierty, *Macromolecules, 18,* 388 (1985).
[23] B. Albert, R. Jerome, P. Teyssié, and D. Bayens-Volant, *J. Polym. Sci., Polym. Chem. Ed., 24,* 2577 (1986).
[24] F. Mikeŝ, H. Morawetz, and K. S. Dennis, *Macromolecules, 17,* 60 (1984).
[25] S. G. Weissberg, R. Simha, and S. Rothman, *J. Res. Natl. Bur. Standards, 47,* 298 (1951).
[26] M. Daoud, J. P. Cotton, B. Farnoux, G. Jannink, G. Sarma, H. Benoit, R. Duplessis, C. Picot, and P.-G. de Gennes, *Macromolecules, 8,* 804 (1975).
[27] J. Jachowicz and H. Morawetz, *Macromolecules, 15,* 828 (1982).
[28] L.-C. P. Chang and H. Morawetz, *Macromolecules, 20,* 428 (1987).
[29] R. M. Siegal, D. H. Johanson, and H. Mark, *J. Polym. Sci., 5,* 111 (1950).
[30] M. J. Richardson, *Proc. Roy, Soc. London, Sec. A, 279,* 50 (1964).

[31] Y. Miyake, Y. Einaga, and H. Fujita, *Macromolecules, 11,* 1180 (1978).
[32] P. J. Flory and W. R. Krigbaum, *J. Chem. Phys., 18,* 1086 (1950).
[33] T. A. Orofino and P. J. Flory, *J. Chem. Phys., 26,* 1067 (1957).
[34] H. Yamakawa, *Modern Theory of Polymer Solutions,* Harper and Row, San Francisco, 1971, p. 169.
[35] O. F. Olaj and K. H. Pelinka, *Makromol. Chem., 177,* 3413 (1976).
[36] P.-G. de Gennes, *J. Chem. Phys., 55,* 572 (1971).
[37] J. Klein and B. J. Briscoe, *Proc. Roy. Soc. London, Ser. A, 365,* 53 (1979).
[38] T. Y.-J. Shiah and H. Morawetz, *Macromolecules, 17,* 792 (1984).
[39] M. Antonietti, J. Contandin, R. Grütter, and H. Sillescu, *Macromolecules, 17,* 798 (1984).
[40] F. E. Bailey, R. D. Lundberg, and R. W. Callard, *J. Polym. Sci., A-2,* 845 (1964).
[41] A. D. Antipina, V. Yu. Baranovsky, I. M. Papisov and V. A. Kabanov, *Vysokomol. Soedin., A-14,* 941 (1972).
[42] I. M. Papisov, V. Yu. Baranovsky, I. Sergeeva, A. D. Antipina, and V. A. Kabanov, *Vysokomol. Šoed., A-16,* 1133 (1974).
[43] V. Yu. Baranovsky, A. A. Litmanovich, I. M. Papisov, and V. A. Kabanov, *Eur. Polym. J., 17,* 969 (1981).
[44] G. Weber and D. J. R. Laurence, *Biochem. J., 56,* XXXI (1954).
[45] H.-L. Chen and H. Morawetz, *Eur. Polym. J., 19,* 923 (1983).
[46] B. Bednář, Z. Li, Y. Huang, L.-C. P. Chang, and H. Morawetz, *Macromolecules, 18,* 1829 (1985).
[47] B. Bednář, H. Morawetz, and J. A. Shafer, *Macromolecules, 17,* 1634 (1984).
[48] J. C. Leyte and M. Mandel, *J. Polym. Sci., A-2,* 1879 (1964).
[49] U. P. Strauss and G. Vesnaver, *J. Phys. Chem., 79,* 1558, 2426 (1975).
[50] U. P. Strauss and M. S. Schlesinger, *J. Phys. Chem., 82,* 1627 (1978).
[51] B. Bednář, H. Morawetz and J. A. Shafer, *Macromolecules, 18,* 1940 (1985).

Herbert Morawetz was born in Prague in 1915. He studied chemical engineering at the University of Toronto obtaining a M.A.Sc. in 1944. In 1945 he came to the U.S.A. where he worked until 1949 as a research chemist for the Bakelite Co. He received a Ph.D. in chemistry at the Polytechnic Institute of Brooklyn (now the Polytechnic Inst. of New York) in 1950 and has been since 1951 a member of its Chemistry Department. From 1971 to 1981 he was Director of the polymer research Institute and in 1981 he was named Institute Professor at the Polytechnic. In 1980 Case Western University named him Centennial scholar, and in 1985 he received the Award in Polymer Chemistry of the American Chemical Society.

Dr. Morawetz's scientific publications deal with various problems in the physical chemistry of polymers, particularly polyelectrolyte behavior, molecular association, polymerization in the crystalline state, polymer reactivity, polymer compatibility and rates of conformational transitions of polymers and their low molecular weight analogs. In 1965 he published a monograph "Macromolecules in Solution" and a revised edition of this book was published in 1975. During the past four years he has written a history of polymer science which was published in March 1985 under the title "Polymers—The Origins and Growth of a Science".

REACTIONS OF HYDROPHILIC POLYMERS

C. G. OVERBERGER
Department of Chemistry and the Macromolecular Research Center, University of Michigan, Ann Arbor, Michigan 48109

SYNOPSIS

Nucleic acids contain several characteristic structural features: a chiral center attached adjacent to the base, a stereoregular, hydrophilic poly(phosphate-ribose) backbone which is negatively charged at neutral pH, and a helical secondary structure arising from conformational preferences enforced by base stacking and inter-chain hydrogen bonding. After the structure and conformational properties of the nucleic acids were elucidated, chemists began to ask whether these features could be built into totally synthetic systems. Early work focused on the polymerization of modified nucleosides, and then advanced to more structurally remote analogs such as polymers of N-vinyl derivatives of the bases. Throughout this work people were concerned with several major aspects. First, did the synthetic polymers possess conformational order resembling polynucleotides—did they show base stacked conformations in solution? Second, was it possible for these polymers to undergo base pairing interactions with suitable complementary polymers? If the answers to these questions was yes (and often times it has been), then could the polymers exhibit some type of activity when introduced into a biological system? A number of these polymers have shown either in vitro or in vivo activity against viruses or tumors and some have been shown to affect either protein synthesis or the replication mechanisms of viruses.

Several synthetic polynucleotide analogs were prepared having hydrophilic polymer backbones such as polyethyleneimine, poly(vinylamine), and poly(dehydroalanine) as polymer backbones in place of the phosphate-ribose backbones, and chiral nucleic acid base derivatives such as 2-(thymin-1-yl)propionic acid, 3-(thymin-1-yl)butyric acid, and 3-(adenin-9-yl)butyric acid as pendants through amide bond formation. The thyminyl pendants were resolved by using conventional methods. A basic resolving agent such as quinine or brucine was used to react with the acids to form the diastereomers which were subjected to fractional crystallization to separate both enantiomers. In the case of the optical resolution of 3-(adenin-9-yl)butyric acid, its ammonium salt, ammonium 3-(adenin-9-yl)butyrate was used to react with brucine at a refluxing water-methanol mixture to generate a stable salt by releasing ammonia gas. The salt was subjected to fractional crystallization to obtain the partially resolved enantiomers. The pure enantiomers were obtained by the successive crystallization procedure based on the fact that racemic mixtures have different solubility from their corresponding enantiomers, and a partially resolved enantiomer can be treated as a combination of a pure enantiomer and racemic mixture.

The preparation of the polynucleotide analogs was carried out by grafting the pendants onto the polymer backbone through amide bond formation using either a direct coupling agent or an active ester method. Mild reaction conditions were employed to prevent racemization. Free radical polymerization of dehydroalanine containing nucleic acid bases was carried out to obtain quantitatively grafted polymers. When the unprotected 3-(adenin-9-yl)butyric acid was used as a pendant for the grafting reaction, the active ester of the acid was prepared by reacting with N-hydroxysuccinimide in the presence of dicyclohexylcarbodiimide as a dehydrating agent at low temperature. Because of the selectivity of this type of active ester, the active ester of the unprotected adenine derivative could be isolated. The active ester reacted with polyethylenimine at room temperature to obtain the polynucleotide analogs.

When poly(vinylamine) was used as the polymer backbone, a pretreatment on the polymer was employed. Since poly(vinylamine) cannot be isolated from an aqueous solution without crosslinking, most grafting reactions of poly(vinylamine) were carried out in aqueous solution. Since our active ester reacted with water, poly(vinylamine) had to be isolated from water for further reaction. When aqueous poly(vinylamine) was mixed with more than six equivalents of N hydroxysuccinimide, the polymer complex could be isolated from water and dissolved in dimethyl sulfoxide. The resulting polymer complex could react with active ester to obtain the polynucleotide analogs or react with acetic anhydride to obtain acetylated poly(vinylamine) for a tacticity study on the poly(vinylamine).

The amide bond increases structural order through its partial double bond character, resulting in restricted rotation of the pendants. A study of the model compounds using UV and NMR spectroscopy has revealed considerable evidence for polynucleotide-like order conformations in these polymers. Significant base stacking was observed in most of these systems. The optically active structures were examined by means of circular dichroism, the first nucleic acid models without a phosphate-ribose backbone which could be so examined. The poly(ethylenimine) graft containing thymine displayed exciton coupling, characteristic of the highly ordered arrangement of the bases in a polynucleotide.

A program of biological testing of these compounds has begun and so far has shown very encouraging results. In most of the cases where the polymers are soluble enough to be tested, antiviral activity has been observed during cell culture tests against a number of viruses of interest in human diseases. Current work involves structural modification of the polymers and further biological tests so that more information on their mechanism of action can be obtained.

The imidazole group is an important nucleophile in catalytic systems. It was found by attaching imidazole to polymer backbones the rate of esterolysis of activated esters could be increased manyfold over monomeric models. Some more recent work on water soluble polymers can be explained with reference to cooperative, hydrophobic and electrostatic effects in the hydrolyses.

INTRODUCTION

Nucleic acids are responsible for the storage and transmission of information necessary to classify living systems. They contain several characteristic structural features: a chiral center adjacent to the base, a stereoregular, hydrophilic poly(phosphate-ribose) backbone, and a helical secondary structure arising from conformational preferences enforced by base stacking and interchain hydrogen bonding. The physical and chemical properties of nucleic acids have not been well rationalized with respect to their complicated structures. The complexity of the sugar alone produces many conformational possibilities. In addition to further research on the nucleic acids, chemists have sought to synthesize simpler polymer models of the nucleic acids in order to better understand their behavior and obtain structure-function relationships. We were concerned with several major aspects. First, did the synthetic polymers possess conformational order resembling polynucleotides—did they show basestacked orientations in solutions? Second, was it possible for these polymers to undergo basepairing interactions with suitable complementary polymers? Could the polymers exhibit some type of activity when introduced into a biological system?

Our synthetic program turned toward the study of polynucleotide analogs, both for their value as models of nucleic acids and for their biological properties. Considerable attention has been given to these analogs; this includes both the homopolymers with poly(phosphate-ribose) backbone, most closely resembling natural nucleic acids, as well as nucleic acid models where the backbone is other than poly(phosphate-ribose) and/or the pendant groups have been modified [1–4]. There have been some studies on the biological activity of the nucleic acid models with backbone and/or pendant group modification [5]. As an example of one of the more well-studied systems, poly(9-vinyladenine) has been found to inhibit the replication of Friend leukemia virus in mice.

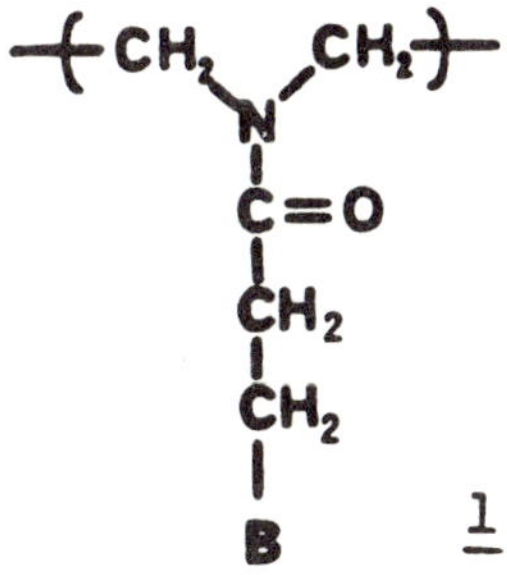

1

B = adenin-9-yl
thymin-l-yl

Our interest in synthetic nucleic acid models began with our experience in making compound 1. These linear PEIs containing nucleic acid base grafts showed the ability to interact by complementary hydrogen bonding with each other and with other polynucleotide analogs such as poly U [6, 7]. The polymers also gave evidence of significant base stacking. In view of these results and the promising biological activity observed in other polynucleotide analogs, we decided to further develop this strategy. Our goals were to increase structural order in the polymers, maximize water solubility, and elucidate the relationship between macromolecular structure and biological activity. Our overall approach combines several advantages:

1. Carboxylic acid derivatives with α- or β-nucleic acid substituents were synthesized. These derivatives were resolved using resolving agent and fractional crystallization procedures, thereby allowing only one enantiomer to be attached to the polymer backbone. In addition, these derivatives are amenable to the methods used in preparing polypeptide amide bonds.

2. Hindered rotation of the nucleic acid base pendant about the amide bond formed during grafting promotes an ordered, base-stacked structure of the type found in such synthetic polymers as polyadenylic acid (poly A) and polyuridylic acid (poly U). The *trans* conformation of the amide bond is greatly favored; this was demonstrated by NMR (nuclear magnetic resonance) studies of model compounds.

3. If a suitable amine-containing vinyl monomer or its precursor is available, the amide bond formation described above can produce optically active nucleic base monomers amenable to the many methods of vinyl polymerization.

4. Many backbone and pendant modifications are possible while retaining most of the above advantages to the retention of order. Macromolecular structure can thus be tailored in response to the results of biological testing, if such information is available.

RESULTS AND DISCUSSION

Preparation of Pendants

Two general methods were used to synthesize the pendants as demonstrated in Scheme 1 [8, 9] and Scheme 2 [10–12]. The hypoxanthine derivative was prepared by treatment of the adenine derivative with hydrochloric acid in a sodium nitrite aqueous solution [13].

OPTICAL RESOLUTION OF PENDANTS

2-(Thymin-l-yl)propionic Acid

One enantiomer of 2-(thymin-l-yl)propionic acid can be obtained easily by the reported procedure [8]. However, the other enantiomer is very difficult to obtain. A modified procedure [14] (Scheme 3) was used to obtain both enantiomers in good yields. The (−) enantiomer was obtained by crystallizing the salt for three times as in the original procedure. The filtrates were combined and the resolving agent, quinine, was removed by passing the mixture through

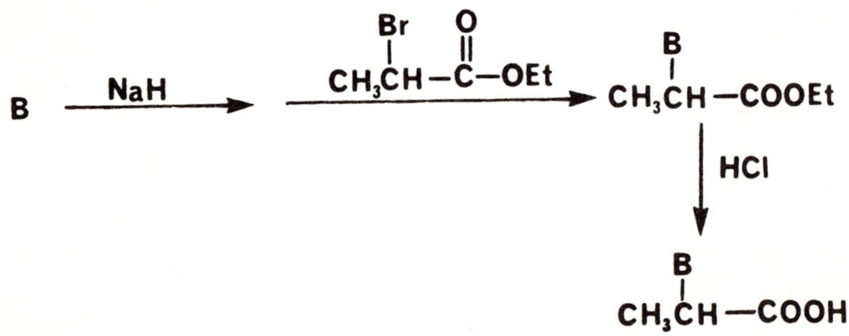

B = thymin-l-yl; adenin-9-yl; uracil-l-yl; cytosin-l-yl

Scheme 1. Preparation of pendants.

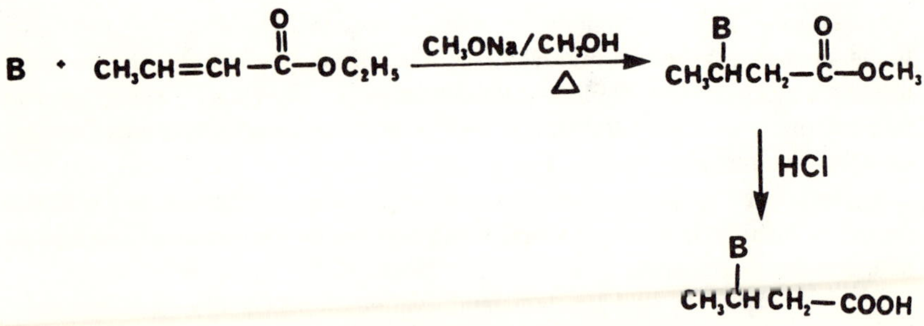

B = thymin-l-yl; adenin-9-yl; uracil-l-yl; cytosin-l-yl

Scheme 2. Preparation of pendants.

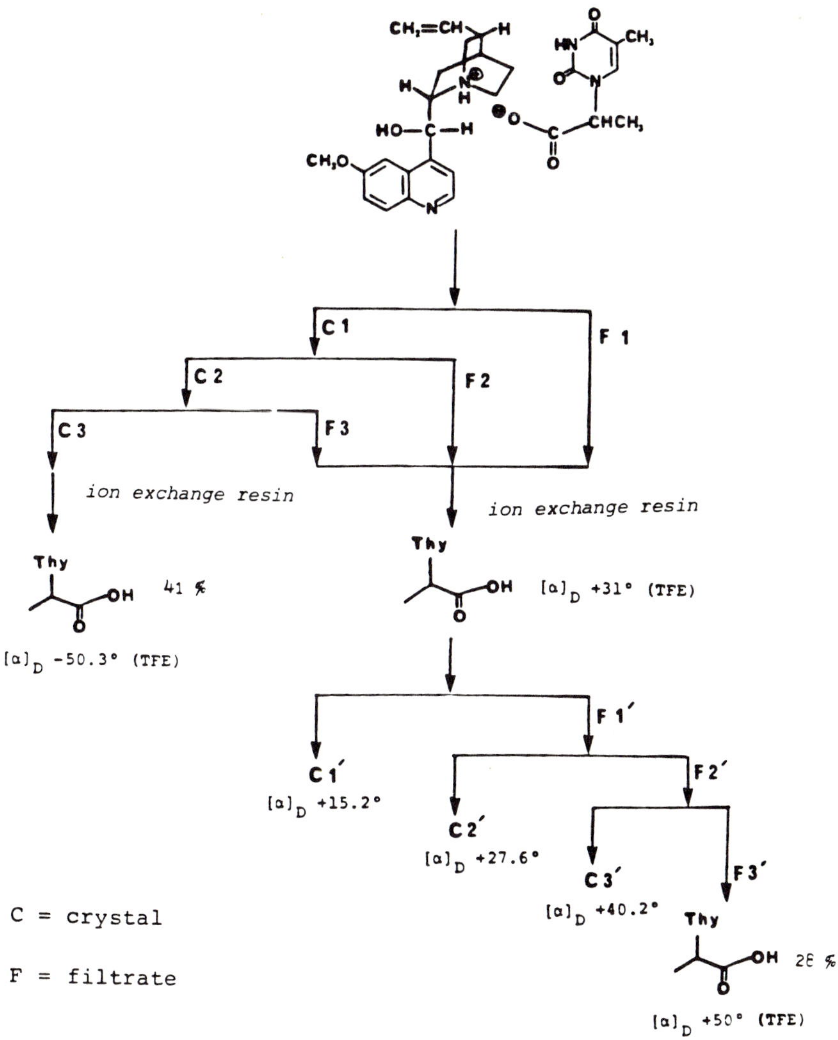

Scheme 3. Optical resolution of 2-(thymin-l-yl)propionic acid.

an ion exchange resin column. The partially resolved (+) enantiomer was obtained. A racemic mixture has a different solubility from a pure enantiomer. A partially resolved enantiomer can be treated as a combination of a racemic mixture and one pure enantiomer. The partially resolved (+) enantiomer was subjected to crystallization from absolute ethanol. The pure enantiomer has higher solubility than its racemic mixture; thus the pure (+) enantiomer can be obtained in the final filtrate.

3-(Adenin-9-yl)butyric Acid

The attempted optical resolution of the unprotected adenine derivative was reported [8]. A stable salt could not be formed. The difficulty is that the bifunctionality of the compound causes the formation of a polymeric ionic species. This problem was overcome by employing the ammonium salt of 3-(adenin-9-yl)butyric acid instead of its free acid. The carboxylic group was protected by the formation of ammonium salt. As demonstrated in Schemes 4(a) and 4(b), ammonia gas is released upon refluxing with the basic resolving agent; a stable salt was obtained. The salt was subjected to fractional crystallization using a 1:4 water-acetone mixture as the solvent. The crystals and filtrates were combined separately. After brucine was removed by adjusting the pH to 4.0, the partially resolved enantiomers were obtained. They had specific rotations of +20.3° and −18.4°C, respectively. The partially resolved adenine pendant was crystallized from absolute ethanol. The crystals showed lower optical rotation than the filtrates. This result implied that the pure enantiomer had higher solubility than the racemic mixture. Thus the pure enantiomers were enriched in the filtrates [11]. The trace amount of brucine was removed by crystallization from 95% ethanol.

3-(Thymin-l-yl)butyric Acid

This acid was resolved by using brucine as the resolving agent and a 1:2 water-acetone mixture as the fractional crystallization solvent [10]. The same procedure as Scheme 4(b) was employed to obtain the partially resolved enantiomers. However, for this acid the pure enantiomers had lower solubilities in absolute ethanol than their racemic mixture. Thus, the pure enantiomers were obtained by crystallizing the partially resolved enantiomers until the optical rotation remained constant.

GRAFTING METHOD

Grafting of the optically active pendants onto the hydrophilic polyamines was carried out by the formation of an amide bond. The amide bond increases structural order by virtue of its partial double bond character, resulting in some restricted rotation of the pendants. Techniques useful in peptide synthesis have been employed here for the grafting procedure. In general, there are two types of methods for this purpose. The first is a direct coupling method which employs a coupling agent in the presence of one equivalent of base to form an amide bond. Some examples of a coupling agent are shown in Scheme 5. One disadvantage of this direct coupling method is that a base is required for the reaction. The second method may be named the active ester method. Scheme 6 demonstrates this method and some examples of the *N*-hydroxy compounds. The reaction of an acid with a *N*-hydroxy compound using DCC as the dehydrating agent generates an active ester. The isolated active ester reacts with amine at room temperature to generate the amide. The reactivity of a given active ester, R—CO—X (prepared from the *N*-hydroxy compound, HX), toward amines depends markedly upon the nature of the leaving group X. Ueda [15] and Lu [16] and colleagues demonstrated that the active esters have selectivity toward both aliphatic and aromatic amines. The active esters of

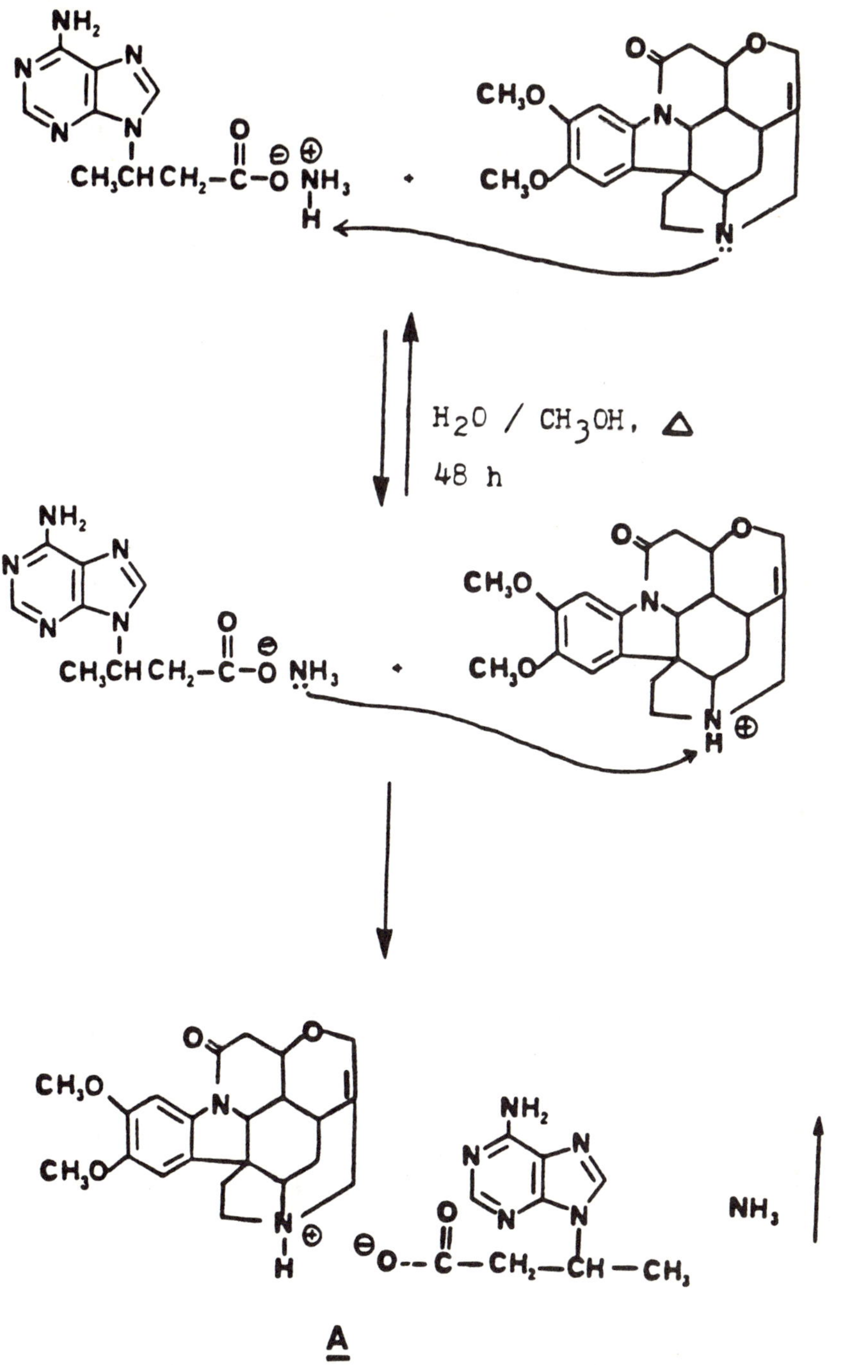

Scheme 4(a). Formation of diastereomeric salt of unprotected adeninyl pendant.

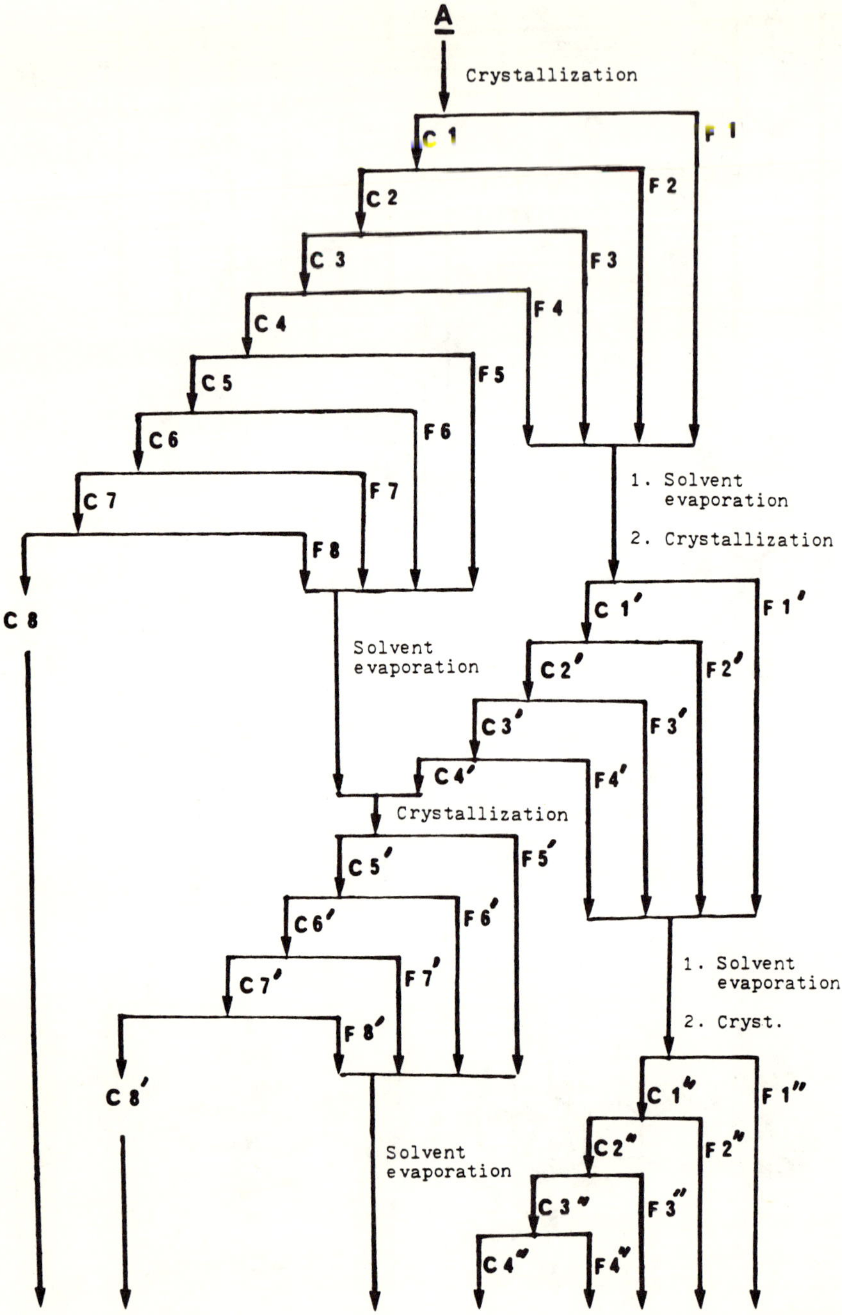

N-hydroxysuccinimide do not react with aniline at room temperature; on the other hand, the active esters of the more acidic *N*-hydroxy compounds such as 3-hydroxy-4-oxo-3,4-dihydro-1,2,3-benzotriazole react with aniline at room temperature to form an amide bond. We employed *N*-hydroxysuccinimide

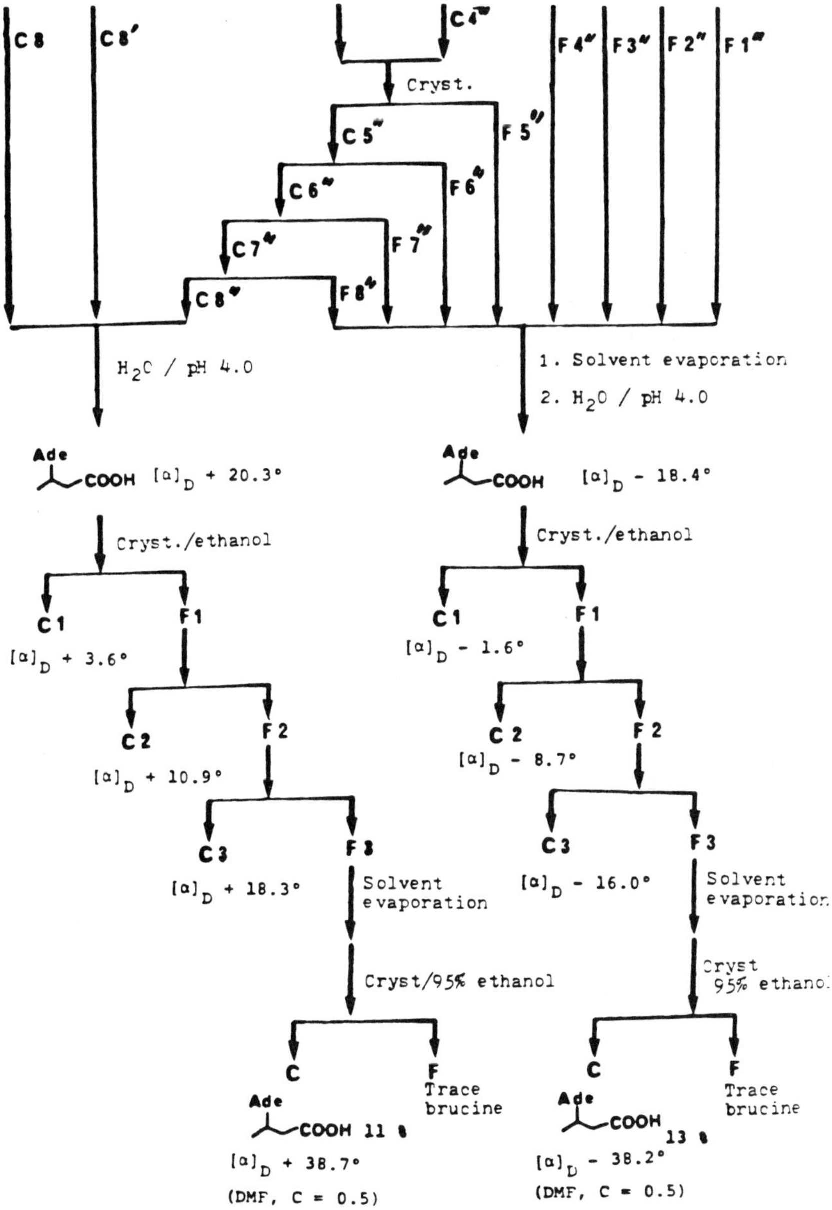

Scheme 4(b). Optical resolution of 3-(adenin-9-yl)butyric acid.

for the successful preparation of the active esters of the unprotected adenine pendants.

POLYETHYLENIMINE ANALOGS

Our most studied system uses linear polyethylenimine (PEI) as the backbone. Linear PEIs of several molecular weights have been used for grafting α-nucleic

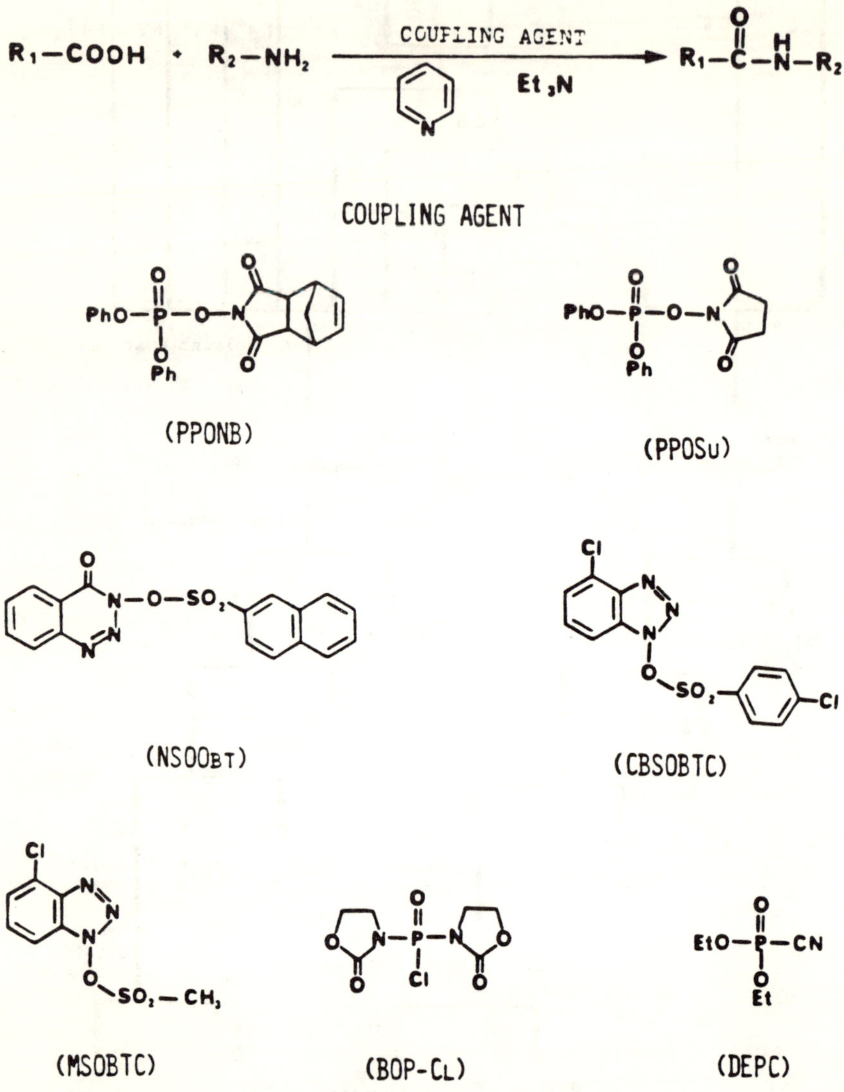

Scheme 5. Direct coupling method of formation of amide bonds.

acid base-substituted pendants. These are described in Scheme 7. The PEIs were obtained by cationic polymerization of 2-substituted-2-oxazolines. Polymer *2* is an optically active compound. This has allowed us to examine the system using not only ultraviolet (UV) spectroscopy (hypochromicity studies) but also circular dichroism (CD) [17]. The CD spectrum of the dimer model compound is shown in Figure 1. The B_{2u} band is caused by the base chromophore. A strong dependence of the exciton splitting of the B_{2u} band on temperature was observed for polymer *2* (Fig. 2). The splitting disappeared if a 2-proline spacer group was placed between the backbone and the base moiety [18]. The

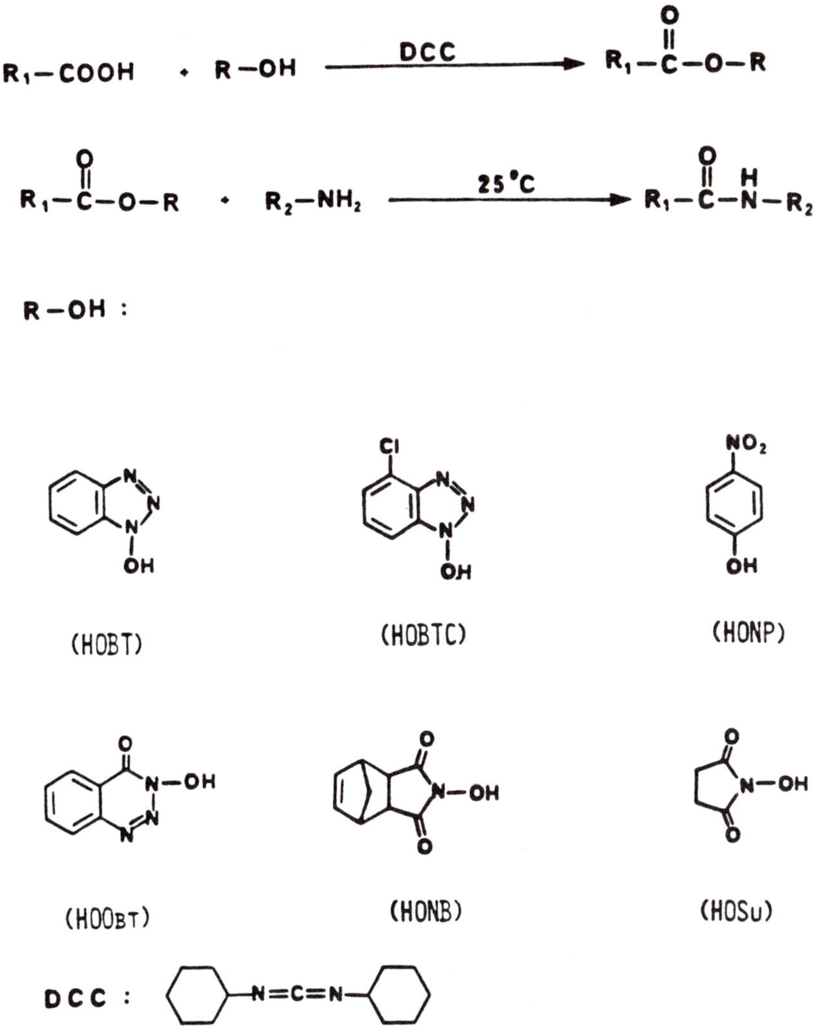

Scheme 6. Active ester method of formation of amide bonds.

UV spectra of this system in water indicate significant hypochromicity (36.8%). From the CD and UV studies we are able to rationalize that the polymer has some conformational order with significant base stacking.

Significant hypochromicity (45.4%) has been observed for the corresponding racemic adenine graft polymer *3(a)*. The conclusion from the hypochromicity studies is that polymers *2–4* have more conformational order than the single strand poly A and poly U. A combination of *2* and *3(a)* in neutral aqueous solution resulted in marked hypochromic effects at certain mixing ratios (mole fraction *3(a)* equals 0.8 and 0.3). In acidic aqueous solution, only a small deviation from Beer's law was observed. Mixing *3(a)* (racemic) and *2* in chloroform/ethanol (1/1, v/v) also resulted in a hypochromic effect; the latter reached a

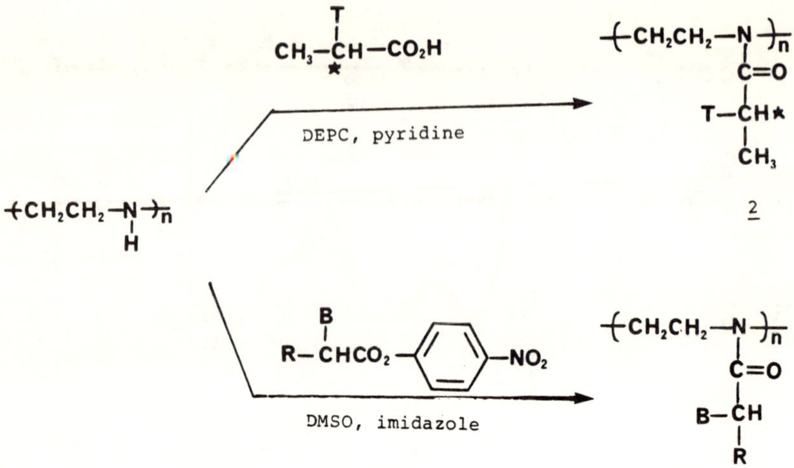

R = CH_3; (B = A, AcA, T, U) *3a*, *3b*, *3c*, *3d*
R = $(CH_3)_2CH$; (B = T) *4*

T = thymin-l-yl; U = uracil-l-yl; A = adenin-9-yl
AcA = N^6-acetyladenin-9-yl
DEPC = diethylphosphoryl cyanide

Scheme 7. Polynucleotide analogs derived from PEI.

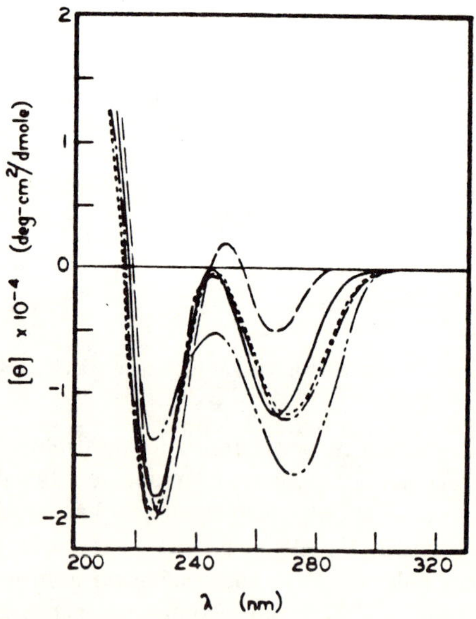

FIG. 1. CD curves of D − (−)T in various solvents; [D − (−)T], $2 \times 10^{-4}M$: (−)TFE, (− − −) EtOH, (− − −) pure H_2O, (− − −) HCl (0.1N), (− − −) NaOH (0.1N).

maximum when there were equal moles of *2* and *3(a)*. The same combination in trifluoroethanol resulted in no hypochromic effect. Mixtures of *2* with the corresponding monomer and dimer models of *3(a)* in neutral aqueous solution

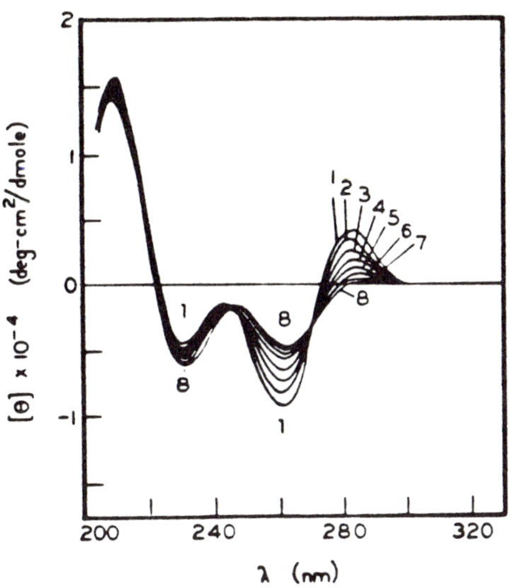

FIG. 2. Influence of temperature on the cotton effect of P − (−)T in neutral aqueous solution (1) 2°C, (2) 15°C, (3) 40°C, (4) 50°C, (5) 60°C, (6) 70°C, (7) 80°C, (8) 90°C: [P − (−)T], $2 \times 10^{-4}M$ residue [11].

gave only a slight deviation from Beer's law. In the chloroform/ethanol solvent system, a combination of the monomer models of *3(c)* (racemic) and *3(a)* gave no significant deviation from Beer's law. The data indicate that base pairing has occurred between *2* and *3(a)* or *3(c)* and *3(a)* so as to enhance base stacking [19].

Uracilyl polymer *3(d)* and polymer *4*, a derivative of 3-methyl butyric acid, have been synthesized and examined spectroscopically [20]. The bulky R group of the latter favors *trans* amide bonds but may also sterically impede base stacking. NMR study shows conformational populations, whether R = CH_3 or $(CH_3)_2CH$, and hypochromicities are comparable. The hypochromicities were dependent on pH of the solutions. Studies in DMSO/D_2O solutions showed increased base stacking in the more aqueous solvents; for polymer *3(d)* with n = 100, a marked increase was noted. Polymer *3(d)* and its monomer model also interact with poly A in *tris* buffer solutions.

The optically active 3-(adenin-9-yl)butyric acid and 3-(thymin-l-yl)butyric acid were grafted onto PEI in 100% grafting efficiency by the active ester method [11, 14]. Scheme 8 shows the reactions employed. Studies on these optically active polymers by UV, CD, and NMR spectroscopy are underway.

POLY(VINYLAMINE) DERIVATIVES

We have synthesized poly(vinylamine) containing optically active (+) or (−)-2-(thymin-l-yl)propionyl groups as the grafted pendants and the related monomer and dimer models [21]. The poly(vinylamine)s were prepared by hydrolysis of poly(*N*-vinylacetamide) [22] or poly(*N*-vinyl-*t*-butyl-carbamate) [23]

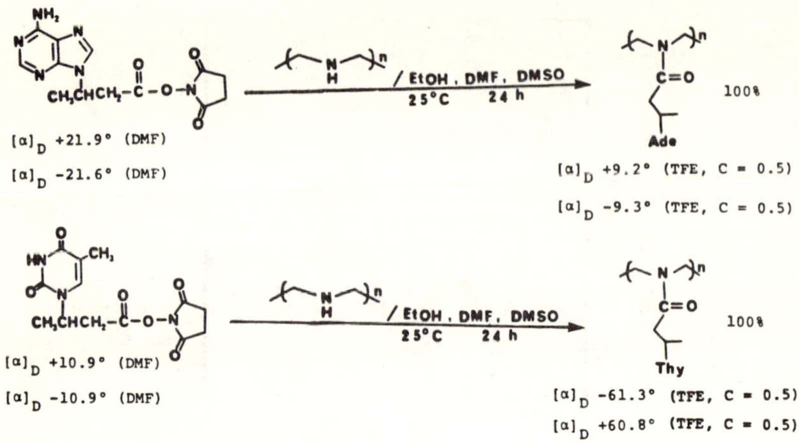

5 6 7 8

Scheme 8. Grafting of polyethylenimine by the active ester method.

which were obtained by free radical polymerization of the corresponding monomers. These compounds have been studied in detail by UV, CD, and NMR spectroscopies [24]. Exciton coupling is not observed in the polymers; hypochromicity in H_2O is 27.9% and 22.4% for the analogs prepared from the two kinds of poly(vinylamine) backbone. The CD spectrum of one of the dimer models displayed extremely solvent-dependent exciton coupling. The efficiency of grafting pendants onto PVAm was greatly improved by employing the PVA-6HOSu complex instead of free poly(vinylamine). Free base form of poly(vinylamine) cannot be isolated from water without crosslinking; however, the amino group of poly(vinylamine) can be protected by 6 equivalent of a weak acid, *N*-hydroxysuccinimide (HOSu). After water was removed, PVA-6HOSu was isolated. This species was reacted with the active esters of 3-(thymin-l-yl)butyric acid and 3-(adenin-9-yl)butyric acid to yield completely grafted polymers [10, 25]. The reactions are demonstrated in Scheme 9. Studies on these optically active polymers by UV, CD, and NMR spectroscopies are currently in progress.

POLY(DEHYDROALANINE) (PDA) DERIVATIVES

As the molecular weight of the PEI and PVAm derivatives described above is increased, water solubility decreases. Therefore, one way to modify these compounds is to attach hydrophilic groups to the polymer backbone. We have effectively prepared a carboxylated version of the PVAm polymers by free radical polymerization of dehydroalanines, *9–12* [26, 27]. These monomers are made by coupling the nucleic acid base derivative with β-chloroalanine methyl ester hydrochloride using *N,N′*-dicyclohexylcarbodiimide (DCC) in DMF (dimethyl formide). Dehydrochlorination of the product is effected using triethylamine; both dehydrochlorination and hydrolysis to the acid occur in aqueous NaOH. Polymerization of these monomers gives water-soluble products with inherent viscosities of up to 0.87 dL/g (DMF, 0.5 g/dL). The hypochromicities of the

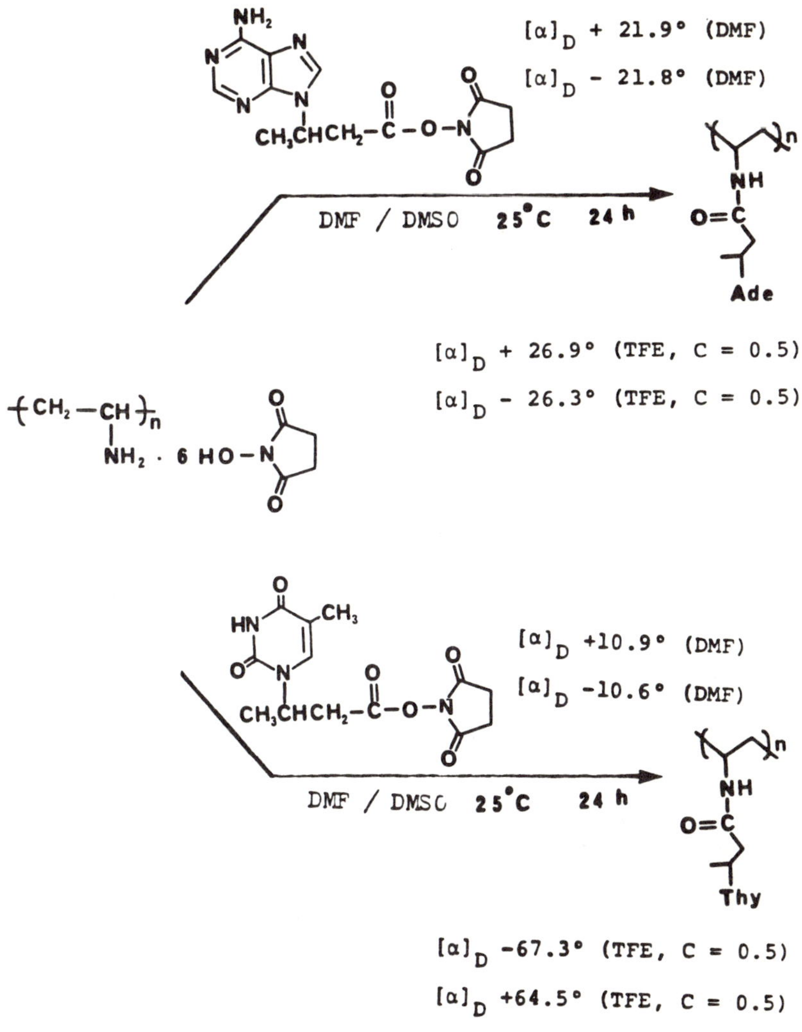

Scheme 9. Grafting of poly(vinylamine) by the active ester method.

pyrimidine polymers are on the order of 10% versus the monomer models, while the purine polymers give slightly higher values. The optically active polymer *16* (R = H) exhibits a CD spectrum which is very dependent on salt concentration, but much less dependent on the degree of neutralization. This behavior is paralleled in the viscometric titrations of the polymers. Further information on the conformational behavior of these types of polymers might be obtained by examining copolymers. A 360 MHz ^{1}H NMR examination of several copolymers of optically active monomers *12* with *10* in DMSO at 60°C revealed upfield shifts of the pyrimidine ring methyl resonance as well as the resonances of the methyl ester and methyl group adjacent to the chiral center. Such effects of the purine ring current are also seen in homopolymer *14* while

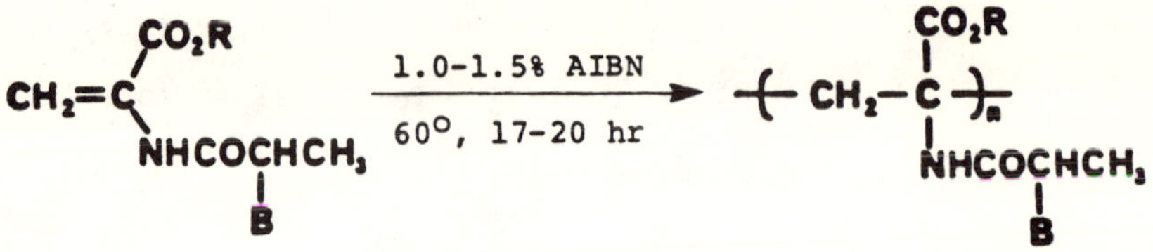

9-12   13-16

Monomer	B	R	Polymer
9	T, U	H, CH_3	*13*
10	AcA	CH_3	*14*
11	A	H	*15*
12	(−)T	H, CH_3	*16*

Scheme 10. Polynucleotide analogs derived from PDA.

the spectrum of *13* resemble their monomer models. These results suggest that pyrimidine units do interact with adjacent purine units. Such interactions were difficult to detect by circular dichroism; the spectra of the copolymers were similar to homopolymers.

BIOLOGICAL ACTIVITY OF THE POLYNUCLEOTIDE ANALOGS

We have begun a screening program for these polymers against a number of viruses which are of interest in human diseases. These polynucleotide analogs have been tested *in vitro* for biological activity. This involves growing cells which are known to support the growth of the viruses of interest and treating them with the viruses and the polymer or control solutions. After a certain period, the amount of activity of the viruses was measured. Table I demonstrates the results of the testing of some of the polymers. Further study will be carried out to understand the correlation between structural features and antiviral activity.

We are grateful to Professor H. F. Maassab for the antiviral activity studies of the polymers.

TABLE I
Antiviral Activity of Some Polynucleotide Analogs

Polymer	Measles	HSV-II	RSV	Influenza A	Influenza B
13 (B = U)	MI	MI	+	+	+
13 (B = T)	+	+	+	0	+
3a	MI	+	0	0	0
7	+	+	+	0	0

Abbreviations: HSV-II = herpes simplex virus type II; RSV = respiratory syncytial virus; 0 = no activity; MI = marginal activity; + = positive activity.

REFERENCES

[1] Z. A. Schomshtein and S. A. Giller, *Chem. Heterocycl. Compd.*, *12*, 27 (1976).
[2] A. S. Jones, *Int. J. Biolog. Macromolecules*, *1*, 194 (1979).
[3] K. Takemoto and Y. Inaki, *Adv. Polym. Sci.*, *41*, 1 (1981).
[4] S. Hoffman, *Z. Chem.*, *7*, 241 (1979).
[5] J. Pitha, *Adv. Polym. Sci.*, *50*, 7 (1983).
[6] C. G. Overberger and Y. Inaki, *J. Polym. Sci.*, *Chem. Ed.*, *17*, 1739 (1979).
[7] C. G. Overberger, Y. Inaki, and Y. Nambu, *J. Polym. Sci.*, *Chem. Ed.*, *17*, 1759 (1979).
[8] C. G. Overberger and Y. Morishima, *J. Polym. Sci.*, *Polym. Chem. Ed.*, *18*, 1247 (1980).
[9] A. G. Ludwick and C. G. Overberger, *J. Polym. Sci.*, *Polym. Chem. Ed.*, *20*, 2123 (1982).
[10] C. G. Overberger and C. C. Chen, *J. Polym. Sci.*, *Polym. Lett. Ed.*, *23*, 345 (1985).
[11] C. G. Overberger and C. C. Chen, *J. Polym. Sci.*, *Polym. Lett. Ed.*, in press.
[12] V. P. Chu, Ph.D. dissertation, University of Michigan, Ann Arbor, MI 1984.
[13] M. J. Lan, Ph.D. dissertation, University of Michigan, Ann Arbor, MI 1985.
[14] C. C. Chen and C. G. Overberger, *Am. Chem. Soc. Polymer Preprints*, *26(2)*, 188 (1985).
[15] M. Ueda, K. Okada, and Y. Imai, *J. Polym. Sci.*, *Polym. Chem. Ed.*, *14*, 2665 (1976).
[16] C. X. Lu, P. Liu, and C. Hu, *J. Polym. Sci. Polym. Chem. Ed.*, *19*, 2091 (1981).
[17] C. G. Overberger and Y. Morishima, *J. Polym. Sci.*, *Polym. Chem. Ed.*, *18*, 1267 (1980).
[18] C. G. Overberger and Y. Morishima, *J. Polym. Sci.*, *Polym. Chem. Ed.*, *18*, 1433 (1980).
[19] Y. Morishima and C. G. Overberger, *J. Macromol. Sci. Chem.*, *A13(5)*, 573 (1979).
[20] A. G. Ludwick and C. G. Overberger, *J. Polym. Sci.*, *Polym. Chem. Ed.*, *20*, 2139 (1982).
[21] C. G. Overberger and S. Kikyotani, *J. Polym. Sci.*, *Polym. Chem. Ed.*, *21*, 525 (1983).
[22] D. J. Dawson, R. D. Gless, and R. E. Wingard, Jr., *J. Am. Chem. Soc.*, *98*, 5996 (1976).
[23] A. R. Hughes and T. St. Pierre, *Macromol. Syn.*, *6*, 32 (1977).
[24] C. G. Overberger and S. Kikyotani, *J. Polym. Sci.*, *Polym. Chem. Ed.*, *21*, 541 (1983).
[25] C. C. Chen, Ph.D. dissertation, University of Michigan, Ann Arbor, MI 1985.
[26] K. A. Brandt and C. G. Overberger, *Nouv. J. Chim.*, *6*, 673 (1982).
[27] C. G. Overberger, K. A. Brandt, S. Kikyotani, and A. G. Ludwick, *Polym. Bull.*, *5*, 481 (1981).

Charles G. Overberger was born in Barnesboro, Pennsylvania, a small town in the bituminous coal area, in 1920. He graduated from Pennsylvania State University with a B.S. degree in 1941; entered the Graduate School of the University of Illinois in the fall of 1941, and obtained the Ph.D. degree in 1944 under the direction of Professor Carl S. Marvel.

As a graduate student he was a Teaching Assistant, a Research Assistant and a University Fellow. From 1944 to 1946 he was a Research Associate at the University of Illinois with Professor Carl S. Marvel. From 1946 to 1947 he held a duPont postdoctoral fellowship at the Massachusetts Institute of Technology with Professor A. C. Cope.

In 1947 he was appointed as an Assistant Professor of Organic Chemistry at the Polytechnic Institute of Brooklyn. In 1950 he was promoted to Associate Professor and to full Professor in 1951. He was Associate Director of the Polymer Research Institute from 1951 to 1955. From 1955 to 1964, he assumed the duties of Head of the Department of Chemistry. From February, 1963, to February, 1964, he was Acting Vice President for Research at the Polytechnic Institute of Brooklyn. From February, 1964 to February, 1967, he was Dean of Science and Director of the Polymer Research Institute. In February, 1967, he became Chairman of the Department of Chemistry at The University of Michigan and in 1968 he helped establish and became Director of the Macromolecular Research Center. From 1972 to 1983 he was Vice President for Research at The University of Michigan.

He holds honorary degrees of Doctor of Science from Holy Cross College, 1966; Long Island University, 1968; and Doctor Honoris Causa, Federal University of Rio de Janeiro, 1979.

Dr. Overberger has four children: Erica, Carla, Charles Thomas and Ellen Ann.

Dr. Overberger's principal researches are in the fields of synthetic organic reaction mechanisms and polymer chemistry. More specifically, they are in the areas of ionic polymerication, reactions of free radicals in solution, the effects of structure on the decomposition of radical formers, particulary cyclic azo compounds, synthetic enzyme analogs, polymers with mercaptan side chains, and optically active polymers.

SYNTHESIS, SPECTROSCOPY AND SIMULATIONS—A NOVEL APPROACH TO DRUG DESIGN

MURRAY GOODMAN
University of California, San Diego, La Jolla, California 92093

Our laboratories have been involved in the design of novel bioactive molecules. To this end we have combined synthesis, spectroscopy, computer simulations and studies of biological activity. In this lecture I wish to cite two examples of our work: one in the area of enkephalins and the second in catecholamine chemistry.

In order to understand the relationship between chemical structures of enkephalins and their selectivities in interaction with receptors in the brain, we undertook to synthesize constrained analogs of enkephalins. Thus, we have prepared a series of cyclic 14-membered ring containing enkephalins (structures **I**, **II**, and **III**): **I**: TFA H-Tyr-c[-D-A_2bu-Gly-Phe-Leu-] **II**: TFA H-Tyr-c[-D-A_2bu-Gly-gPhe-mLeu-] **III**: TFA H-Tyr-c[-D-A_2bu-gGly-mPhe-Leu-] and measured their biological activities using the guinea pig ileum (μ-receptor) and mouse vas deferens (δ-receptor) (Table I). The μ-receptor reflects opiate activities from morphine-like molecules while the δ-receptor is a measure of opiate activities of enkephalins. These tests were carried out in collaboration with Dr. Peter Schiller at the University of Montreal. At the same time we have under-

TABLE I
Biological Activity of Cyclic Enkephalin Analogs

Compound	Potencies in Guinea Pig Ileum Assay (μ-receptor)[a, b]	Potencies in Mouse vas Deferens Assay (δ-receptor)[a, b]
I	11.4 ± 1.6	0.035 ± 0.003
II (L) slower eluting	163 ± 21	1.47 ± 0.60
II (D) faster eluting	9.64 ± 0.74	0.767 ± 0.258
III (D) faster eluting	0.144 ± 0.022	0.00137 ± 0.00014
III (D) slower eluting	0.0638 ± 0.0075	<0.000025

[a] Potencies relative to [Leu^5]enkephalin (=1).
[b] Mean of three determinations ± SEM.

taken a program to study the conformations of these molecules by high resolution nuclear magnetic resonance spectroscopy and molecular dynamics simulations. From the analogs available and our conformational studies, we have proposed a preliminary model for the structural requirements for the opiate activities of the peptides at the μ- and δ-receptors in the brain.

In like manner we have undertaken the synthesis of catecholamine congener derivatives and conjugates. These molecules involve modifications of isoproterenol. We have measured the biological activity of these compounds using the cyclic AMP accumulation in an S49 lymphoma cell line in collaboration with Dr. Kenneth Melmon of Stanford University. Both compounds are much more potent than isoproterenol in this *in vitro* test. Conformational analysis of these molecules has also been undertaken. Preliminary results from molecular dynamics simulations indicate that specific folding are required for the design of super agonists. Using synthesis together with computer simulation we have devised a fundamental approach for the design of biologically active catecholamine congener derivatives and conjugates.

Murray Goodman was born on July 6, 1928.

PRESENT POSITION

Professor of Chemistry
University of California, San Diego La Jolla, CA 92093.

EDUCATION

University of California, Berkeley; Ph.D., 1952.

POSITIONS

Atomic Energy Commission Research Assistantship, University of California, Berkeley 1949–1952.
Research Associate, MIT 1952–1955.
National Research Council Fellow, Cambridge University 1955–1956.
Assistant Professor of Chemistry, Polytechnic Inst. of Brooklyn 1956–1960.
Associate Professor of Chemistry, Polytechnic Inst. of Brooklyn 1960–1964.
Professor of Chemistry, Polytechnic Inst., of Brooklyn 1964–1971.
Director, Polymer Research Inst., Polytechnic Inst. of Brooklyn 1967–1971.
Professor of Chemistry, University of California, San Diego 1971–.
Chairman, Department of Chemistry, University of California, San Diego 1976–6/30/81.

PROFESSIONAL ACTIVITIES

Editor, Journal Biopolymers, 1971–.
Member, Steering Committee for Human Reproduction, WHO, 1974–.
Chairman, Medicinal Chemistry A Study Section, NIH, 1972–1974; member 1970–1974.
Member, U.S. National Committee for IUPAC, 1980–.
Chairman, Section C, American Association for the Advancement of Science.

OTTO VOGL

Otto Vogl was born on November 6, 1927 in Traiskirchen, Austria. He became naturalized in 1959. Hc is married to the former Jane Cunningham of Sag Harbor, NY, and has two children, Eric and Yvonne.

EDUCATION

Ph.D., University of Vienna (Austria), 1950.
University of Michigan (Research Associate), 1953–1955.
Princeton University (Research Associate), 1955–1956.

EMPLOYMENT

University of Vienna (Austria) (Instructor in Analytical Chemistry), 1948–1955.
E. I. DuPont de Nemours & Co. (Plastics & Central Research Dept.), 1956–1970.
University of Massachusetts (Professor of Polymer Science and Engineering), 1970–1983.
Professor Emeritus, University of Massachusetts, 1983–.
Polytechnic Institute of New York (Herman F. Mark Professor of Polymer Science), 1983–.

GUEST PROFESSORSHIPS

Kyoto University (Japan), Department of Synthetic Chemistry, 1968 and 1980.
Osaka University (Japan), Department of Polymer Chemistry, 1968.
Royal Institute of Technology (Sweden), Department of Polymer Technology, 1971.
University of Freiburg (F.R.G.), Institute of Macromolecular Chemistry, 1973.
University Louis Pasteur, Strasbourg (France), 1976.
University of Berlin, Institute of Organic Chemistry, 1977.
Technical University Dresden (G. D.R.), Department of Chemical Technology, 1982.

AWARDS

Fullbright Fellowship, 1976.
Humboldt Award, F.R.G., 1977.
Senior Scientist Fellowship, JSPS, Japan, 1980.
Dr. rer. nat., honoris causa, University of Jena, 1983.
Foreign Member, Academy of Sciences (G.D.R.), 1983.
Distinguished Service Award, ACS, Division of Polymer Chemistry, Inc., 1985.
Chemical Pioneer Award, American Institute of Chemists, 1985.

RESEARCH INTERESTS

Aldehyde polymerization, optically active polymers, ring opening polymerization, solid state polymerization, polymerization at very high pressures, NMR spectroscopy of polymers, regular copolyamides, physiologically active polymers, polymer degradation and stabilization, polymerizable and polymeric stabilizers.

PROFESSIONAL ACTIVITIES

National Academy of Sciences: Chairman, NRC Committee on Macromolecular Chemistry, 1977–1979; Member, ad hoc Committee for National Needs in Polymer Science and Engineering, 1980–1981.
NSF: International Division: Advisory Committee, 1981–.
ACS, Division of Polymer Chemistry: Chairman, 1974.

ACS, Connecticut Valley Section: Chairman, 1974.

Secretariat for Macromolecules: General Secretary, 1976.

General Chairman: Polymer Chemistry, ACS-JCS Meeting in Honolulu, 1979.

Gordon Research Conferences: Polymers, Chairman, 1981.

PUBLICATIONS

Author and Coauthor of over 300 papers and patents, eight books and several book chapters.

ADVISORY OR EDITORIAL BOARD

Macromolecules (1966–1978); Journal of Polymer Science, Polymer Chemistry Edition; Macromolecular Synthesis; Journal of Macromolecular Science, Chemistry; Macromolecular Science (Reviews); Polymer (London); Die Makromolekulare Chemie; Monatshefte fur Chemie; Polymer News; Acta Polymerica; Valchemist (1973–1976); Del. Chem. Bulletin (1968–1970); Polymer Contents.

CONGRATULATORY MESSAGES

The following are excerpts from telegrams and letters sent to Professor Herman F. Mark in honor of his 90th birthday celebration.

April 24, 1985

"On behalf of all Americans, whose lives are touched daily by the miracle of polymers — whether they know it or not — congratulations on your 90th birthday."

The White House

May 2, 1985

"Your 90th birthday gives me the occasion to express my high esteem for your life's work in polymer chemistry for which you are internationally recognized as a groundbreaking scientist."

Dr. Rudolf Kirchschlaeger
President, Republic of Austria
(translated from the German)

May 6, 1985

". . . I am grateful for this opportunity to add my voice to the chorus of praise and tribute that you are receiving at this International Symposium . . . !

". . . you began by founding the very first polymer research institute right here in New York 44 years ago . . . the Polytechnic Institute of New York . . . stands as a monument to your vision and as proof of your genius."

Edward I. Koch
Mayor, City of New York

May 1985

". . . I offer my heartiest congratulations to the internationally famous chemist, researcher and teacher on the occasion of his birthday celebration and I convey to him the best wishes of his native city and hope that the coming years will bring him much happiness and enjoyment."

Helmut Zilk
Mayor, Vienna
(translated from the German)

May 1985

". . . my heartiest congratulations and good wishes on your 90th birthday.

"The University of Vienna has always greatly admired your scientific work and considers itself fortunate to be affiliated with you."

Dr. Hans Tuppy
Rector, University of Vienna
(translated from the German)

April 25, 1985

"Heartiest congratulations . . .accompanied by many good wishes for the coming years . . .

"We would like to take this opportunity to express our sincere thanks for the close ties which have existed since the 1920's between you and our Society and which have endured throughout the difficult times and events until today . . . "

Heinz A. Staab
President, Max-Planck-Gessellschaft
for the Advancement of the Sciences
(translated from the German)

April 20, 1985

"On this auspicious occasion to celebrate your 90th birthday, I, on behalf of the Chemical Society of Japan, would like to extend our heartfelt congratulations to you."

Prof. Dr. Teiji Tsuruta
President, the Chemical Society of Japan

April 30, 1985

"This happy event . . . provides us with yet another occasion to express our profound gratitude and admiration for your enormous contributions to science."

Erich Bloch
Director, National Science Foundation

Published by John Wiley & Sons, Inc.

April 22, 1985

The Chancellor of the Republic of Austria, who according to Article 64 Paragraph 1 of the Austrian Constitution carries out the functions of the President, hereby certifies that he has awarded to

Professor Emeritus Dr. HERMAN F. MARK

the GRAND DECORATION OF HONOR IN GOLD WITH STAR for services to the Republic of Austria.

Dr. Fred Sinowatz
(translated from the German)

Grand Decoration of Honor in Gold with Star for Services to the Republic of Austria

Plate I

Herman F. Mark receiving Medal from Thomas Klestil, Austrian Ambassador

Plate II

Otto Vogl, Edward Teller, Herman F. Mark

Mary E. Curtis, Otto Vogl, Herman F. Mark, Andrew H. Neilly

Plate III

Ephraim Katchalski-Katzir, Jane Vogl, Thomas Klestil

Herman F. Mark, Otto Vogl, Takeo Saegusa

Plate IV

Herman F. Mark, Karl Schlögl

Helga Winkler - Campagna, Otto Vogl, Thomas Klestil, Dr. and Mrs. Gerhard Pflueger, Peter Marboe

Plate V

George Bugliarello, President, Polytechnic University

Bengt Rånby

Helga Winkler-Campagna, Austrian Counsul General

Plate VI

Herman F. Mark

Linus Pauling

Plate VII

Otto Vogl

John P. McTague

Ellis Fields

Plate VIII

Linus Pauling

C.H. Bamford

Paul Flory

Max Perutz

Plate IX

Elkan Blout

Ephraim Katchalski-Katzir

Bruce Merrifield

Plate X

Karl Schlögl

Günther Wilke

Hans Mark

Plate XI

L. Guy Donaruma

George Smets

Bengt Rånby

Plate XII

Piero Pino

Baogong Qian

Naoya Ogata

Norbert Bikales

Plate XIII

Eli Pearce

James O'Donnell

Helmut Ringsdorf

Plate XIV

Frederick Eirich

Herbert Morawetz

Charles G. Overberger

Murray Goodman

Plate XV

Plaza Hotel, May 6, 1985

Plate XVI